T0187585

Handbook of Flavor Characterization

FOOD SCIENCE AND TECHNOLOGY

A Series of Monographs, Textbooks, and Reference Books

EDITORIAL BOARD

Senior Editors

Owen R. Fennema   University of Wisconsin–Madison
Y. H. Hui   Science Technology System
Marcus Karel   Rutgers University (emeritus)
Pieter Walstra   Wageningen University
John R. Whitaker   University of California–Davis

Additives   **P. Michael Davidson**   University of Tennessee–Knoxville
Dairy science   **James L. Steele**   University of Wisconsin–Madison
Flavor chemistry and sensory analysis   **John H. Thorngate III**   University of California–Davis
Food engineering   **Daryl B. Lund**   University of Wisconsin–Madison
Food lipids and flavors   **David B. Min**   Ohio State University
Food proteins/food chemistry   **Rickey Y. Yada**   University of Guelph
Health and disease   **Seppo Salminen**   University of Turku, Finland
Nutrition and nutraceuticals   **Mark Dreher**   Mead Johnson Nutritionals
Phase transition/food microstructure   **Richard W. Hartel**   University of Wisconsin–Madison
Processing and preservation   **Gustavo V. Barbosa-Cánovas**   Washington State University–Pullman
Safety and toxicology   **Sanford Miller**   University of Texas–Austin

Additional Volumes in Preparation

Industrialization of Indigenous Fermented Foods: Second Edition, Revised and Expanded, *edited by Keith H. Steinkraus*

Genetic Variation in Taste Sensitivity, *edited by John Prescott and Beverly J. Tepper*

Handbook of Food Analysis: Second Edition, Revised and Expanded: Volumes 1, 2, and 3, *edited by Leo M. L. Nollet*

Handbook of Flavor Characterization

Sensory Analysis, Chemistry, and Physiology

edited by

Kathryn D. Deibler

Cornell University
Geneva, New York, U.S.A.

Jeannine Delwiche

The Ohio State University
Columbus, Ohio, U.S.A.

CRC Press
Taylor & Francis Group
Boca Raton London New York

CRC Press is an imprint of the
Taylor & Francis Group, an **informa** business

CRC Press
Taylor & Francis Group
6000 Broken Sound Parkway NW, Suite 300
Boca Raton, FL 33487-2742

First issued in paperback 2019

© 2004 by Taylor & Francis Group, LLC
CRC Press is an imprint of Taylor & Francis Group, an Informa business

No claim to original U.S. Government works

ISBN-13: 978-0-8247-4703-9 (hbk)
ISBN-13: 978-0-367-39502-5 (pbk)

This book contains information obtained from authentic and highly regarded sources. Reasonable
efforts have been made to publish reliable data and information, but the author and publisher
cannot assume responsibility for the validity of all materials or the consequences of their use. The
authors and publishers have attempted to trace the copyright holders of all material reproduced in
this publication and apologize to copyright holders if permission to publish in this form has not
been obtained. If any copyright material has not been acknowledged please write and let us know so
we may rectify in any future reprint.

Except as permitted under U.S. Copyright Law, no part of this book may be reprinted, reproduced,
transmitted, or utilized in any form by any electronic, mechanical, or other means, now known or
hereafter invented, including photocopying, microfilming, and recording, or in any information
storage or retrieval system, without written permission from the publishers.

For permission to photocopy or use material electronically from this work, please access www.
copyright.com (http://www.copyright.com/) or contact the Copyright Clearance Center, Inc.
(CCC), 222 Rosewood Drive, Danvers, MA 01923, 978-750-8400. CCC is a not-for-profit organization
that provides licenses and registration for a variety of users. For organizations that have been
granted a photocopy license by the CCC, a separate system of payment has been arranged.

Trademark Notice: Product or corporate names may be trademarks or registered trademarks, and
are used only for identification and explanation without intent to infringe.

Visit the Taylor & Francis Web site at
http://www.taylorandfrancis.com

and the CRC Press Web site at
http://www.crcpress.com

Preface

Many challenges facing flavor research and how they can be dealt with are discussed in this handbook by scientists from around the world. Flavor analysis continues to evolve as new techniques and insights are developed. Innovative and multidisciplinary approaches are being used to tackle the challenges associated with flavor analysis. Psychologists, physiologists, geneticists, and sensory specialists are now working together with chemists to uncover the mysteries of flavor.

Although the term *flavor* refers to all aspects of food that are detected during consumption, this book primarily is concerned with odor and aroma. A flavor compound is a stimulant that activates a sensory receptor, producing a perception; e.g., some flavor compounds in a lemon activate receptors on the olfactory epithelium, resulting in the impression of "lemon." An individual may have a hedonic response (degree of liking) to a food that is the result of his or her flavor perception modified by emotions and memories. A comprehensive study of flavor considers the stimuli, the receptors, the processing of a sensory signal, and the hedonic response of individuals. Challenges related to the analysis of a flavor and its perception and interpretation by humans are addressed throughout this handbook.

CHALLENGES RELATED TO THE HUMAN RECEPTOR (SENSORY)

Human psychology contributes to the flavor experience and can be evaluated by various sensory analysis techniques described in Part I. Huge variations, both between panelists and within a single panelist, may be experienced in sensory analyses, even with stringently designed experiments. Properly designing experiments that measure specific characteristics requires consideration of many aspects. Taste and tactile signals interact with the aroma reception signal to affect flavor perception (Chapters 9, 12, 24, and 25). Time-intensity sensory methods add the dimension of time to perception (Chapters 7, 9, and 11).

CHALLENGES RELATED TO THE STIMULI (CHEMICAL)

The nature of flavor compounds creates challenges for analysis. Aroma compounds must be volatile. They are usually present at very low concentrations in foods. Despite the fact that hundreds of volatile compounds are often present in a food, only a few may be odor-active. Gas chromatography has been an invaluable tool for separation and subsequent identification of volatile compounds. Concentration of flavor chemicals is often necessary since the compounds are usually present at low levels. Some methods of sample preparation are described in this handbook, including solid-phase microextraction (see Chapters 16, 20–22, 30, and 31), sorptive stir bar extraction (Chapter 32), absorption on a porous polymer (Chapters 21, 22, and 27), super-critical CO_2 extraction (Chapter 22), simultaneous steam distillation (Chapter 31), accelerated solvent extraction (Chapter 35), simultaneous distillation extraction (Chapters 21 and 31), and direct gas injection with cryofocusing (Chapter 20). Sampling conditions are considered in Chapters 20, 23, and 24, and comparisons of some chemical detector sensitivities are made in Chapters 18, 23, and 27–29.

Hundreds of volatile compounds may be present in a food system. Identifying which of those compounds contribute to flavor requires consideration of their odor thresholds and the ratios of the compounds present, as well as physiological considerations. To selectively identify compounds that could potentially contribute to the flavor, humans must be used as detectors. Part III covers gas chromatography olfactometry (GCO), which has been used for over 35 years and gives a direct link between chemical and sensory analyses. Methods for determining relative aroma potency of compounds have been well established (CharmAnalysis^{TM}, aroma extraction dilution analysis (AEDA), odor activity values). Focusing on the small subset of volatile compounds with the highest odor potency has proved useful in comparing products under different aging conditions (Chapter 34), and identifying impact chemicals during grinding of coffee (Chapter 16), in orange essence oil (Chapter 14), and in different tomato cultivars (Chapter 13). Identification of the most potent odorants in various extracts is discussed, along with in vivo methods that allow the monitoring of preidentified volatile compounds during food consumption. Nonaroma constituents of foods can influence the degree to which a particular compound volatilizes (flavor release), thus affecting the ratio of volatilized compounds. The effects on flavor release of fat (Chapters 10, 11), protein

(Chapters 7, 11), polysaccharides (Chapter 12), and orange pulp (Chapter 30) also are discussed.

In addition, the chemical formation of aroma compounds for the synthesis of desired flavors and prevention of off-flavors is considered. Chapters 15, 21, 26, 33, and 34 discuss how light and heat can affect the formation of both desirable and undesirable flavors. Microorganisms and aging are key in off-flavor formation in dairy products, wine, and grain degradation. Potent flavor compounds are released during the processing of coffee.

CHALLENGES RELATED TO HUMAN PHYSIOLOGICAL CHARACTERISTICS

The effect of human physiological processes on aroma perception is a rapidly growing area of interest. The dynamics of aroma presentation can influence the ratio of compounds available for reception on the olfactory epithelium (Chapters 3, 24). A single flavor compound can activate more than one olfactory receptor. Chapter 9 covers quantifying perception of mixtures of flavor compounds activating multiple receptors with various intensities—a multivariate issue clearly requiring multidisciplinary approaches. The complexity and diversity of human genomics add another dimension to the enigma of flavor analysis, addressed in Chapters 4, 5, and 6. Incorporation of measurement of physiological effects into flavor research assists in accounting for human variation, and is discussed in Chapter 10.

ACKNOWLEDGEMENT

Financial support for compiling this handbook was made possible through industry sponsorship from International Flavors & Fragrances, Inc.; Pepsi-Cola Company; Frito Lay, Inc.; McCormick & Co., Inc.; Givaudan Flavors Corporation; Firmenich Inc.; and Takasago International Corporation. Organizational and financial support was provided by the Agricultural and Food Chemistry Division of the American Chemical Society.

All chapters included in this handbook were peer reviewed. Many thanks to those who contributed their time and consideration to these reviews.

Kathryn D. Deibler
Jeannine Delwiche

Contents

Contents

Part IV.   Comparisons of Techniques, Methods, and Models

Contents

Part VI. Compounds Associated with Flavors

Handbook of Flavor Characterization

1

Methods, Approaches, and Caveats for Functionally Evaluating Olfaction and Chemesthesis

Charles J. Wysocki and Paul Wise
Monell Chemical Senses Center, Philadelphia, Pennsylvania, U.S.A.

> Only human sensory data provide the best models for how consumers are likely to perceive and react to food [*and aroma*] products in real life.*

I. INTRODUCTION

What occurs in the cranium can only be inferred. When a perfectly enjoyable meal is consumed or an exhilarating walk in a blooming garden includes the pleasures of fragrances carried on the wind, only the individual can experience such sensory pleasures. Although advances in brain imaging have made significant clinical impact, the biomedical community cannot yet relay the emotions, the feelings, or even the basic sensory components of such experiences to an outside observer. Hence, investigators wishing to quantify, or even approximate, these experiences must rely, at least in part, on the psychophysical approach.

Human psychophysics has a long history, beginning with the works of Gustav Fechner in the 19th century [2], with considerable emphasis on evaluating function in vision [3], audition [4], somatosensation [5], and proprioception [6]. The chemical senses, although historically less well studied, have not been neglected [7,8]. Through the appropriate use of

---

* From Ref. 1. Italic added.

psychophysics, investigators can obtain considerable information about what the individual experiences when an odor is smelled. Individual sensitivity to odorants can be measured; methods to quantify odorant intensity and pleasantness are available; and insights into odor quality and acceptance, both quite personal experiences, can be obtained. After laying a foundation describing how chemoreception works in the nose, this chapter briefly summarizes some of the approaches that have been used to extract information about personal experiences with odors from individuals willing to serve as subjects in experiments. Mindful that experiments are extremely artificial, we proceed to make inferences about the joys of good meals and garden walks based upon data that have been collected in such situations. At present, it is the best approach available.

A distinction must be made between psychophysics and physics, or better yet, in the case of the chemical senses, chemistry. Psychophysics attempts to extract information that is contained within the psychological realm of the individual. In so doing, it has established a vocabulary that separates the physical from the psychological construct but also provides the bridge that links the two. As individuals communicate across disciplines, the importance of appreciating subtleties in definitions becomes critical to interdisciplinary growth. These distinctions in definitions are not trivial. They recognize the import of the physical sciences while maintaining the integrity of the individual perceiver. Examples are included in Table 1.

Table 1 Comparisons Among Physical, Psychophysical, and Psychological Terms

Concept	Chemistry	Psychophysics	Psychological construct
Stimulus	Molecule/ compound	Odorant/irritant	Odor/aroma
Complex stimulus	Mixture	Mixture of odorants	Odor/aroma
Smell	Odorant molecule or mixture	Odorant	Odor/aroma
Quantity	Concentration	Intensity	Strength
Hedonics	Not defined	Scale ratings	Pleasantness
Quality	Substance	Adjectival ratings	Identity or association

II. DISSECTING ODORANTS BY THE "PATHWAYS" THEY TAKE TO THE BRAIN

Although the proceeding allusions to odor experiences suggest pleasant experiences, all too often individuals are faced with unpleasant or even irritating odors. These take many forms and have many sources, e.g., the presence of other humans; agricultural activities; biological degradation, including spoilage of foods; industrial by-products; even natural defense mechanisms, e.g., of skunks. When unpleasant odors become quite strong, some people are *irritated* by their presence. Often the *irritation* is psychological in nature. This form of irritation should be distinguished from pure sensory irritation.

Sensory irritation in the mouth is easily described by example. After eating a hot pepper individuals experience the residual burn. For some the burn is pleasant, whereas for others it may be extremely unpleasant. Regardless, the sensory experience of irritation is caused by capsaicin, the active ingredient in hot peppers that produces the burn.

Analogous events occur in the nose when many odorants interact with the epithelia therein. The odor component is conveyed via the odorant's interaction with molecular receptors on olfactory sensory neurons (first cranial nerve) in the olfactory epithelium [9]; irritation is initiated by interactions with receptors or other mechanisms that stimulate the trigeminal (fifth cranial) nerve [10]. The term *chemesthesis* has been applied to distinguish this sensory experience from olfaction or, in the oral cavity, taste [11,12]. Importantly, chemesthesis is a bodywide experience. It is only on some portions of the head, e.g., eyes, nose, mouth, and some other facial areas, where information is conveyed by the trigeminal nerve (Fig. 1).

A. Olfaction

The 10–20 million olfactory sensory neurons in the human nose are confined to a relatively small patch of tissue located high in the nasal cavity. When odorants, e.g., carvone in Fig. 2, are deposited within the mucus covering the distal ends of the olfactory receptor cells, they interact with some of the membrane-bound, G-protein-coupled receptors [9]. This interaction initiates a transduction process that converts physicochemical information in the odorant, e.g., its structure or other attributes, into electrical energy that is conveyed in the form of pulses (action potentials) along olfactory axons to the brain.

The life cycle of an olfactory receptor cell is quite interesting. The sensory cell is a neuron, but a short-lived one: most die within 60 days of formation. It begins from a cellular division of a basal cell (Fig. 2). Through

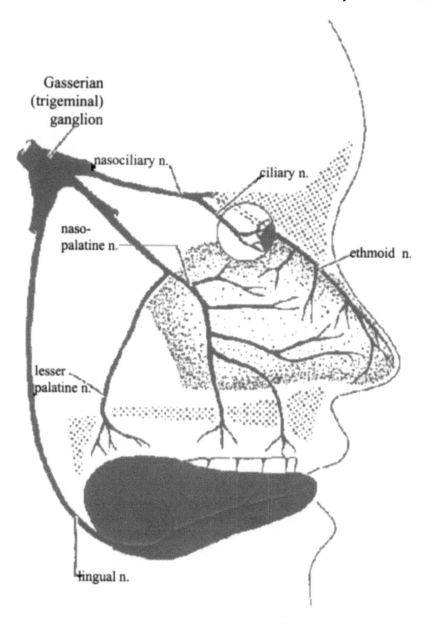

Figure 1 Representation of innervation of the facial regions by branches of the trigeminal (fifth cranial) nerve. Although nasal, oral, and some facial chemesthesis is conveyed via the fifth cranial nerve, chemesthesis is bodywide and is communicated to the brain by nerves other than the trigeminal.

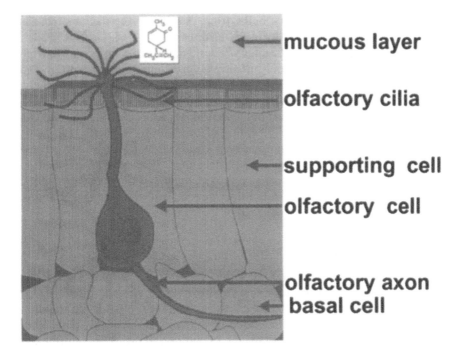

mucous layer

olfactory cilia

supporting cell

olfactory cell

olfactory axon
basal cell

Figure 2 Depiction of some components of the vertebrate olfactory epithelium in the nose. Odorants, e.g., carvone, deposit themselves in the mucous layer and interact with molecular receptors in the membrane of cilia of the olfactory receptor cells. Subsequent to intracellular signal transduction events, action potentials are sent via the olfactory axons to the olfactory bulbs in the brain. Supporting cells provide physical and physiological support for the olfactory neurons. Undifferentiated basal (stem) cells are the source of new supporting and olfactory receptor cells.

a process not yet fully understood [13], the dividing basal cell directs replacement of supporting cells or sensory cells. Should a neuron develop, it sends an axon to the brain, where it makes connections in the olfactory bulb. The connectivity is not random. Each sensory neuron expresses one olfactory receptor gene [14], or at least a very limited number of genes [15]. Axons from sensory cells expressing the same molecular receptor coalesce and converge at the same destination within the olfactory bulb, viz., two olfactory glomeruli: one located in the medial zone and one in the lateral zone. These regions are interconnected and form the basis for a spatial map that apparently codes odor quality [16]. Although each sensory cell lives only a fraction of the lifetime of a typical vertebrate, the pattern of innervation of the olfactory bulbs from the sensory epithelium remains static [17], unless disturbed by injury or disease [18].

Farther upstream in the olfactory system, the projection neurons from the olfactory bulb (mitral and tufted cells) distribute themselves to myriad destinations [19], which in turn project elsewhere and, in most cases, send a return link to the originating location [19]. These distributed projection pathways allow for numerous influences on behavior: e.g., the odorant can be spoken about, presumably because of connections to cortex [19]; odors may influence mood and emotion, even at subconscious levels [20], because of diverse connections to limbic structures [19]; physiological, e.g., autonomic and neuroendocrine, responses may be modulated by odors, because olfaction communicates rather directly with the hypothalamopituitary gonadal axis [21].

B.   Chemesthesis

Chemesthesis is a combination of forms, viz., chemical stimuli that activate subcomponents of somesthesis, one's ability to locate stimulation of the skin or underlying tissues. Different classes of neurons provide information about various aspects of stimulation of the body surface, e.g., pain, heat, cooling, light touch, and pressure [22]. In addition to providing information about the stimulus modality, e.g., pain versus cooling, information about its location also is extracted. For example, a needle stick is readily localized. Although with less accuracy, heating and cooling also can be localized. Chemicals too can stimulate some of these subsystems. In the extreme, strong acids or caustic fluids can cause injury. Milder forms of stimulation also are readily detected, especially in mucous tissue, where chemical compounds have easier access to stimulate receptors; surrounding cells (which may release compounds that activate local neurons [23], including within the nose [24]); or perhaps neuronal membranes directly. Importantly, as anyone who has had a large dose of capsaicin can report, regions of mucous tissue that are not innervated by the trigeminal nerve can be stimulated by the irritant molecule, although some time after eating the meal.

In the nose, however, the trigeminal nerve provides information to the brain about chemical irritation (as well as many other sensory attributes, e.g., temperature, humidity, and physical changes). Initially, odorants that stimulate chemesthesis must deposit themselves in the mucus. Although Fig. 3 indicates such deposition in the region of the olfactory epithelium, this can take place anywhere within the nasal cavity.

At present, some specific receptors that are associated with irritation, and the genes that code for their expression, e.g., the burn of capsaicin and

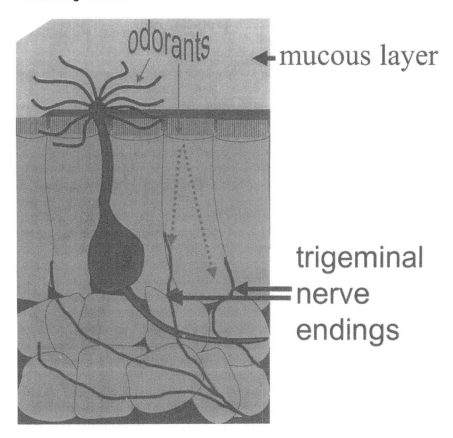

Figure 3 Schematic of stimulation of the trigeminal nerve in the nose by odorants/ irritants. After depositing themselves in the mucus, irritants penetrate the epithelium to contact receptors or free nerve endings of the trigeminal nerve.

the coolness of menthol, have been identified (Refs. [25] and [26], respectively). These receptors are presumably expressed on endings of the trigeminal nerve (and elsewhere).

Once these and other receptor processes are stimulated, individuals report any one or a mix of the following: itchy, tickling, scratchy, prickling, furry, stinging, pungent, painful, cool, cold, peppery, warm, hot, burning, or perhaps other attributes. Although suggestive, it is extremely subjective for individuals to determine when an odorant stimulus that is increasing in concentration becomes a true sensory irritant by relying solely upon an adjective base to describe it.

III. DISTINGUISHING BETWEEN SENSORY AND COGNITIVE IRRITATION

All sensation and perception are psychological; however, as defined herein and elsewhere [27,28], irritation derived from the nose can take two forms. A strong odor can be irritating because the individual does not like it. This could be an example of pure cognitive irritation, provided there were sufficient evidence that the odorant did not stimulate incoming chemesthetic pathways. Alternatively, an odorant could provide an odor and be a true sensory irritant; ammonia provides a good example. After opening a bottle of ammonia or an ammonia-containing cleaning product at arm's length, one can appreciate its odor by gently wafting the air in the direction of the nose. By slowly moving the opening of the bottle to the nose, the odor of the ammonia becomes stronger. At some point, an ensuing sniff brings on the characteristic "kick" in ammonia—chemesthetic pathways from the nose have become activated. Here, too, interpreting when the "kick" begins is subjective. A more objective measure of chemesthetic onset would be preferred. If the ammonia were to be provided to one side of the nose only, localizing the "kick" to the side of the nose receiving stimulation would ensure that the individual was utilizing chemesthetic information. As pain does, low-level chemical stimulation of the trigeminal nerve provides spatial information about the locus of stimulation: the individual is able to determine which nostril is being stimulated.

If an individual is stimulated by an odorant that is not an irritant in one nostril only, the side of stimulus delivery cannot be determined with a level of accuracy above that of chance [28]. For example, a strong concentration (3.2%) of phenylethyl alcohol (Fig. 4), vanillin, or lyral (data not shown for the latter two) in the vapor phase presented unilaterally cannot be localized. Apparently, in the vapor phase, these compounds fail to stimulate chemesthesis adequately. Ammonia and many other volatile organic compounds (VOCs) stimulate both olfaction (to provide an odor) and chemesthesis; hence, for any individual, there must be some concentration below which these dual-mode VOCs are undetectable, a higher concentration at which an odor is apparent (but not recognized), a higher-still concentration at which the odor is clearly identified (recognition), and a concentration, typically higher, at which localization in the nose becomes apparent. By using established psychophysical procedures (see later discussion), it then becomes possible to establish three thresholds for most odorous VOCs, viz., an odorant detection threshold, an odor recognition threshold, and a localization threshold. The first is defined as the lowest concentration at which an individual can reliably determine the presence of an odorant. The second is the concentration at which the individual can

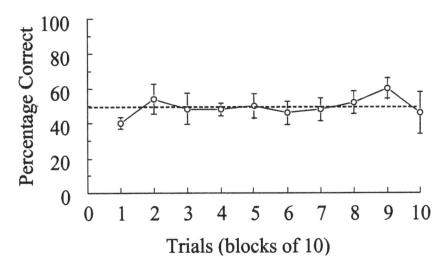

Figure 4 Over 100 trials, individuals were *trained* to localize phenylethyl alcohol (PEA) when it was presented in a single nostril; none succeeded. These results support the interpretation that PEA does not adequately stimulate the trigeminal nerve to provide spatial information about stimulus location (side of the nose). (From Ref. [28].)

reliably identify the odor. The third is the concentration at which an individual can reliably determine which nostril is being stimulated when an odorant/irritant is presented monorhinally (unilaterally).

IV. CHARACTERISTICS OF LOCALIZATION

Localizing the irritation of an odorant is a straightforward task for a subject in an experiment. Typically, the individual must choose, left or right, which nostril received the stimulus. If the stimulus is an odorant [rather than a non odorous irritant, e.g., carbon dioxide (CO_2)], a smell will be detected on each trial, and it will be perceived as coming from the nose in toto; however, the task is not to smell the odor, but to report the side of stimulation, and that cannot be done by relying solely upon olfactory information [28].

Using a two-alternative (left vs. right), forced-choice, modified-staircase procedure (see Sec. V), this task generates reliable thresholds for l-butanol (test-retest, $r = 0.76$; data for other irritants not shown). Furthermore, interindividual variation for intranasal chemesthetic thresholds

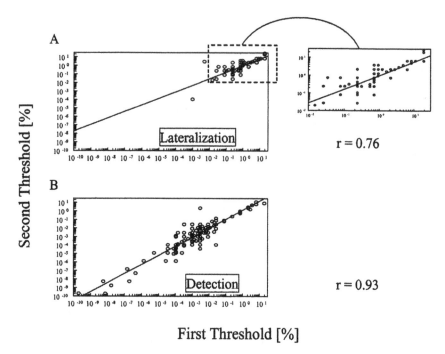

Figure 5 Results of two threshold measures for localization (A, lateralization) and detection (B) of butanol performed in the same test session. For direct comparisons of thresholds, the axes, $\log_{10}$ of the percentage of butanol in solution, are the same in A and B (note the greater variation in detection thresholds than in lateralization thresholds). The magnification of A captures the 1000-fold range in which most localization thresholds occurred. (From Ref. [28].)

(as defined by the localization task; Fig. 5a) is not as extreme as noted for olfactory detection thresholds (Fig. 5b).

V. MEASURING OLFACTION AND CHEMESTHESIS

Many psychophysical approaches have been proposed to obtain quantitative estimates of individual percepts. Some studies focus on measurements of threshold sensitivity [29]; others focus on the use of suprathreshold stimuli [30]; both can be important in determining overall impact. Interests also lie in hedonic evaluations, odor quality and odor (aroma) acceptance in the academic (understanding the workings of the brain), marketing (understanding how to influence and motivate the consumer), and corporate (understanding how to increase the profit margin) settings. At least in part, psychophysical approaches provide data that address some of these issues.

A. Sensitivity

Thresholds are important springboards from which other data may be obtained. Establishing thresholds for odorants has had a history of its own, and different methods have yielded different results. Leafing through the volume of odorant thresholds published by Devos and colleagues [31] provides an excellent exercise in method-related variation. These authors have attempted to standardize thresholds by statistically manipulating results obtained by various methods across numerous laboratories over many decades. What becomes clear from the numerous investigations is the variation inherent in such endeavors. In part, this variation is attributable to different methods used to establish detection thresholds and the criterion established by the investigators. Approaches include tasks requiring "yes-no" or forced-choice answers; methods include the use of constant stimuli or adaptive procedures.

What is the minimal concentration that elicits a perceptible odor? The question implies two discrete sensory states: odor present (above threshold) or odor absent (below threshold). Some theories propose the existence of sensory thresholds of this type (e.g., [2,32]). Most evidence supports another class of theories, in which sensation can assume any value along a continuum rather than one of two (or a few) discrete states [33–35]. In virtually all literature on odor sensitivity, the term *threshold* has no necessary relationship to the internal state of an organism; rather, it refers to the concentration that produces some criterion level of performance on a given task. Accordingly, one must consider thresholds in light of the methods used to obtain them. Important considerations include (a) the task subjects perform and (b) the method experimenters use to select concentrations used during a session.

1. Tasks

a. "Yes-No". In a "yes-no" task, subjects receive a single sample (i.e., dilution of an odorant). They respond "yes" if the sample elicits a sensation, and "no" otherwise (Fig. 6; idealized data). "Yes-no" often serves in adaptive procedures (i.e., ascending or descending series of concentrations) but can also serve in procedures that use constant stimuli (see later discussion). The task has a subjective component since individuals must establish, from their own experience, appropriate criteria by which to judge whether or not they smelled an odor. In short, personal criterion can

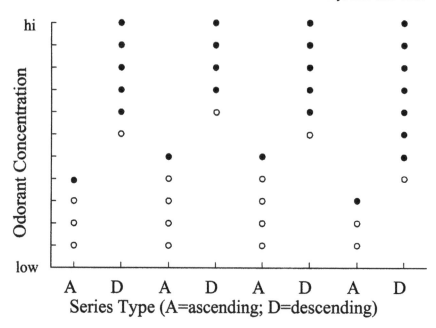

Figure 6 Idealized data representing the method of limits. Solid circles indicate a correct choice. Note that the method can be used with either ascending or descending concentrations.

become confounded with sensory acuity. For example, some might respond "yes" at the slightest hint of a sensation, perhaps giving the impression of greater sensitivity than they truly possess. Others might say, "yes" only after they clearly recognize an odor (subjects typically require higher concentrations to identify than to detect). Therefore, proportion of "yes" responses need not provide a monotonic measure of sensitivity. In some procedures, experimenters assess a criterion by interspersing odorless blanks (solvent only), sometimes called "catch trials." Methods exist to adjust proportion of "yes" responses ("correct for guessing"), but these procedures tacitly accept the validity of thoroughly refuted theories [33,34]. A later section describes an alternative approach.

   b. *"Forced-Choice"*. In forced-choice (FC), experimenters do not attempt to determine directly whether subjects smell an odorant solution (do not rely on what subjects say they smell) but, rather, determine whether subjects can distinguish the odorant apart from one or more blanks (determine what subjects can do). For example, in two-alternative FC (2-AFC), subjects receive an odorant and a blank (in random order) and must decide which sample contains the odorant. Subjects may also receive an odorant plus two (3-AFC) or more (*n*-AFC) blanks, two odorants plus three blanks, etc. In all cases, subjects must identify the sample or samples that contain an odorant. If subjects fail to perceive an odor, they must guess.

It may seem strange to demand that subjects select a sample, even if they report no sensation. In many situations, psychologists frown on demands that can guide subjects to an outcome. In this case, however, the structure of the task makes it less susceptible to bias. Unlike "yes-no" without catch trials, experimenters can score correctness objectively: The subject either picks the odorant solution or mistakenly picks a blank. Bias can still play a role (i.e., as a default in 2-AFC, a subject might select the second sample received in the face of even slight uncertainty); however, such bias proves relatively easy to detect and usually exerts smaller effects on estimates of sensitivity. Therefore, for a given FC task, proportion correct usually qualifies as a monotonic measure of sensitivity.

Because FC proves less susceptible to bias, most experimenters favor FC over "yes-no" for most applications. Situations in which response latency is of interest constitute important exceptions. Chamber studies that require long exposures might constitute other exceptions; e.g., one could do three trials of "yes-no" in the time it takes to do one trial of 3-AFC. One might also use a "yes-no" design for technical reasons, i.e., if an apparatus requires some time to switch between presentation of odorants and blanks.

2. Methods

Investigators wishing to establish a threshold can choose from an approach that is amenable to their needs. Typically, these include at least two alternatives.

a. Method of Constant Stimuli. In the method of constant stimuli, the experimenter chooses, in advance, (a) the range of concentrations to present, (b) the difference between successive concentrations, and (c) the number of trials per concentration. Generally, all trials occur in random order.

Researchers summarize FC results by plotting proportion correct versus concentration. Urban [36] called this a *psychometric function* (Fig. 7). The function ranges from chance, or no ability to choose the correct sample, e.g., 0.50 for 2-AFC or 0.33 for 3-AFC, to perfect ability to choose (1.00 for all tasks). Fitted functions (typically sigmoid, or S-shaped) help researchers estimate the concentration that corresponds to a criterion proportion correct (usually halfway between chance and perfect).

For "yes-no" tasks that include no blanks, one can plot proportion of "yes" responses at each concentration and calculate the concentration that

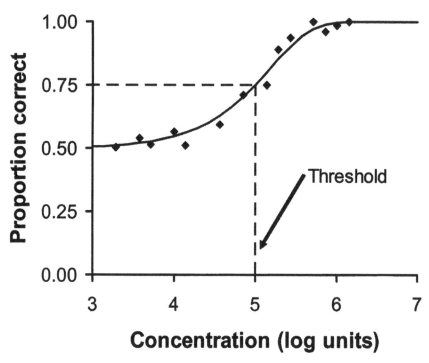

Figure 7 A hypothetical psychometric function for a two-alternative forced-choice (2-AFC) task. Experimenters typically define threshold as halfway between chance and perfect performance (for 2-AFC, 75% correct). Since functions rarely include a point that has a y value of exactly 0.75, experimenters often interpolate by fitting S-shaped functions. Others have described various functions used and procedures for fitting them to empirical date (e.g., [131–133]). The solid curve in the figure represents a cumulative Gaussian, which has the longest history [134]. Using the fitted function, we find that threshold equals 5 log units.

results in a criterion level of detection (usually the probability of a "yes" response = 0.50, since detection functions range from 0 to 1). Thresholds measured this way include effects of response bias. If a "yes-no" design includes blanks ("catch trials"), signal detection theory [37,38] provides statistical machinery to derive a measure of sensitivity unaffected by bias (d'; see Fig. 8). One can then plot d' versus concentration and calculate a threshold (a fixed level of performance, perhaps $d' = 0.95$, which corresponds to 75% correct in a 2-AFC task).

Method of constant stimuli has an important advantage: it yields an entire psychometric function. The shape and slope of the function can

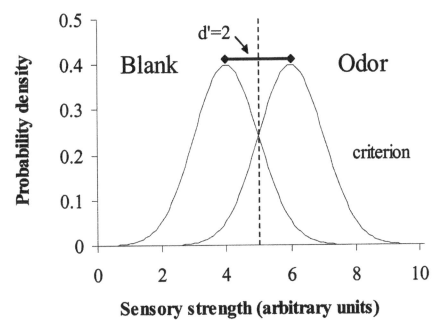

Figure 8 The decision space for signal detection theory, which has received support both in general [33,34,41] and in the detection of odors [55,135]. According to one common version of the theory, both blank and odorant give rise to Gaussian distributions of sensory strength with equal variance. The subject responds *yes* if an observed value of sensory strength exceeds some criterion (dashed vertical line), and *no* otherwise. The area under each distribution to the right of criterion corresponds to the probability the observer will respond *yes* to a given stimulus: correct responses (*hits*) for odorants and incorrect responses (*false alarms*) for blanks. Empirical estimates of these probabilities allow one to calculate the distance between the means of the two distributions in units of their common standard deviation, termed d'; d' (which equals 2 in this case) remains constant as criterion changes (see Ref. [41] for an excellent overview).

have theoretical interest [39]. The function also can have practical interest, since it allows researchers to calculate the likelihood that a person or group will report an odor at any concentration. Such information might, for example, allow a manufacturer to do a complete cost-risk analysis by computing the cost to scrub factory emissions to reach a range of concentrations.

    Method of constant stimuli has an important drawback as well: it requires a great deal of time. First, the method of constant stimuli may require extensive pilot work, since experimenters must fix stimulus parameters in advance. Second, because subjects receive all stimuli an equal number of times, runs include a large number of trials. The large number of trials proves especially difficult in olfaction, in which relatively rapid adaptation and slow recovery impose long intervals between trials. For this reason, few olfactory laboratories collect complete psychometric functions. Regardless, this protocol can be used to generate detection, recognition, or localization thresholds.

    b.   *Adaptive Methods.* Adaptive methods attempt to locate the region close to threshold and present a number of trials in this region (see Refs. [40,41] for reviews). Of the many techniques available (e.g., [42–46]), only two have seen common use in olfactory experiments [47], viz., some variant of a staircase method and the ascending method of limits (using forced choice).

    Olfactory researchers (e.g., [48–50]) often use some variant of the Wetherill and Levitt staircase method [51]; cf. [52], usually with a 2-AFC or 3-AFC task. Testing begins at the best available estimate of a subject's threshold, i.e., the mean of previous measurements, and performance on recent trials determines concentration after that. One common rule (others are possible) follows: increase concentration by one step (typically two- or three-fold) after an incorrect response; present the same concentration again after a single correct response; and decrease concentration after two successive correct responses (one-up, two-down rule). Concentrations above threshold lead to more correct responses (the result is a decrease in concentration), whereas concentrations below threshold lead to more incorrect responses (and result in increasing concentrations). Eventually, for 2-AFC, the method begins to present stimuli close to the concentration that would result in correct responses 71% of the time ("tracks" the 71% correct point of the psychometric function; other rules track different points). If concentration changes in a direction opposite to the previous change (increases after a decrease or a decreases after an increase), the event is called a *reversal*. A run consists of a predetermined number of reversals (often five or six). The experimenter notes the concentration at which reversals occur and averages them (often disregarding the first reversal) to estimate threshold. The number of reversals can sometimes influence the threshold [53]. In modified staircase procedures, other subtle criteria also may be imposed (e.g., [49]). Here, too, this protocol can be used for detection, recognition, or localization thresholds in either two- or three-alternative, forced-choice tasks (see Fig. 9 for idealized results).

    Olfactory researchers also favor ascending method of limits (e.g., [54–59]). Most employ a 2-AFC or 3-AFC task [60]. Testing begins with a very low concentration. After an error, the experimenter presents the next higher concentration. After a correct response, the experimenter presents the same

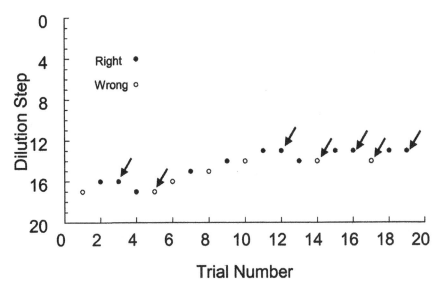

Figure 9 Idealized data from a modified staircase procedure. Note the stimulus movement (increasing or decreasing), depending upon the response of the subject. Arrows indicate reversal points. Typically, the first reversal is ignored and the remaining reversals are averaged to form the threshold.

concentration again [American Society for Testing and Materials (ASTM) standard E 679-91 [60] is an exception; correct responses are not verified before an increase in concentration is mandated]. Testing stops when the subject achieves a criterion number of successive correct responses (typically five in a row) at a given concentration. That concentration serves as a threshold estimate. ASTM E 679-91 recommends terminating "with two or more plusses," but generating sufficient trials in each subject to estimate the threshold: i.e., "a panelist continues to take the test until there is no doubt by that person of the correctness of the choice" [60, p. 3]; in Fig. 10 we have been conservative and extended the number of correct in a series to five; the total number of series to estimate the threshold may therefore be lower than a criterion of two "pluses"). Unlike most staircase procedures, ascending method of limits tracks 100% correct. Thus, this approach would tend to produce slightly higher thresholds than most staircase procedures [47].

B. Intensity and Hedonics

The task is straightforward, but the approach is quite variable. Scales can be used to obtain estimates of odor (or irritation), strength (intensity), and

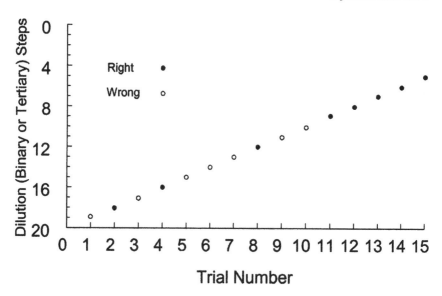

Figure 10 Idealized data depicting threshold determination using American Society for Testing and Materials (ASTM) procedure E 679-91. The ASTM protocol stipulates that testing cease when two or more pluses have been recorded. The method depicted in this figure required five correct answers before cessation of testing.

pleasantness/unpleasantness (hedonic value). Category scales (Likert-type or similar) are common (Figs. 11 and 12 provide some examples; an example that has been in use for many years in our laboratory, especially to generate estimates of odor hedonics, is included in Fig. 13). Such scales typically assure only ordinal (rank-order) data. For this reason, nonparametric statistics provide the most orthodox treatment of data from category scales (although, in most situations, parametric statistics offer more power and prove remarkably robust against false-positive results).

Generating ratio-level (linear vector with a defined zero-point) data, which can be safely analyzed with parametric statistics, is an enviable goal. Magnitude estimation can provide such data. In this procedure, subjects are trained to use numbers to estimate the quantity of the stimulus. In one common technique (modulus-free magnitude estimation), subjects assign a number (any that seems reasonable to them) to the first stimulus and assign proportional numbers to subsequent stimuli (i.e., a stimulus that smells half as strong receives a number half as large; a stimulus that smells twice as strong receives a number twice as large; zero may represent no sensation). Subjects should be cautioned that choosing a very low number for a strong

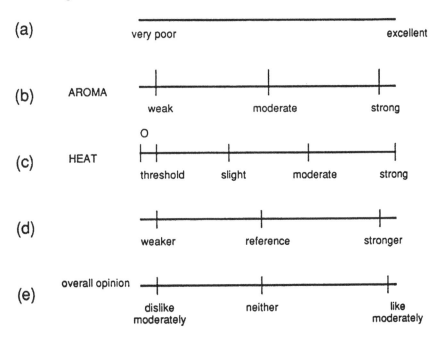

Figure 11 Examples of categorical scales that can be used to generate ordinal data. Subjects would either circle or check the appropriate response. (From Ref. [1].)

stimulus could impose limits on the use of the scale. Some investigators allow subjects to sample the range of stimulus strengths before collecting data to minimize this likelihood. Care should be taken to ensure that context does not bias the results: weak and strong samples should be equally represented in the series of trials. Once data are collected, individual values are normalized for analysis [61].

Not all subjects are comfortable with the use of numbers, and often a subject encounters difficulties using magnitude estimation. If data are collected from such a subject, they often must be must be discarded once it becomes clear that the subject cannot reliably generate magnitude estimates. Magnitude estimation also is subject to context effects: a moderately intense stimulus is embedded among many weak stimuli, it will be rated as more intense than it would have been had the same stimulus been embedded among a set of stimuli having similar intensities. Furthermore, if the same stimulus is embedded within a set of stimuli that are more intense, its rating of intensity will be lowered (e.g., see Chapter 8 in this volume and Ref. [63]).

Green and coworkers [64], noting the difficulties inherent in magnitude estimation, proposed a different metric for use by subjects who participate

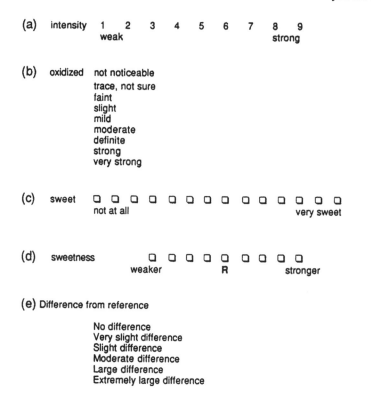

(a) intensity 1 2 3 4 5 6 7 8 9
                  weak   strong

(b) oxidized not noticeable
              trace, not sure
              faint
              slight
              mild
              moderate
              definite
              strong
              very strong

(c) sweet ▢ ▢ ▢ ▢ ▢ ▢ ▢ ▢ ▢ ▢ ▢ ▢ ▢ ▢ ▢
           not at all   very sweet

(d) sweetness   ▢ ▢ ▢ ▢ ▢ ▢ ▢ ▢ ▢
                  weaker   R   stronger

(e) Difference from reference

              No difference
              Very slight difference
              Slight difference
              Moderate difference
              Large difference
              Extremely large difference

(f) Hedonic Scale for children

Super Good   Really Good   Good   Maybe Good or Maybe Bad   Bad   Really Bad   Super Bad

Figure 12 Examples of linear scales that can be used to generate ordinal data. Although linear, scales of this type typically generate data that are compressed; individuals tend not to use the extremes. (From Ref. [1].)

in experiments in which psychophysical estimates of stimulus intensity are to be obtained. They proposed the labeled magnitude scale (LMS), which is a verbal scale based upon personal sensory experiences (see Fig. 14). Initially, the scale was not developed for use in olfaction; however, subsequent work validated its use with odors [65]. Unfortunately, as does magnitude

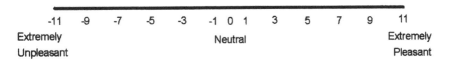

Figure 13 Example of a scale that can be used to generate hedonic ratings, e.g., perceptual unpleasantness/pleasantness of an odor or irritant.

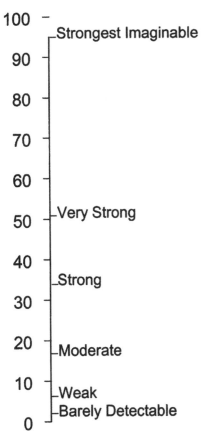

Figure 14 Labeled magnitude scale (LMS); adapted from Ref. [11] used to quantify perception of stimulus intensity. Numerical values on the left and their associated hash marks are not visible to the subjects. They are included herein for reference. With proper instructions, the LMS produces data linearly proportional to data from magnitude estimation [63,64]. Since magnitude estimation provides ratio-level data [140,141]; cf. [142], agreement with LMS implies that the LMS also provides ratio-level data.

estimation (upon which the LMS is based), the LMS suffers from context effects [63].

C. Quality

Scholars have attempted to measure odor quality in a variety of ways [66]. Others have provided thorough reviews (e.g., [67–69]). We provide a brief overview of four major techniques: categorization, profiling, ratings of similarity, and discrimination.

1. Categorization

Various authors have attempted to group odors according to qualitative resemblance (e.g., [7,70–72]). Some hoped categories would represent "primary" odors (analogous to the red, green, and blue primaries of additive color mixing) and searched for molecular common denominators within categories (which might appeal to corresponding receptors). That categorizers failed to identify such primaries comes as no surprise in light of current knowledge (mammals express hundreds of receptor proteins, all broadly tuned [73]).

   Categorization could potentially specify quality relative to other odorants (odorants in the same category should smell more similar than odorants in different categories) or independently (according to category membership). Unfortunately, different authors categorize the same odorants quite differently. Since categorization, usually based at least in part on an author's own nose, qualifies as subjective [69], it was never clear who, if anyone, had created an accurate scheme. Without a valid and universally accepted scheme, the value of categorization as measurement is limited.

2. Odor Profiling

In one style of odor profiling, subjects evaluate odors against a list of verbal descriptors [74–80]. For example, in the scheme of Dravnieks [81], subjects receive a list of 146 descriptors, e.g., cardboardlike, fishy, leatherlike, sauerkraut-like. For a given test odorant, subjects rate, on a scale from 0 to 5, the applicability of all descriptors. Ratings of applicability form a quality "profile" of the test odorant (Fig. 15). Subjects differ greatly both in the kind and in the number of descriptors they apply to a given odor (e.g., [78,79,81]. However, profiles achieve considerable stability when researchers (a) average across subjects and purge aggregate profiles of infrequently used descriptors [78,79,81] or (b) train sensory panels in use of descriptors [82].

   In another style of profiling, subjects evaluate odors against reference chemicals [83–86]. Use of chemical references may circumvent the problem

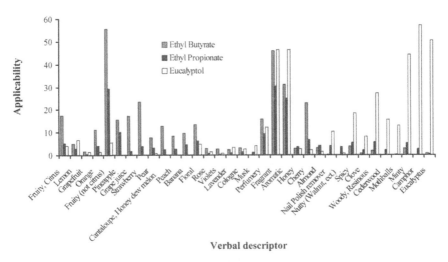

Figure 15 Partial odor profiles obtained using Dravnieks's [81] system of descriptors. Descriptors (some edited for brevity) appear along the x axis. The y axis represents applicability of descriptors to three odors. Gray bars represent a profile for ethyl butyrate ("tutti-fruity"). Black bars represent a profile for ethyl propionate, another, homologous ester. Profiles for the two similar-smelling esters both include many fruit-related descriptors (particularly "Fruity, not Citrus"). They also have "Fragrant" and "Aromatic" in common. White bars represent a profile for eucalyptol, an odorant that smells quite different from the two esters. As the profiles of the esters do the profile for eucalyptol contains the descriptors "Fragrant" and "Aromatic." However, the profile differs from those of the esters in the lack of fruit-related descriptors and the addition of "Minty," "Camphor," and "Eucalyptus."

of lack of agreement regarding meanings of verbal descriptors. At any rate, subjects using such schemes can show impressive agreement (e.g., [83,87–89]; cf. [85]).

With both styles, the obtained profile represents an operational definition of odor quality, the validity and utility of which depend on the choice of descriptors/references. No one knows how many descriptors a truly general system would include, but even Dravnieks's 146 fail to distinguish among some qualitatively distinct stimuli [80,90]. In practice, investigators must often tailor systems to the exact specifications of the substances they evaluate. Specialized systems exist for diverse products, including wine, Scotch whisky, and drinking water [91–93].

Profiling has produced few data of use to scientists engaged in basic research on the processing of odor quality [69]. However, in more applied settings, the reliability of average profiles allows industry researchers to describe the quality of odors with a standard vocabulary [82,88,94]. Further, descriptors that represent specific ingredients, products of chemical reactions, and so on, can provide valuable information to those working in quality control or product development. In general, if one needs to know exactly *how* two odors differ, some form of profiling scheme probably offers the best solution. If one wants to know *how much* two odors differ, matters are less clear, since estimates of similarity will depend not only on the descriptors one chooses, but also on the method by which one compares profiles [95].

   a.   Ratings of Similarity/Dissimilarity.   Various experimenters have taken a direct approach to measurement of qualitative similarity by asking subjects to estimate similarity or dissimilarity between pairs of odors on one or another scale (e.g., [75,96–105]). The technique may have intuitive appeal but has not proved useful for most purposes.

   Multidimensional analyses of ratings of similarity typically yield a dominant hedonic dimension: pleasant odors cluster on one side of an "odor space" and unpleasant odors cluster on the other ([75,97,99,102–104,106]; cf., [96]). Analyses often yield an intensity dimension as well, and occasionally one or two others that are unimportant in statistical terms and defy clear interpretation.

   Some authors [69] suggest that, faced with a difficult and unfamiliar task, subjects resort to judgments of pleasantness as a way to convey some useful information. Beyond this, rating the similarity between two odors, which could vary along many perceptual dimensions, might be analogous to rating the similarity between a tennis ball and a basketball. Some subjects find them similar on the basis of shape. Others find them dissimilar on the basis of size. Hedonic valence might survive in average data because it alone plays an important role in the judgments of all subjects. At any rate, since (a) pleasantness can change with pairing to an aversive stimulus, context, state of hunger versus satiety, or even mere exposure [67,102,107,108], and (b) qualitatively diverse odors can smell equally pleasant [107], a measure that primarily reflects relative pleasantness might provide an impoverished or even distorted representation of qualitative similarity.

   b.   Discrimination.   Discrimination, a second approach to measurement of qualitative similarity, does not require subjects to decide how similar odorants smell, thereby requiring subjects to choose for themselves appropriate criteria by which to judge similarity, but only to decide whether or not odorants differ [69,93]. In perhaps the most common paradigm, subjects receive three odors. Two serve as "standards." Subjects must decide which standard matches a third odor [109–113]. Other methods can also serve (e.g., [114–116]). For all methods, a greater number of errors indicate greater similarity. For most techniques, models exist to convert proportion correct to a linear metric of similarity [35].

Unlike categorization, profiling, and ratings of similarity, discrimination relies on tests of performance rather than reports of mental content. In this sense, discrimination qualifies as a more objective technique [69]. Unfortunately, discrimination has two important drawbacks. First, since many trials are required to obtain stable estimates of performance, discrimination proves time-consuming. Automated testing might help alleviate this problem. Second, once odors differ to a certain degree, performance reaches an asymptote beyond which discrimination fails to resolve greater differences. For example, many people might achieve identical (near-perfect) performance in discriminations of strawberry versus lemon and in discriminations of strawberry vs. rotten fish, even though strawberry presumably resembles lemon more than it does rotten fish. Researchers might extend the range of differences that discrimination resolves by using new dependent variables. For example, response time tends to decrease as discrimination becomes easier, even past the point of asymptotic performance (see [48]). Researchers might also make discriminations more difficult by manipulating the stimulus [113].

For now, discrimination seems most appropriate in two situations. First, discrimination is particularly appropriate in determining whether two odorants differ at all, rather than how much they differ. The study of enantiomers, molecules of identical composition that are structural mirror opposites, qualifies as a good example (e.g., [109,111,117–120]). Second, discrimination is particularly appropriate in the study of small differences in quality. The study of the effects of small and systematic changes in molecular properties qualifies as a good example [121–124].

VI. CAVEATS

With any psychophysical approach, there are limitations. In the chemical senses, considerable variation is encountered because of individual differences.

A. Olfaction

1. Variation in Olfactory Sensitivity and Qualitative Assessments

Some variation in olfaction can be explained by other factors, e.g., genetic influence, effects of aging, modification by environmental factors; however, investigators can expect to encounter significant variability (cf. [125]). Data in Fig. 5B provide an excellent example.

For approximately 140 individuals (50% of each biological sex) ages 20 to 90 years, the range of detection thresholds for l-butanol spanned over 11 orders of magnitude. Importantly, there was very respectable test-retest reliability ($r = 0.93$; Fig. 5B).

As mentioned, at suprathreshold levels subjects differ greatly both in the kind and in the number of verbal descriptors they apply to a given odor [78,79,81]. Some differences probably stem from the way subjects use labels, but some may reflect true individual differences in perceived quality [126]. To determine exactly how the "odor spaces" of two subjects might differ, some test of performance (a task based on discrimination) might serve best [69].

2. Variation Due to Genetic Contributions

Specific anosmia is an inability of an otherwise healthy individual, with an intact sense of smell, to experience an odor when others readily detect such. Amoore and Steinle made many contributions to this area of investigation, and interested individuals are directed to their review [127]. Care should be taken when encountering such individuals in olfactory research: some specific anosmias may be dynamic—individuals can first appear as anosmic to a specific odorant, and that may even have an apparent genetic influence [128]— however, with repeated experience with the odorant the individual may become sensitized to the odor [129].

Complications arise if the subject is incapable of detecting the stimulus. In this situation care must be taken to recognize false positive findings. "Thresholds" can be generated when it is physically impossible to do so. Monte Carlo simulations of these procedures indicate a rather high probability of assigning a threshold to an individual when none should be. In the modified staircase procedure previously described, the probability of a false-positive result on any single series of trials, when the subject cannot smell the odorant, approximates 0.40.

3. Variation Due to the Effects of Aging

Investigators must pay special attention to age effects in olfaction. With advancing age there is, in general, a decline in olfactory function [49]. Indeed, much of the variation in Fig. 5B results from the contributions of age. When subjects are grouped by decade of age, this effect is readily visualized (Fig. 16).

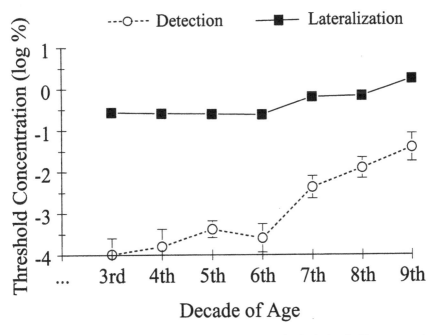

Figure 16 Average detection and localization (lateralization) thresholds, expressed as $\log_{10}$ of the percentage in solution, $\pm$ standard error of the mean (s.e.m.) (those for lateralization are buried within the marker and cannot be seen), across the age range of subjects, grouped by decade. (From Ref. [28].)

4. Variation Due to Environmental Exposures

Individuals exist within an environment, and its impact upon the chemical senses becomes obvious with simple testing. Exposure to an odorant results in adaptation to the odorant: i.e., individuals lose sensitivity to the odorant often to the point where they can no longer perceive its presence. This is a common, everyday experience. What many individuals fail to appreciate are the broader implications of such changes in olfaction. For example, exposure to one odorant may affect sensitivity to other odorants, whether the additional odorant smells similar to the first odorant or not. This is an example of cross-adaptation (see Fig. 17).

The effects of adaptation may be short-lived, as seen in the recovery phase in Fig. 17; however, this may not always be the case. Individuals who experience long-term exposures to odorants, e.g., in the workplace, may

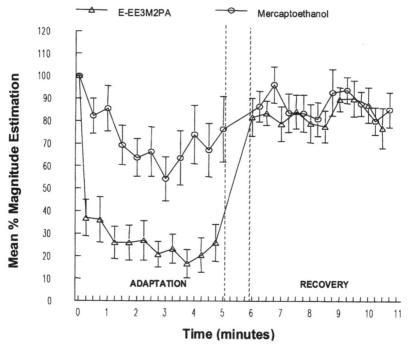

Figure 17 During 5 min subjects were exposed to the E-isomer of the ethyl ester of 3-methyl-2-pentenoic acid (E-EE3M2PA), which has a floral odor. Every 15 sec they were asked to rate the intensity of either E-EE3M2PA or mercaptoethanol (a sulfurous malodor) across 11 min (except for the 1 min between adaptation to E-EE3M2PA and recovery from adaptation). Relative to estimates of perceived intensity that were obtained before adaptation, the perceived intensity of E-EE3M2PA decreased during repeated exposures to the odorant (adaptation). There also was a significant decrease in the perceived intensity of mercaptoethanol during exposure to E-EE3M2PA (cross-adaptation).

develop a form of adaptation (and perhaps cross-adaptation) that is resistant to rapid recovery ([130]; also see Fig. 18).

B. Chemesthesis

1. Variation in Chemesthetic Sensitivity

Much less is known about individual variation in chemesthetic sensitivity. Using CO_2 as an intranasal chemesthetic stimulus, individual psychometric functions for sensitivity can be generated: two such functions are contained

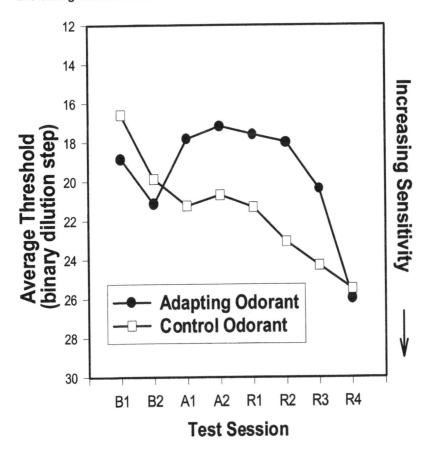

Figure 18 Subjects were tested for sensitivity to two different odorants in two weekly baseline tests (B1 and B2). Roughly half of the subjects were then exposed to one odorant, which was diffused in their living environment; the remainder were similarly exposed to the other. Adaptation exposures lasted for 2 weeks. During this time subjects were tested for sensitivity to the two odorants (at A1 and A2). Subjects showed elevated thresholds to the exposed odorant (solid circles), relative to unexposed odorant. After test A2, the odorant was removed from the living environment. Subjects returned for sensitivity tests at weekly intervals (R1–R4) during the recovery phase. Full recovery from long-term adaptation required 3–4 weeks [139].

within Fig. 19 (unpublished data). These examples demonstrate significant individual differences, which can be seen in a much larger sample of people.

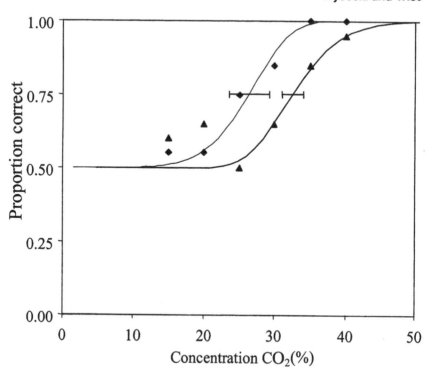

Figure 19 Psychometric functions for detecting the irritation from carbon dioxide (CO_2) in the nose obtained from two individuals. Horizontal bars at 0.75 indicate the 95% confidence interval at the localization threshold. These two individuals are significantly different from each other. Across many individuals, there are broad differences in sensitivity to CO_2.

This variation appears to be much less extreme than that observed in olfaction (compare Fig. 5a and 5b).

2. Variation Due to the Effects of Aging

As noted in olfaction, with advancing age there is a decline in chemesthetic sensitivity within the nose. These effects can be seen in Fig. 16.

3. Variation Due to Environmental Exposures

Much more striking than individual differences in sensitivity are the differences across individuals in their response to longer-duration puffs of CO_2. Some individuals exhibit rapid adaptation to the CO_2, whereas others appear to be resistant to adaptation (unpublished). The mechanisms underlying such differences remain to be elucidated.

Long-term exposure to a chemesthetic stimulus also affects sensitivity to the agent. Individuals who work with cellulose acetate and are exposed to acetone were studied. These workplace exposures shifted chemesthetic thresholds in the direction of decreased sensitivity, but the effects appeared to be specific to acetone; thresholds for l-butanol were unaffected [27].

VII. CONCLUSIONS

A. Summary

For both olfaction and intranasal chemesthesis, one can determine thresholds (detection for olfaction and localization for chemesthesis) in various fashions. For concentrations that are above the detection threshold, but below the localization threshold, individuals should experience an odor. If they find the odor *irritating*, one can safely assume that this percept is cognitive, rather than sensory. True sensory irritation would be expected at or above the localization threshold.

Intensity estimates for olfactory and chemesthetic stimuli also can be obtained. The ability to quantify sensory strength provides data that can be used to assess overall strength of the odor or sensory irritant.

Individuals also can provide hedonic evaluations of odorant and chemesthetic stimuli. Typically without difficulty, one can recall experiencing odors that are quite pleasant and, at the other end of the continuum, odors that are quite offensive. Importantly, during any odorant/irritant exposure, these psychological percepts can be quantified. Typically, data collected from the psychological realm are closely linked to physical measures; e.g., they are concentration dependent. Pleasant smelling odors, however, can become offensive with increasing concentration. Likewise, some odorants that are offensive at moderate concentrations may be neutral or even have a positive hedonic valence at concentrations that just exceed detection threshold.

Irritant is often used to describe odorants that also stimulate the trigeminal nerve in the nose; however, this word has negative connotations. Indeed, some people may reach unwarranted conclusions. For example, they may believe that they have been exposed to a chemical toxin (perhaps better called a *cognitoxin*) when they have not. There is a flip side to this argument: many times individuals enjoy irritation that accompanies stimulation by chemical stimuli, e.g., in the nose, menthol or mint (cool irritation), acetic acid (vinegar; sharp or tingling), or wasabi (the kick in the nose); and in the mouth, CO_2 in carbonated drinks or capsaicin and zingerone from hot peppers and ginger, respectively.

Odor quality has its origins in the nose, and specific anosmias can make an overall contribution. For complex mixtures, an individual who lacks the ability to perceive an important note may not appreciate what others describe as a superb wine. Alternatively, if an individual cannot detect some malodorants, what is offensive to most others may not even register as disagreeable. To this latter point, some individuals appear not to detect the full mixture in the spray of a skunk. Indeed, some people, including one of the authors of this chapter, find the odor pleasant.

Beyond the nose, considerable cognitive "baggage" can affect how one perceives odor quality and hedonics. For example, if one believes that the smell from butyric acid is of personal origin, it is easy to conclude that the odor arose from *sweaty socks*; however, if told that the odor is from a food source, one could easily conclude that Parmesan cheese was being smelled.

Olfactory and chemesthetic acceptance and pleasantness undoubtedly shift with experience. Indeed, with repeated exposures young children can overcome novelty associated with new foods and will begin to ingest much more than their counterparts who have not had such exposures [136]. Exposure-induced shifts also have been reported for strictly olfactory stimuli [107,137].

B. Applications

For both olfaction and chemesthesis, sensitivity, and any differences in sensitivity, where they exist, can be determined in populations of interest. Furthermore, by establishing olfactory detection and chemesthetic localization thresholds, one can then know when an odorant becomes an irritant. These quantitative measures then allow discrimination between cognitive and sensory irritation.

In some instances or in specific target populations, some chemosensory irritation may be positive. Unfortunately, little is known about how chemesthetic and olfactory percepts interact [138] to form the overall sensory impression. Armed with psychophysical protocols, however, investigators can learn how various stimuli interact and can grow to appreciate the influence of myriad factors that contribute to product development.

ACKNOWLEDGMENTS

Funding to support this chapter was provided in part by the NIH, P50 DC 00214 and T32 DC 00014, and by the Life-Oriented Software Laboratory, Muroran Institute of Technology, Muroran, Japan. While preparing this contribution, CJW lived through one earthquake and one very powerful typhoon in housing provided, in part, by Muroran Institute of Technology.

REFERENCES

1. HT Lawless, H Heymann. Sensory Evaluation of Food, Principles and Practices. New York: Chapman & Hall, 1998, p 17.
2. G Fechner. Elemete der Psychophysik. Leipzig: Breitkopf & Harterl, 1860.
3. J Armington, J Krauskopf, eds. Visual Psychophysics and Physiology: A Volume Dedicated to Lorrin Riggs. New York: Academic Press, 1978.
4. WA Yost, AN Popper, RR Fay, eds. Human Psychophysics. New York: Springer-Verlag, 1993.
5. L Kruger, ed. Pain and Touch. San Diego: Academic Press, 1996.
6. BJ Noble, RJ Robertson. Perceived Exertion. Champaign, IL: Human Kinetics, 1996.
7. H Zwaardemaker. Die Physiologie des Geruchs. Leipzig: Wilhelm Engelmann, 1895.
8. GK Beauchamp, LM Bartoshuk, eds. Tasting and Smelling: Handbook of Perception and Cognition, 2nd ed. San Diego: Academic Press, 1997.
9. NE Rawson, G Gomez. Cell and molecular biology of human olfaction. Microsc Res Tech 58: 142–151, 2002.
10. T Hummel, A Livermore. Intranasal chemosensory function of the trigeminal nerve and aspects of its relation to olfaction. Int Arch Occup Environ Health 75: 305–313, 2002.
11. BG Green. Measurement of sensory irritation of the skin. Am J Contact Dermat 11: 170–80, 2000.
12. B Bryant, WL Silver. Chemesthesis: The common chemical sense. In: TE Finger, WL Silver, D Restrepo, eds. Neurobiology of Taste and Smell, 2nd ed. New York: Wiley-Liss, 2000, pp 73–100.
13. MD Hayward, JL Morgan. The olfactory system as a model for the analysis of the contribution of gene expression to programmed cell death. Chem Senses 20: 261–269, 1995.
14. P Mombaerts. The human repertoire of odorant receptor genes and pseudogenes. Annu Rev Genomics Hum Genet 2: 493–510, 2001.
15. NE Rawson, J Eberwine, R Dotson, J Jackson, P Ulrich, D Restrepo. Expression of mRNAs encoding for two different olfactory receptors in a subset of olfactory receptor neurons. J Neurochem 75: 185–195, 2000.
16. L Belluscio, C Lodovichi, P Feinstein, P Mombaerts, LC Katz. Odorant receptors instruct functional circuitry in the mouse olfactory bulb. Nature 419: 296–300, 2002.
17. SM Potter, C Zheng, DS Koos, P Feinstein, SE Fraser, P Mombaerts. Structure and emergence of specific olfactory glomeruli in the mouse. J Neurosci 21: 9713–9723, 2001.

18. RM Costanzo. Rewiring the olfactory bulb: Changes in odor maps following recovery from nerve transection. Chem Senses 25: 199–205, 2000.
19. MT Shipley, M Ennis. Functional organization of olfactory system. J Neurobiol 30: 123–176, 1996.
20. S Jacob, LH Kinnunen, J Metz, M Cooper, MK McClintock. Sustained human chemosignal unconsciously alters brain function. Neuroreport 12: 2391–2394, 2001.
21. CJ Wysocki. Neurobehavioral evidence for the involvement of the vomeronasal system in mammalian reproduction. Neurosci Biobehav Rev 3:301–341, 1979.
22. JH Martin. Receptor physiology and submodality coding in the somatic sensory system. In: ER Kandel, JH Schwartz, eds. Principles of Neural Science, 2nd ed. New York: Elsevier, 1985, pp 301–315.
23. EW Boddeke. Involvement of chemokines in pain. Eur J Pharmacol 429: 115–119, 2001.
24. R Eccles. Pathophysiology of nasal symptoms. Am J Rhinol 14: 335–338, 2000.
25. MJ Caterina, D Julius. The vanilloid receptor: A molecular gateway to the pain pathway. Annu Rev Neurosci 24: 487–517, 2001.
26. DD McKemy, WM Neuhausser, D Julius. Identification of a cold receptor reveals a general role for TRP channels in thermosensation. Nature 416: 52–58, 2002.
27. CJ Wysocki, P Dalton, MJ Brody, HJ Lawley. Acetone odor and irritation thresholds obtained from acetone-exposed factory workers and from control (occupationally unexposed) subjects. AIHA J 58: 704–712, 1997.
28. CJ Wysocki, BJ Cowart, T Radil. Nasal-trigeminal chemosensitivity across the adult life-span. Percept Psychophys 65: 115–122, 2003.
29. JE Amoore. Specific anosmia and the concept of primary odors. Chem Senses Flav 2: 267–281, 1977.
30. RL Doty, P Shaman, SL Applebaum, R Giberson, L Sikorsky, L Rosenberg. Smell identification ability: Changes with age. Science 226: 1441–1443, 1984.
31. M Devos, F Patte, J Rouault, P Laffort, LJ Van Germert. Standardized Human Olfactory Thresholds. Oxford, England: IRL Press, 1990.
32. HR Blackwell. Psychophysical thresholds: Experimental studies of methods of measurement. University of Michigan Engineering Research Institute Bulletin, 36. Ann Arbor: University of Michigan Press, 1953.
33. JA Swets. Form of empirical ROC's in discrimination and diagnostic tasks: Implications for theory and measurement of performance. Psychol Bull 99: 181–198, 1986.
34. NA MacMillan, DC Creelman. Response bias: Characteristics of detection-theory, threshold-theory, and "nonparametric" indexes. Psychol Bull 107: 401–413, 1990.
35. DM Green, TG Birdsall. Detection and recognition. Psych Rev 85: 192–206, 1978.

36. FM Urban. The application of statistical methods to problems of psychophysics. Philadelphia: Psychological Clinic Press, 1908.

37. WP Tanner, JA Swets. A decision-making theory of visual detection. Psychol Rev 61: 401–409, 1954.

38. JA Swets, WP Tanner, TG Birdsall. Decision processes in perception. Psychol Rev 68: 301–340, 1961.

39. DM Green, RD Luce. Parallel psychometric functions from a set of independent detectors. Psychol Rev 82: 483–486, 1975.

40. LO Harvey, Jr. Efficient estimation of sensory thresholds with ML-PEST. Spatial Vis 11: 121–128, 1997.

41. NA MacMillan, DC Creelman. Detection theory: A user's guide. Cambridge: Cambridge University Press, 1991.

42. A Wald. Sequential Analysis. New York: Wiley, 1947.

43. MM Taylor, CD Creelman. PEST: Efficient estimates on probability functions. J Acoust Soc Am 41: 782–787, 1967.

44. JL Hall. Hybrid adaptive procedure for estimation of psychometric functions. J Acoust Soc Am 69: 1763–1769, 1981.

45. CS Watson, DG Pelli. Quest: A Bayesian adaptive psychometric method. Percept Psychophys 33: 113–120, 1983.

46. R Madigan, D. Williams. Maximum likelihood procedures in two-alternative forced-choice: Evaluation and recommendations. Percept Psychophys 42: 240–249, 1987.

47. MR Linschoten, LO Harvey, Jr, PM Eller, BW Jafek. Fast and accurate measurement of taste and smell thresholds using a maximum-likelihood adaptive staircase procedure. Percept Psychophys 63: 1330–1347, 2001.

48. PM Wise, WS Cain. Latency and accuracy of discriminations of odor quality between binary mixtures and their components. Chem Senses 25: 247–265, 2000.

49. E Pribitkin, MD Rosenthal, BJ Cowart. Prevalence and etiology of severe taste loss in a chemosensory clinic population. Eastern Section Meeting of the Triological Society, Philadelphia, January 26, 2002.

50. JF Gent, WS Cain, LM Bartoshuk. Taste and smell measurement in a clinical setting. In: HL Meiselman, RS Rivlin, eds. Clinical Measurement of Taste and Smell. New York: Macmillan, 1986, pp 107–116.

51. GB Wetherill, H Levitt. Sequential estimation of points on a psychometric function. Br J Math Psychol 18: 1–10, 1965.

52. G Kobal, K Palisch, SR Wolf, ED Meyer, KB Huttenbrink, S Roscher, R Wagner, T Hummel. A threshold-like measure for the assessment of olfactory sensitivity: The "random" procedure. Eur Arch Otorhinolaryngol 258: 168–172, 2001.

53. RL Doty, DA McKeown, WW Lee, P Shaman. A study of the test-retest reliability of ten olfactory tests. Chem Senses 20: 645–656, 1995.

54. WS Cain, JF Gent, FA Catalanotto, RB Goodspeed. Clinical evaluation of olfaction. Am J Otolaryngol 4: 252–256, 1983.

55.   RL Doty, P Snyder, G Huggins, LD Lowry. Endocrine, cardiovascular, and psychological correlates of olfactory sensitivity during the human menstrual cycle. J Comp Physiol Psychol 95: 45–60, 1981.

56.   P Gagnon, D Mergler, S Lapare. Olfactory adaptation, threshold shift and recovery at low levels of exposure to methyl isobutyl ketone (MIBK). Neurotoxicology 15: 637–642, 1994.

57.   JE Commetto-Muñiz, WS Cain. Relative sensitivity of the ocular trigeminal, nasal trigeminal, and olfactory systems to airborne chemicals. Chem Senses 20: 191–198, 1995.

58.   S Nordin, C Murphy, TM Davidson, C Quinonez, AA Jalowayski, DW Ellison. Prevalence and assessment of qualitative olfactory dysfunction in different age groups. Laryngoscope 106: 739–744, 1996.

59.   B. Moll, L Klimek, G Eggers, W Mann. Comparison of olfactory function in patients with seasonal and perennial allergic rhinitis. Allergy 53: 297–301, 1998.

60.   ASTM. E 679-91: Standard practice for determination of odor and taste thresholds by a forced-choice ascending concentration series method of limits. West Conshohocken, PA: ASTM, 1991 (reapproved 1997).

61.   JD Pierce, Jr, CJ Wysocki, EV Aronov. Mutual cross-adaptation of the volatile steroid androstenone and a non-steroid functional analog. Chem Senses 18: 245–256, 1993.

62.   S Nordin. Context effects, reliability, and internal consistency of intermodal joint scaling. Percept Psychophys 55: 180–189, 1994.

63.   HT Lawless, J Horne, W Spiers. Contrast and range effects for category, magnitude and labeled magnitude scales in judgements of sweetness intensity. Chem Senses 25: 85–92, 2000.

64.   BG Green, GS Shaffer, MM Gilmore. Derivation and evaluation of a semantic scale of oral sensation magnitude with apparent ratio properties. Chem Senses 18: 683–702, 1993.

65.   BG Green, P Dalton, BJ Cowart, G Shaffer, K Rankin, J Higgins. Evaluating the "labeled magnitude scale" for measuring sensations of taste and smell. Chem Senses 21: 323–334, 1996.

66.   PM Wise. Latency and accuracy of same-different discriminations of odor quality between binary mixtures and their components. PhD dissertation, University of California at San Diego, 2000.

67.   WS Cain. Olfaction. In: RC Atkinson, RJ Hernstein, G Lindzey, RD Luce, eds. Stevens' Handbook of Experimental Psychology: Perception and Motivation. Vol 1. New York: Wiley, 1988, pp 409–459.

68.   KJ Rossiter. Structure-odor relationships. Chem Rev 96: 3201–3240, 1996.

69.   PM Wise, MJ Olsson, WS Cain. Quantification of odor quality. Chem Senses 25: 429–443, 2000.

70.   H Henning. Der Geruch. Leipzig: Johann Ambrosius Barth, 1916.

71.   JE Amoore. The stereochemical theory of olfaction. 1. Identification of the seven primary odours. Proc Sci Seet Toilet Goods Assoc 37: 1–12, 1962.

72. JE Amoore. The stereochemical theory of olfaction. 2. Elucidation of the stereochemical properties of the olfactory receptor sites. Proc Sci Sect Toilet Goods Assoc 37: 13–23, 1962.

73. K Mori, H Nagao, Y Yoshihara. The olfactory bulb: Coding and processing of odor molecule information. Science 286: 711–715, 1999.

74. FJ Pilgrim, HG Schutz. Measurement of quantitative and qualitative attributes of flavor. National Academy of Sciences-National Research Council Symposium: Chemistry of Food Flavors. Washington, DC: National Academy Press, 1957, pp 47–58.

75. M Yoshida. Studies in psychometric classification of odors (5). Jpn Psychol Res 6: 145–154, 1964.

76. HR Moskowitz, CD Barbe. Profiling of odors components and their mixtures. Sens Processes 1: 212–226, 1977.

77. JM Coxon, AM Gregson, RG Paddick. Multidimensional scaling of perceived odour of bicyclo [2.2.1] heptane, 1,7,7-trimethylbicyclo [2.2.1] heptane and cyclohexane derivitives. Chem Senses Flav 3: 431–441, 1978.

78. A Dravnieks, FC Bock, M Tibbets, M Ford. Comparison of odors directly and through odor profiling. Chem Senses Flav 3: 191–220, 1978.

79. A Dravnieks. Odor quality: Semantically generated multidimensional profiles are stable. Science 218: 799–801, 1982.

80. DG Laing, ME Willcox. Perception of components in binary odour mixtures. Chem Senses 7: 249–264, 1983.

81. A. Dravnieks. Atlas of Odor Character Profiles. Vol. 61. Philadelphia: American Society for Testing and Materials, 1985.

82. H Stone, JL Sidel. Sensory Evaluation Practices, 2nd. ed. Orlando, FL: Academic Press, 1993.

83. HG Schutz. A matching-standard method for characterizing odor qualities. Ann N Y Acad Sci 116: 517–526, 1964.

84. JE Amoore, D Venstrom. Correlations between stereochemical assesments and organoleptic analysis of odorous compounds. In: T Hayashi, ed. Olfaction and Taste II. Oxford: Pergamon, 1967, pp 3–17.

85. M Yoshida. Psychometric classification of odors. Chem Senses Flav 1: 443–464, 1975.

86. E Polak, D Trotier, E Baliguet. Odor similarities in structurally related odorants. Chem Senses Flav 3: 369–380, 1978.

87. EG Boring. A new system for the classification of odors. Am J Psychol 40: 345–349, 1928.

88. ED Crocker. Comprehensive method for the classification of odors. Proc Sci Sect Toilet Goods Assoc 6: 1–3, 1946.

89. RH Wright, KM Michels. Evaluation of infrared relations to odor by a standards similarity method. Ann NY Acad Sci 116: 535–551, 1964.

90. MA Jeltma, EW Southwick. Evaluation and applications of odor profiling. J Sens Stud 1: 123–136, 1986.

91. AC Noble, J Buechsenstein, EJ Leach, JO Schmidt, PM Stern. Modification of a standardized system of wine aroma terminology. Am J Enol Viticult 38: 143–146, 1987.

92. JS Swan, SM Burtles. Quality control of flavour by the use of integrated sensory analytical methods at various stages of Scotch whisky production. In: H van der Starre, ed. Olfaction and Taste VII. London: IRL Press, 1980, pp 451–452.

93. RL Doty. Psychophysical measurement of odor perception in humans. In: DG Laing, RL Doty, W Breipohl, eds. The Human Sense of Smell. Berlin: Springer-Verlag, 1991, pp 95–151.

94. GV Civille, B Lyon. ASTM Lexicon Vocabulary for Descriptive Analysis. Philadelphia: American Society for Testing and Materials, 1996.

95. P Callegari, J Rouault, P Laffort. Odor quality: From descriptor profiles to similarities. Chem Senses 22: 1–8, 1997.

96. T Engen. Psychophysical scaling of intensity and quality. Ann NY Acad Sci 116: 504–516, 1964.

97. M Yoshida. Studies in psychometric classification of odors (4). Jpn Psychol Res 6: 115–124, 1964.

98. KB Doving, AL Lange. Comparative studies of sensory relatedness of odors. Scand J Psychol 8: 47–51, 1967.

99. MH Woskow. Multidimensional scaling of odors. In: N Tanyolaç, ed. Theories of Odor and Odor Measurement. Istanbul: Bebek, 1968, pp 147–188.

100. B Bergland, U Bergland, T Engen, G Ekman. Multidimensional analysis of twenty-one odors. Scand J Psychol 14: 131–137, 1973.

101. A Dravnieks. A building-block model for the characterization of odorant molecules and their odors. In: WS Cain, ed. Odors: Evaluation, Utilisation, and Control. Ann NY Acad Sci 237: 144–163, 1974.

102. HR Moskowitz, CL Gerbers. Dimensional salience of odors. In: WS Cain, ed. Odors: Evaluation, Utilisation, and Control. Ann N Y Acad Sci 237: 1–16, 1974.

103. SS Schiffman. Characterization of odor quality utilizing multidimensional scaling techniques. In: HR Moskowitz, CB Warren, eds. Odor Quality and Chemical Structure. Washington, DC: American Chemical Society, 1981, pp 1–19.

104. SS Schiffman. Mathematical approaches for quantitative design of odorants and tastants. In: C Warren, J Walradt, eds. Computers in Flavor and Fragrance Research. Washington, DC: American Chemical Society, 1984, pp 33–50.

105. D Kurtz, TL White, M Hayes. The labeled dissimilarity scale: A metric of perceptual dissimilarity. Percept Psychophys 62: 152–161, 2000.

106. M Yoshida. Studies in psychometric classification of odors (6). Jpn Psychol Res 14: 70–86, 1972.

107. WS Cain, F Johnson. Lability of odor pleasantness: Influence of mere exposure. Perception 7: 459–465, 1978.

108. IL Bernstein, MM Webster. Learned taste aversions in humans. Physiol Behav 25: 363–366, 1980.
109. FN Jones, D Elliot. Individual and substance differences in the discriminability of optical isomers. Chem Senses Flav 1: 317–321, 1975.
110. B Eskenazi, WS Cain, K Friend. Olfactory functioning in temporal lobectomy patients. Neuropsychologia 24: 553–562, 1986.
111. CA Hormann, BJ Cowart. Olfactory discrimination of carvone enantiomers. Chem Senses 18: 13–21, 1993.
112. RA de Wijk, WS Cain. Odor quality: Discrimination versus free and cued identification Percept Psychophys 56: 12–18, 1994.
113. MJ Olsson, WS Cain. Psychometrics of odor quality discrimination. Chem Senses 25: 493–499, 2000.
114. JER Frijters. Sensory difference testing and the measurement of sensory discriminability. In: JR Piggot, ed. Sensory Analysis of Foods, 2nd ed. London: Elsevier Applied Science, 1988, pp 117–140.
115. B Rousseau, A Meyer, M O'Mahony. Power and sensitivity of the same-different test: Comparison with triangle and duo-trio methods. J Sens Stud 13: 149–173, 1998.
116. J Bi, DM Ennis. Statistical models for the A-Not A method. J Sens Stud 16: 215–237, 2001.
117. ET Theimer, MR McDaniel. Odor and optical activity. J Soc Cosmet Chem 22: 15–26, 1970.
118. L Friedman, JG Miller. Odor incongruity and chirality. Science 172: 1044–1046, 1971.
119. T Hummel, C Hummel, E Pauli, G Kobal. Olfactory discrimination of nicotine-enantiomers by smokers and non-smokers. Chem Senses 17: 13–21, 1992.
120. ML Laska, R Tuebner. Olfactory discrimination ability for ten pairs of enantiomers. Chem Senses 24: 161–170, 1999.
121. ML Laska, D Freyer. Olfactory discrimination ability for aliphatic esters in squirrel monkeys and humans. Chem Senses 22: 457–465, 1997.
122. M Laska. Olfactory discrimination ability for aromatic odorants as a function of oxygen moiety. Chem Senses 27: 23–29, 2002.
123. M Laska, P Tuebner. Olfactory discrimination ability for homologous series of aliphatic alcohols and aldehydes. Chem Senses 24: 263–270, 1999.
124. M Laska, S Trolp, P Tuebner. Odor structure-activity relationships compared in human and non-human primates. Behav Neurosci 113: 998–1007, 1999.
125. M Kendal-Reed, JC Walker, WT Morgan, M LaMacchio, RW Lutz. Human responses to propionic acid. I. Quantification of within- and between-participant variation in perception by normosmics and anosmics. Chem Senses 23: 71–82, 1998.
126. DA Stevens, RJ O'Connell. Individual differences in thresholds and quality reports of human subjects to various odors. Chem Senses 16: 57–67, 1991.

127. JE Amoore, S Steinle. A graphic history of specific anosmia. In: CJ Wysocki, MR Kare, eds. Chemical Senses. Vol 3. Genetics of Perception and Communication. New York: Marcel Dekker, 1991, pp 331–351.

128. CJ Wysocki, GK Beauchamp. Ability to smell androstenone is genetically determined. Proc Natl Acad Sci USA 81: 4899–4902, 1984.

129. CJ Wysocki, K Dorries, GK Beauchamp. Ability to perceive androstenone can be acquired by ostensibly anosmic people. Proc Natl Acad Sci USA 86: 7976–7978, 1989.

130. P Dalton, CJ Wysocki. The nature and duration of adaptation following long-term odor exposure. Percept Psychophys 58: 781–792, 1996.

131. RR Bush. Estimation and evaluation. In: RD Luce, RR Bush, E Galanter, eds. Handbook of Mathematical Psychology. Vol 1. New York: Wiley, 1963, pp 429–469.

132. LO Harvey, Jr. Efficient estimation of sensory thresholds. Behav Res Methods Instrum Comput 18: 623–632, 1986.

133. B Treutwein, H Strasburger. Fitting the psychometric function. Percept Psychophys 61: 87–106, 1999.

134. DJ Finney. Probit Analysis, 3rd ed. Cambridge: Cambridge University Press, 1971.

135. G Semb. The detectability of the odor of butanol. Percept Psychophys 4: 335–340, 1968.

136. LL Birch, L McPhee, BC Shoba, E Pirok, L Steinberg. What kind of exposure reduces children's food neophobia? Looking vs. tasting. Appetite 9: 171–178, 1987.

137. P Dalton, N Doolittle, PA Breslin. Gender-specific induction of enhanced sensitivity to odors. Nat Neurosci 5: 199–200, 2000.

138. WS Cain, C Murphy. Interaction between chemoreceptive modalities of odour and irritation. Nature 284: 255–257, 1980.

139. PH Dalton. Unpublished observations.

140. SS Stevens. The psychophysics of sensory function. Am Scient 48: 226–253, 1960.

141. SS Stevens. Psychophysics. New York: Wiley, 1975.

142. NH Anderson. Foundations of Information Integration Theory. New York: Academic Press, 1981.

2

Sensory Analysis and Analytical Flavor Chemistry: Missing Links

Susan E. Ebeler
University of California, Davis, California, U.S.A.

I. INTRODUCTION

Significant progress has been made in the past several decades in our understanding of the flavor chemistry of foods and beverages (reviewed in Ref. [1]). Analytical chemistry has advanced from methods that provided information only about compounds present at high concentrations to the current application of sophisticated extraction, separation, and detection techniques that provide information about component concentration and identity at nanogram per liter (ng/L) levels. Simultaneously, sensory analysis has evolved to a science that applies the current knowledge of behavioral research, psychology, and statistics to provide sensitive, accurate, and precise information about the sensory properties of foods and beverages. In addition, our ability to link analytical and sensory information has been advanced through application of numerous multivariate statistical analyses.

The impact of these advances on our understanding of wine flavor has recently been reviewed [2–5]. One of the most promising areas of current study is the use of gas chromatography mass spectrometry (GCMS) and gas chromatography olfactometry (GCO) combined with sensory aroma recombination studies to identify a small subset of aroma compounds that can mimic the aroma of varietal wines. These techniques have been successfully used to identify important aroma components of Gewürztraminer, Scheurebe, and Grenache rosé wines [6–9].

However, to date, these studies have not fully taken into account the subtle variability in wine aromas that can occur within a variety as a function of processing conditions, vintage year, geographical location, climate, and viticultural practices. Furthermore, these studies cannot predict consumer perception or preferences of the wines on the basis of chemical or sensory properties. Using examples from recent studies, this chapter explores some of the factors that are limiting our ability to correlate analytical and sensory information. We also discuss some approaches being used to provide the missing links needed to quantify accurately and to describe complex food flavors.

II. UNDERSTANDING FLAVOR OF COMPLEX MIXTURES

Most food and beverage flavors are extremely complex and arise from the combination of a number of chemical components. For example, the distinctive varietal flavor of most wines is not due to a single impact compound but to the combination of several components, most of which are not unique to a single grape variety. Wine flavor is further complicated by the characteristic that wines of a single variety can also have distinctly different sensory attributes. For example, in a 2001 study, Yegge [10] showed that Chardonnay wines from a single vintage but from different producers were quite variable in their flavor and aroma profiles: some wines were dominated by oak aromas and others by peach/apricot or other fruit-related aromas, and some had more floral characteristics (Fig. 1). In similar studies, Fischer and associates [11] have shown that sensory properties of Riesling wines can vary significantly, even within the same vineyard designation. Heymann and Noble [12] and Guinard and Cliff [13] have shown that judges can distinguish among wines of different geographic origin on the basis of their aroma properties. In some cases, chemical identification of trace aroma compounds (e.g., methoxyisobutylpyrazine, wine lactone, 3-mercapto-1-hexanol) and variations in ratios of individual components present at higher levels may account for some of the observed variations in sensory properties of these wines. However, in many cases, analysis of wine composition alone cannot fully account for observed variabilities in sensory properties [5,14,15].

One factor limiting our ability to predict the flavor of mixtures on the basis of chemical composition is an incomplete understanding of the complex processes involved in the sensory perception of these mixtures. For example, although the relationship between concentration and perceived intensity of single compounds has been well characterized at suprathreshold levels, individual components can mask or enhance the aroma of other

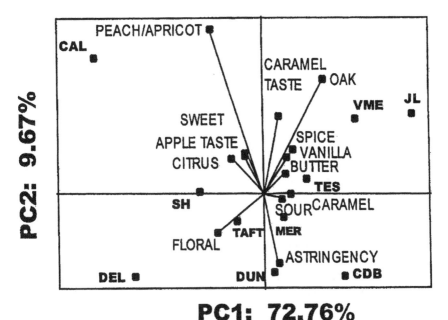

Figure 1 Principal component analysis of descriptive data for 10 Chardonnay wines. Individual wines are indicated in bold capital letters. Mean perceived intensity of each descriptive term is indicated by distance from origin (farther from origin is more intense). Wines associated with specific descriptive terms are located near the descriptive analysis terms. PC1 = principal component 1. (From Ref. [10].)

compounds in ways that are not completely understood [16]. Even at subthreshold levels, compounds may influence the overall perceived aroma. As discussed by Buttery [17], there is an additive effect of detection when a mixture of compounds is present at subthreshold concentrations. In addition, in 2000 Dalton and colleagues [18] showed that taste and smell senses can also interact so that a subthreshold taste combined with a subthreshold odor can be detected.

In addition to intensity, the qualitative character of an individual compound can change as a function of concentration and when it is mixed with other compounds. For example, in wine, the compounds ethylbutyrate and diacetyl, which alone have bubble-gum/fruity and buttery characters, respectively, yield a distinctive butterscotch character when combined. Context can also have a significant effect on perceived odor quality. Lawless and coworkers [19] have shown that the character of the compound dihydromyrcenol is described by sensory panelists as either citruslike or pinelike, depending on the aroma qualities of other compounds presented at the same time. These context effects may influence the perceived aroma of a wine, depending on whether it is evaluated with a group of wines with similar or with distinctly different overall flavor properties. For example, a wine with a distinct fruity character may smell different when evaluated with other fruity wines as compared with wines having strong vegetal or bell pepper characteristics.

Recent advances in neurobiology have shed some light on the biological and physiological processes involved in the perception of odor mixtures. Research in this area has led to isolation of olfactory and taste receptors and their genes, to improved understanding of mechanisms for signal transduction, and to identification of brain regions involved in taste and olfactory perception (reviewed in [20,21]). These studies have shown that odor receptors are broad-banded, that each receptor responds to a variety of chemical stimuli, and that the corresponding patterns of signals sent to the brain allow us to distinguish different odor qualities.

A second area requiring further study is that of the dynamic nature of flavor release in complex food and beverage matrices. There is an increasing understanding that aroma perception is dynamic and that interactions between volatile compounds and matrix components can have a significant impact on flavor volatility and flavor release (reviewed in [22,23]). For example, in wine, ethanol can suppress the volatility of esters so that perception of fruitiness is decreased [24,25]. Similarly, polyphenols, important nonvolatile constituents of red wine in particular, may interact with some flavor compounds, to alter their volatility and flavor release [26–31]. However, the mechanisms of these interactions and their effects on sensory perception are not yet fully characterized.

Analytical methods involving exhaustive extraction of flavor compounds (i.e., liquid/liquid extraction, dynamic headspace) do not take these matrix effects into account. However, new instrumentation and methodologies are yielding improved information on the mechanisms involved in flavor/matrix interactions and the effects on flavor perception. For example, spectroscopic techniques, such as nuclear magnetic resonance (NMR), can provide information on complex formation as a function of chemical environment and have been used to study both intra- and intermolecular interactions in model systems [28,31]. In addition, NMR techniques, initially developed to study ligand binding for biological and pharmaceutical applications, were applied in 2002 to model food systems to screen flavor mixtures and identify those compounds that will bind to macromolecules such as proteins and tannins [32]. Flavor release in the mouth can be simulated with analytical tools such as the retronasal aroma simulator (RAS) developed by Roberts and Acree [33]. These release cells can provide quantitative information on release of volatiles from complex matrices under controlled conditions of temperature, shear rate, saliva addition, and gas flow [34]. Actual analysis of volatile release in the mouth or nose space can be accomplished by sampling exhaled air directly into the source of an atmospheric pressure ionization mass spectrometer (APIMS) [35]. The breath analysis technique follows the time course of release for different flavor compounds from a food or beverage matrix while taking into account matrix composition and structure effects as well as enzymatic, chemical, and physical changes that occur during chewing [36–39]. By simultaneously recording sensory perception during the breath analysis, a direct correlation and modeling of the chemical release profiles to the perceived aroma may be possible. The complementary information obtained from all of these analytical techniques can be used to predict better and to optimize flavor release and flavor perception in foods and beverages.

Finally, there has been recent interest in developing analytical tools that will provide a more integrated approach to flavor analysis. In general, these techniques rely on pattern recognition and chemometrics to distinguish subtle differences in complex spectra or instrument responses produced by the different products. For example, NMR spectra and Fourier transform infrared (FTIR) spectra obtained over a range of wavelengths have been used to distinguish between alcoholic beverages produced under different processing conditions or from different grape varieties [40–42]. Similarly, electronic noses that utilize an array of polymeric-based chemical sensors with differential responses to individual compounds or classes of compounds have been used to screen and classify products with off-odors [43,44]. GCMS without chromatographic separation has also been proposed as a "sensor" that can monitor changes in odor quality of food products [45,46]. Although these techniques cannot fully mimic all of the factors influencing odor perception discussed previously, they do hold promise for rapid screening of off-odors, optimization of processing conditions, and preliminary characterization or classification of products based on chemical properties.

III. LINKING SENSORY AND ANALYTICAL MEASUREMENTS WITH CONSUMER PREFERENCES

The complexities of consumer behavior were recently demonstrated by Noble and coworkers in a study of red wines obtained from a number of producers [2,47]. Sensory descriptive analysis using trained panelists showed that the wines could be readily distinguished and had sensory properties ranging from fruity, to vegetative, to leather, to sour, to astringent. When typical (untrained) consumers rated their preferences for these wines, the most preferred wine was one shown by descriptive analysis to have strong fruity characteristics, whereas the least preferred wine had the least amount of fruity character. However, preference for other wines did not appear to be related to either fruity character (i.e., some wines with strong fruity characteristics were not preferred by the consumers) or any other sensory attributes. This study shows that even if sensory perception could be completely correlated to analytical measurements, prediction of consumer preferences and perceptions from this information alone would still not be possible.

Some individual differences in perception and preferences may be the result of intrinsic or genetic factors. For example, Buettner [48] observed that differences in salivary enzyme activity among individuals can significantly influence flavor perception and persistence of odor-active thiols and esters in the mouth. Wang and associates [49] have shown strong links among obesity, brain regions associated with the sensory processing of food, and receptors for dopamine, a neurotransmitter associated with satisfaction and pleasure. In addition, a genetic basis for preferences to alcoholic beverages has also been postulated by Ishibashi and colleagues [50]. Individual differences in ability to metabolize and detoxify ethanol, controlled by the enzyme aldehyde dehydrogenase 2 (ALDH2), were related to acquired preferences for alcoholic beverages. These studies, made possible by advances in molecular biology and application of analytical tools such as magnetic resonance imaging (MRI), are beginning to provide strong evidence that individual genetic and physiological differences can influence perception, eating behavior, and food choices. The sequencing of the human genome now also holds much promise for continued explorations of the complex linkages between genetics and individual behavior.

In addition to these intrinsic or genetic factors, extrinsic factors such as price, label design, and packaging can influence consumer perceptions and preferences. However, Yegge [10] has shown that not all consumers respond similarly to these extrinsic factors. Prior experience and familiarity with a product seem to be important in determining whether sensory or nonsensory factors have a greater impact on an individual consumer's behavior. For example, wine consumers who are knowledgeable about wine and who consume wine on a regular basis seem to be less impacted by these nonsensory variables. An area requiring much further research is the role of context and suggestion in influencing consumer responses to a product. For example, it is not clear how cognitive behavior and perception of flavor attributes change when information about product attributes is provided to consumers in the form of wine label descriptions or "expert" opinions.

IV. CONCLUSIONS

Numerous challenges face scientists in their efforts to link analytical information regarding food and beverage composition to sensory perception and ultimately to consumer preferences. In order to provide these links, an improved understanding of the mechanisms involved in the sensory perception of complex mixtures, consideration of the dynamic nature of flavor release and the impact of matrix interactions, and development of new analytical technologies that provide an integrated approach to flavor analysis will be needed. In addition, continued study of the genetic factors and nonsensory variables that influence perception and behavior is critical for a complete understanding of flavor perception. Food and flavor chemists will look to many fields of science to contribute knowledge in these areas (neurobiology, genetics, psychology, statistics, analytical chemistry, etc.), while the ultimate challenge will be to piece together the links to form a fully integrated picture of food and beverage flavor.

REFERENCES

1. R Teranishi, E Wick, I Hornstein. Flavor Chemistry: 30 Years of Progress. New York: Kluwer Academic, 1999.
2. SE Ebeler. Linking flavor chemistry to sensory analysis of wine. In: R Teranishi, E Wick, I Hornstein, eds. Flavor Chemistry: 30 Years of Progress. New York: Kluwer Academic, New York, 1999, pp 409–421.
3. SE Ebeler. Analytical chemistry: Unlocking the secrets of wine flavor. Food Rev Int 17(1): 45–64, 2001.
4. SE Ebeler, AC Noble. Past and the future: Bucket flavor chemistry to sensochemistry. Am J Enol Viticult 51(5): 205–208, 2000.
5. AC Noble, SE Ebeler. Use of multivariate statistics in understanding wine flavor. Food Rev Int 18(1): 1–20, 2002.
6. H Guth. Identification of character impact odorants of different white wine varieties. J Agric Food Chem 45: 3022–3026, 1997.
7. H Guth. Quantitation and sensory studies of character impact odorants of different white wine varieties. J Agric Food Chem 45: 3027–3032, 1997.
8. H Guth. Comparison of different white wine varieties in odor profiles by instrumental analysis and sensory studies. In: AL Waterhouse, SE Ebeler, eds. Chemistry of Wine Flavor. Washington, DC: American Chemical Society, 1998, pp 39–52.
9. V Ferreira, N Ortin, A Escudero, R Lopez, J Cacho. Chemical characterization of the aroma of Grenache rosé wines: Aroma extract dilution analysis, quantitative determination, and sensory reconstitution studies. J Agric Food Chem 50: 4048–4054, 2002.

10. J Yegge. Influence of sensory and non-sensory attributes of Chardonnay wine on acceptance and purchase intent. Ph. D. Dissertation. University of California, Davis, 2001.

11. U Fischer, D Roth, M Christmann. The impact of geographic origin, vintage and wine estate on sensory properties of *Vitis vinifera* cv. *Riesling* wines. Food Qual Preference 19: 281–288, 1999.

12. H Heymann, AC Noble. Descriptive analysis of commercial Cabernet Sauvignon wines from California. Am J Enol Viticult 38: 41–44, 1987.

13. J-X Guinard, M Cliff. Descriptive analysis of Pinot noir wines from Carneros, Napa, and Sonoma. Am J Enol Viticult 38: 211–215, 1987.

14. AC Noble, RA Flath, RR Forrey. Wine headspace analysis: Reproducibility and application to varietal classification. J Agric Food Chem 28: 346–353, 1980.

15. AC Noble. Analysis of wine sensory properties. In: HF Linskens, JF Jackson, eds. Wine Analysis. New York: Springer-Verlag, 1988, pp 9–28.

16. HT Lawless, H Heymann. Sensory Evaluation of Food: Principles and Practices. Gaithersburg, MD: Aspen Publishers, 1999.

17. RG Buttery. Flavor chemistry and odor thresholds. In: R Teranishi, E Wick, I Hornstein, eds. Flavor Chemistry: 30 Years of Progress. New York: Kluwer Academic, 1999, pp 353–365.

18. P Dalton, N Doolittle, H Nagata, PAS Breslin. The merging of the senses: Integration of subthreshold taste and smell. Nat Neurosci 3(5): 431–432, 2000.

19. HT Lawless, S Glatter, C Hohn. Context dependent changes in the perception of odor quality. Chem Senses 16: 349–360, 1991.

20. G Brand, J-L Millot, D Henquell. Complexity of olfactory lateralization processes revealed by functional imaging: A review. Neurosci Biobehav Rev 25(2): 159–166, 2001.

21. P Given, D Paredes. Chemistry of Taste: Mechanisms, Behaviours, and Mimics. Washington, DC: American Chemical Society, 2002.

22. RJ McGorrin, JV Leland. Flavor-Food Interactions. Washington, DC: American Chemical Society, 1996.

23. DD Roberts, A Taylor. Flavor Release. Washington, DC: American Chemical Society, 2000.

24. AA Williams, PR Rosser. Aroma enhancing effects of ethanol. Chem Senses 6(2): 149–153, 1981.

25. U Fischer. Mass balance of aroma compounds during the dealcoholization of wine: Correlation of chemical and sensory data. Ph. D. Dissertation, Universitat Hannover, Germany, 1995.

26. B King, J Solms. Interactions of volatile flavor compounds in model food systems using benzyl alcohol as an example. J Agric Food Chem 27: 1331–1334, 1979.

27. B King, J Solms. Interaction of volatile flavor compounds with caffeine, chlorogenic acid, and naringin. In: P Schreier, ed. Flavour '81. New York: de Gruyter, 1981, pp 707–716.

28. C Dufour, CL Bayonove. Interactions between wine polyphenols and aroma substances: An insight at the molecular level. J Agric Food Chem 47: 678–684, 1999.

29. C Dufour, I Sauvaitre. Interactions between anthocyanins and aroma substances in a model system: Effect on the flavor of grape-derived beverages. J Agric Food Chem 48: 1784–1788, 2000.

30. H Escalona, L Birkmyre, JR Piggott, A Paterson. Relationship between sensory perception, volatile and phenolic components in commercial Spanish red wines from different regions. J Instit Brew 107(3): 157–166, 2001.

31. D-M Jung, JS de Ropp, SE Ebeler. Study of interactions between food phenolics and aromatic flavors using one-and two-dimensional ^{1}H NMR spectroscopy. J Agric Food Chem 48: 407–412, 2000.

32. D-M Jung, JS de Ropp, SE Ebeler. Application of pulsed field gradient NMR techniques for investigating binding of flavor compounds to macromolecules. J Agric Food Chem 50: 4262–4269, 2002.

33. DD Roberts, TE Acree. Simulation of retronasal aroma using a modified headspace technique: Investigating the effects of saliva, temperature, shearing, and oil on flavor release. J Agric Food Chem 43: 2179–2186, 1995.

34. JR Piggott, CJ Schaschke. Release cells, breath analysis and in-mouth analysis in flavour research. Biomole Eng 17: 129–136, 2001.

35. AJ Taylor, RST Linforth, BA Harvey, A Blake. Atmospheric pressure chemical ionization mass spectrometry for in vivo analysis of volatile flavour release. Food Chem 71(3): 327–338, 2000.

36. MS Brauss, RST Linforth, AJ Taylor. Effect of variety, time of eating, and fruit-to-fruit variation on volatile release during eating of tomato fruits (*Lycopersicon esculentum*). J Agric Food Chem 46(6): 2287–2292, 1998.

37. MS Brauss, B Balders, RST Linforth, S Avison, AJ Taylor. Fat content, baking time, hydration and temperature affect flavour release from biscuits in model-mouth and real systems. Flavour Fragrance J 14(6): 351–357, 1999.

38. MS Brauss, RST Linforth, I Cayeux, B Harvey, AJ Taylor. Altering the fat content affects flavor release in a model yogurt system. J Agric Food Chem 47(5): 2055–2059, 1999.

39. KM Hemingway, MJ Alson, CG Chappell, AJ Taylor. Carbohydrate-flavour conjugates in wine. Carbohydr Polym 38: 283–286, 1999.

40. A Edelmann, J Diewok, KC Schuster, B Lendl. Rapid method for the discrimination of red wine cultivars based on mid-infrared spectroscopy of phenolic wine extracts. J Agric Food Chem 49: 1139–1145, 2001.

41. MA Coimbra, F Goncalves, AS Barros, I Delgadillo. Fourier transform infrared spectroscopy and chemometric analysis of white wine polysaccharide extracts. J Agric Food Chem 50: 3405–3411, 2002.

42. I Duarte, A Barros, PS Belton, R Righelato, M Spraul, E Humpfer, AM Gil. High-resolution nuclear magnetic resonance spectroscopy and multivariate analysis for the characterization of beer. J Agric Food Chem 50: 2475–2481, 2002.

43. P Mielle. "Electronic noses": Toward the objective instrumental characterization of food aroma. Trends Food Sci Technol 7: 432–438, 1996.

44. S Rocha, I Delgadillo, AJ Ferrer Correia, A Barros, P Wells. Application of an electronic aroma sensing system to cork stopper quality control. J Agric Food Chem 46: 145–151, 1998.

45. L Lecanu, V Ducruet, C Jouquand, JJ Gratadoux, A Feigenbaum. Optimization of headspace solid-phase microextraction (SPME) for the odor analysis of surface-ripened cheese. J Agric Food Chem 50: 3810–3817, 2002.

46. RT Marsili. SPME-MS-MVA as an electronic nose for the study of off-flavors in milk. J Agric Food Chem 47: 648–654, 1999.

47. MB Froest, AC Noble. Preliminary study of the effect of knowledge and sensory expertise on liking for red wines. Am J Enol Viticult 53: 275–284, 2002.

48. A Buettner. Influence of human salivary enzymes on odorant concentration changes occurring in vivo. 1. Esters and thiols. J Agric Food Chem 50: 3283–3289, 2002.

49. G-J Wang, ND Volkow, C Felder, JS Fowler, AV Levy, NR Pappas, CT Wong, W Zhu, N Netusil. Enhanced resting activity of the oral somatosensory cortex in obese subjects. Neuroreport 13(9): 1151–1155, 2002.

50. T Ishibashi, S Harada, C Fugii, A Taguchi, T Ishii. Relationship between ALDH2 genotypes and choice of alcoholic beverages. Jpn J Alcohol Stud Drug Depend (Nihon Arukoru Yakubutsu Igakkai Zasshi) 34: 117–129, 1999.

3

When Are Oral Cavity Odorants Available for Retronasal Olfaction?

Bruce P. Halpern
Cornell University, Ithaca, New York, U.S.A.

I. INTRODUCTION

The human oral cavity is potentially connected to the nasal cavity by way of the buccopharynx (oropharynx), pharynx, and nasopharynx [1,2]. Under those circumstances in which this potential connection is open, the air movement of an exhalation that exits from the anterior nares (nostrils) can acquire odorants from the oral cavity and move them through the nasal cavity. If these odorants, while in the nasal cavity, reach the olfactory mucosa at a flow rate and concentration [3,4] that allow penetration to olfactory receptor neurons [5] and activation of these receptors such that sufficient central nervous system (CNS) responses develop, retronasal olfaction may occur. A limitation to the present understanding of retronasal olfaction is the absence of empirical or numerical models of retronasal odorant transport in adult humans. Such models have been published for orthonasal olfaction via the anterior nares [4,6] but are not presently available for retronasal olfaction (experimental airflow and odorant uptake analysis is in progress: PW Scherer, personal communication, October 2002).

Access from the oral cavity to the nasal cavity is necessary for retronasal olfaction, but it is not sufficient. As suggested, it is conceivable that odorants could reach the nasal cavity from the mouth but not arrive at the olfactory epithelium at all, or at least not in a quantity per unit time that would be sufficient to partition into the mucous covering of that sensory epithelium, activate relevant populations of olfactory receptor neurons, and then initiate the type of central nervous system responses that would constitute olfaction.

Several core questions emerge: When is the potential connection between the human oral and nasal cavities open, and when is it closed? Can odorants known to be present in the human oral cavity be detected by physical means in the nasal cavity, and, if so, under what circumstances? As a measure of overall function of retronasal olfaction, with what accuracy can human observers identify odorants originating in the oral cavity?

II. WHEN IS THE CONNECTION BETWEEN THE HUMAN ORAL AND NASAL CAVITIES OPEN, AND WHEN IS IT CLOSED?

Under normal circumstances, during most speech a tight velopharyngeal closure largely or completely seals the potential connection between the oral and nasal cavities [7–9]. Moreover, with appropriate practice, individuals can learn to produce more sustained or more complete velopharyngeal closures [10,11]. During swallows, velopharyngeal closures also normally occur [12]. None of these data are under controversy.

What is in dispute is when the pathway between the oral and nasal cavities is open such that chemicals in the mouth reach the nasal cavity. It has been reported that when 25 mL helium was retained in the oral cavity for 3 min and breathing was done with the lips closed, no helium could be measured in the efflux from the nostrils into a $\sim$94-cm^3 cylinder with the detector 30 cm from the nostrils unless the subjects were either "deliberately exhaling 'mouth-air' through the nose" or making vigorous mouth and tongue movements [13]. Unfortunately, no details were provided concerning in what manner "deliberately exhaling 'mouth-air' through the nose" differed from other exhalations. However, in a related later publication with the same first author [14], "The helium was kept for 1 min in the oral cavity while normal respiration was continued. Then, the air present in the oral cavity was deliberately exhaled through the nose." It appears that in the latter instance, at least, subjects were instructed to retain the helium in the oral cavity and to exhale it upon command. Videofluoroscopic observations indicated that a velopharyngeal closure allowed the helium to be restricted to the oral cavity, and that this closure could be voluntarily ended [14]. This is in agreement with earlier reports that individuals can learn to control and augment velopharyngeal closure voluntarily [10,11]. It is unclear to what extent the reported absence of oral cavity helium in nasal exhalation during normal breathing [13] was due to voluntary velopharyngeal closure [14].

Whether or not a gas bubble held in the mouth appears in nasal exhalations may not be a realistic test of normal access of the oral cavity to the nasal cavity. The presence in air exhaled from the nostrils of a molecule or molecules introduced into the oral cavity in liquid or solid phase might be more relevant. This was studied [13] for the odorant ethyl butanoate [15] (i.e., ethyl butyrate CAS# 105-54-4). Twenty-five milliliters of 8.6 micromolar aqueous ethyl butanoate was held in the mouth for 1 min while the subject was breathing normally during either "modest tongue and mouth movements" or no tongue movement, followed by a swallow upon command with or without "the air present in the oral cavity deliberately . . . exhaled through the nose" (in all cases the exhaled air was passed through anhydrous $CaCl_2$ into a Tenax® filled column that was subsequently eluted with diethyl ether, with analysis for ethyl butanoate done by using high-resolution gas chromatography mass spectrometry [13]. It was reported that in the absence of tongue movement, swallowing, or deliberate exhalation through the nose of the air in the oral cavity, recovered ethyl butanoate over 1 min ranged from 20 to 150 ng, and that the recovered quantity increased by a factor of 100 when swallowing occurred. No numerical data on the effect of "the air present in the oral cavity deliberately . . . exhaled through the nose" without a swallow were provided, although it was stated that results were highly variable, ranging from less than a third of that found with a swallow to almost a third more than the amount of ethyl butanoate recovered with a swallow [13]. Furthermore, no data were offered on effects of tongue movement per se. Nonetheless, the authors concluded that for liquid foods, the mouth can be regarded as a closed system unless swallowing or lowering of the base of the tongue due to tongue movements occurs. This conclusion may be unwarranted. The 20- to 150-ng amounts of ethyl butanoate that were recovered during 1 min of quiet breathing may have exceeded the odor threshold for this molecule. This is the case because the 150 ng could only have appeared during the few seconds of the exhalations. The calculated air phase concentration of ethyl butanoate during exhalations was 7.33 ppb (using 4 sec as the duration of a respiratory cycle and 500 mL as the tidal volume [16,17]). Although this concentration is below the average threshold (the threshold in aqueous phase is approximately 1 ppb [18]; 23 ppb in air phase [19]), individual differences in odorant thresholds are well known. Further, the amounts found with deliberate exhalation, which may be equivalent to ceasing learned velopharyngeal closure, were sometimes equal to or greater than those that immediately preceded or followed a swallow.

Opposite conclusions from those of the preceding report [13] were reached in a study that used various intraoral odorant sources, including the liquid food orange juice and a carbonated beverage [20]. Subjects drank naturally, with no instructions to hold, or not hold, liquids in the mouth, and swallowing was done at the subjects' volition rather than at a command. Breath-by-breath analyses were done with input from a nostril to a modified mass spectrometer (MS-Nose) [21] via an open tube and a perpendicular capillary. Molecular ions known to be present in each liquid were measured. It was observed from these in vivo studies that volatile components of the liquids repeatedly appeared in the nasal cavity during successive exhalations and disappeared only with swallows. This study also examined effects of chewing solid foods, both artificial and semiartificial ones into which known molecules were incorporated and normal foods such as cookies, banana, and a peanut butter sandwich [20]. Processing of these solid foods in a retronasal aroma simulator (RAS), and analyses of the effluent with the MS-Nose, provided information on what molecular ions could be expected in air exhaled from the nasal cavity. The foods were in bite-size pieces and were eaten naturally without constraints. As had been the case with liquids, the known molecular ions in these foods appeared with each exhalation and disappeared after one or two swallows (Fig. 1). This series of experiments [20], as well as earlier ones by some of the authors, demonstrated unequivocally that molecules in foods and beverages, if released into the airspace of the oral cavity, will appear in the nasal cavity during exhalations with the lips closed. Of course, a functional connection between the oral and nasal cavities must have been present.

A different approach to the question of an oral cavity-to-nasal cavity connection is to image the position of the velum in relation to breathing and swallowing. Videofluoroscopy was used to measure velum position before, during, and after swallows of liquids or a paste, and simultaneously a respiratory pressure transducer was used to measure the direction of airflow both during breathing (exhalation or inhalation) and when breathing stopped during a swallow (deglutition apnea) [12]. These direct measurements showed that the air phase of the oral cavity usually had access to the nasal cavity, the connection was more often open than closed at the beginning of deglutition apnea, and always open at the end of deglutition apnea (Fig. 2).

III. WITH WHAT ACCURACY CAN HUMAN OBSERVERS IDENTIFY ODORANTS ORIGINATING IN THE ORAL CAVITY?

Human responses to sensory stimulation with liquids introduced into the oral cavity have been compared with orthonasal (smelling via the anterior nares) olfactory judgments in a number of studies (e.g., [22–27]). Interactions between the oral liquids and orthonasal odorants were often

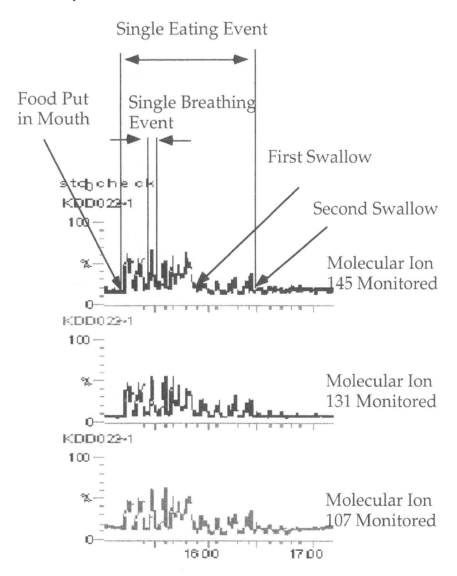

Figure 1 MS-Nose monitoring of ions in the breath of a human while eating an imitation cheese. Molecular ion 145 was ethyl hexanoate; 131 was isoamyl acetate; and 107, benzaldehyde. All had been incorporated into the imitation cheese. (From Ref. [21] with permission.)

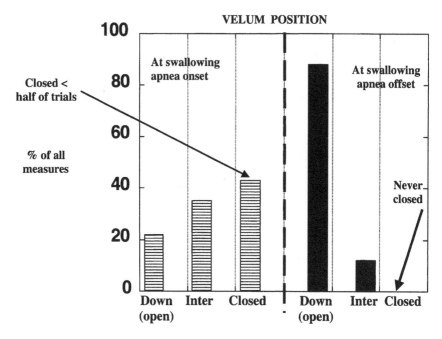

Figure 2 Velum (soft palate) position at the beginning and end of the swallowing (deglutition) apnea that accompanied swallows. Barium-containing liquid or paste allowed videofluoroscopy to characterize the position of the velum as Down (fully open), Inter (intermediate between open and closed), or Closed at the beginning and end of the swallowing apnea. A respiratory pressure transducer (respirodeglutometer) measured both the direction of airflow during breathing (exhale or inhale) and when breathing stopped during a swallow. (From Ref. [12].)

noted. However, if an investigation of solely *olfactory* human responses to oral odorants, i.e., retronasal olfaction per se, is desired, stimuli should be presented to the oral cavity such that only an air phase (vapor phase) access is possible. This can be accomplished either by delivering odorants in vapor form to the oral cavity (e.g., Ref. [28]) or by confining odorants within the oral cavity to a container that prevents direct contact between the odorant and oral cavity tissues but permits free passage of air phase components [29]. These odorant containers (Figs. 3 and 4), designated *odorant presentation containers* (OPCs), are made of high-density polyethylene, fit comfortably into the adult human mouth (or under the anterior nares), and can be discarded after use for one individual with a single concentration of one odorant [29].

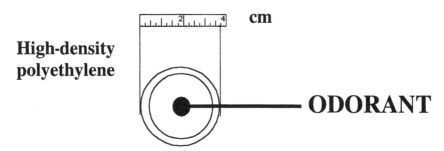

Figure 3   Schematic drawing of the top view of the odorant presentation container (OPC) that is used to present odorants in the oral cavity without taste stimulation (Kodak® gray high-density lid for canisters). An odorant is schematically represented in the center of the OPC.

The OPCs were used to determine the accuracy with which human observers could identify odorants placed in the oral cavity [29]. The observers, after being tested to ensure general olfactory and verbal report competence, first learned to assign arbitrary verbal labels (e.g. one, two, three, four) to the air phase components of oregano powder, ground coffee, garlic powder, or minced chocolate when they were presented orthonasally. Visual observations of the OPC was prevented, and sniffing was not permitted (that is, quiet resting breathing was permitted, but not sniffing, defined as "to inhale forcibly through the nose" [30]). Next, the same odorants in OPCs were presented retronasally, and ability to identify them

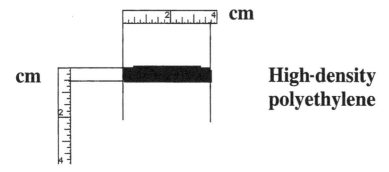

Figure 4   Schematic drawing of the side view of the odorant presentation container (OPC) that is used to present odorants in the oral cavity without taste stimulation.

was measured. Tongue movement was discouraged; lips were closed (subjects were instructed not to move their tongue; with an OPC in place in the mouth and lips closed, little tongue movement was possible); and swallowing did not occur with the OPC in the mouth. Under conditions of natural, quiet breathing, retronasal correct identification of the four odorants was made in an average of 83% of the trials (Fig. 5). This indicated that air phase components of the odorants that were in the oral cavity (a) had access to the nasal cavity, (b) arrived at the olfactory epithelium of the nasal cavity at a flow rate and concentration that allowed penetration to olfactory receptor neurons and activation of these recept, and (c) elicited sufficient central nervous system olfactory responses to provide reliable human olfactory perception. The accuracy of these retronasal identifications was less than that of orthonasal identifications and remained so after direct retronasal learning of arbitrary verbal labels for the odorants but nonetheless was far above chance (Fig. 5).

Human ability to identify retronasal odorants was subsequently confirmed in a series of studies that used intraoral OPCs containing as odorants plant-derived liquid flavors prepared and sold for incorporation in human foods [31–36]. The observers, who were not tested for general olfactory ability, learned and subsequently provided the common, *veridical* name of each odorant, such as *coffee*, *lemon*, *orange*, *oil*. In one of these studies [28], nine different undiluted odorants were presented. When the veridical names were learned retronasally and then tested retronasally, identifications were at least 80% correct for seven of the nine odorants (Table 1). For the other two odorants, lemon and orange, confusions between these two identifications occurred, but the combined *lemon/orange* identifications also exceeded 80% correct. Similar lemon/orange confusions were seen with orthonasal olfaction. Overall, the results of these human psychophysical retronasal olfaction studies demonstrate that with no swallowing, no vigorous tongue movement, and no forced breathing, humans can readily identify by olfaction alone odorants originating in the oral cavity.

IV. CONCLUSIONS

1. The oral cavity in normal healthy adult humans is generally connected to the nasal cavity whenever the individual is not speaking, swallowing, or choosing to produce a velopharyngeal closure. Radiological imaging techniques such as videofluoroscopy demonstrate that this connection is usually open but normally closes during speech or a swallow. Learned closure can also occur.

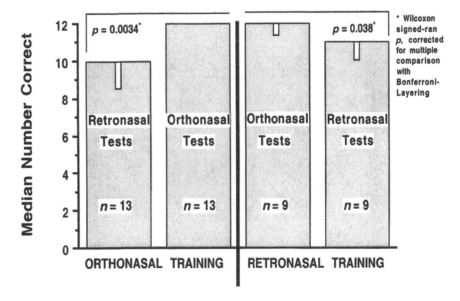

Figure 5 Median number of correct identifications of four odorants presented three times each in random order. Fifteen subjects were trained to identify by assigned number the air phase components of 0.5 to 0.9 g of oregano powder, ground coffee, garlic powder, or minced baking chocolate presented in an odorant presentation container (OPC) via an orthonasal route for 10 sec. After training, odorant identification learning was evaluated by measuring ability to identify correctly to a criterion. The 13 subjects who met the criterion were then tested first with the stimuli presented via a retronasal route (a nose clip was used to prevent any smelling of the odorant in the OPC until the OPC was in the oral cavity, and the lips closed), and second via an orthonasal route. Next the 13 subjects were trained via a retronasal route, evaluated, and the 9 who now met the criterion were tested orthonasally and then retronasally. An average of 10 of the 12 stimuli was correctly identified with the initial retronasal testing (83% correct); 11 of 12, with the second retronasal testing (92% correct). (From Ref. [29].)

2. Air phase odorants derived from liquid or solid foods in the oral cavity reach the nasal cavity during the exhale portion of each natural breathing cycle that occurs during lip closure and normal lingual manipulation or dental chewing of the food. Breath-by-breath in vivo physical measurements with a modified mass spectrometer (e.g., MS-Nose) of exhaled air find the expected food-derived molecular ions during each exhalation. The molecular ion signals become attenuated or absent when the

Table 1 Percentage Veridical Name Identifications of the Air Phase Components of Nine Undiluted Odorants[a]

Odorants	Peanut	Coffee	Lemon	Cinnamon	Banana	Chocolate	Orange	Wintergreen	Oil (canola)
Peanut	88.9	8.3	0.0	0.0	0.0	2.8	0.0	0.0	0.0
Coffee	0.0	86.1	5.6	0.0	0.0	2.8	0.0	0.0	5.6
Lemon	0.0	2.8	63.9	0.0	0.0	2.8	19.4	2.8	8.3
Cinnamon	0.0	2.8	0.0	80.6	2.8	5.6	0.0	8.3	0.0
Banana	0.0	0.0	0.0	0.0	100	0.0	0.0	0.0	0.0
Chocolate	0.0	0.0	5.6	0.0	2.8	88.9	2.8	0.0	0.0
Orange	0.0	0.0	13.9	0.0	5.6	0.0	72.2	2.8	5.6
Wintergreen	0.0	0.0	0.0	2.8	2.8	0.0	2.8	91.6	0.0
Oil (canola)	0.0	0.0	2.8	0.0	0.0	5.6	0.0	5.6	86.1
Odorants	Peanut	Coffee	Lemon	Cinnamon	Banana	Chocolate	Orange	Wintergreen	Oil (canola)

[a] Odorants were presented twice each in random order via the retronasal route in OPC to 18 subjects, after retronasal identification learning. A nose clip was used to prevent any smelling of the odorant in the OPC until the OPC was in the oral cavity, and the lips closed. Correct identification percentages are in cells with gray backgrounds. Overall retronasal identifications were correct on 86% of trials (median). *Source:* Ref. [33].

food is swallowed. Normal adult humans can perceive a wide variety of oral cavity odorants by means of a retronasal route. This retronasal olfaction can occur without taste stimulation, tongue movement, or swallowing. Accuracy of identification of the oral cavity odorants generally ranges between 80% and 90%; 100% accuracy sometimes occurs. The existence of retronasal olfactory perception indicates that air phase components of the odorants in the oral cavity not only reach the nasal cavity but also effectively initiate olfactory processes at peripheral receptor structures and in the CNS.

ACKNOWLEDGMENTS

T. E. Acree, K. D. Deibler, C. A. Pelletier, and the anonymous referee provided valuable information and references; R. B. Darlington, statistical consulting. I thank K. D. Deibler, J. F. Delwiche, C. A. Pelletier, and the anonymous referee for comments on earlier versions of the chapter. It was prepared during support from USDA grant NRI-2001-355503-10102.

REFERENCES

1. V Negus. The Comparative Anatomy and Physiology of the Nose and Paranasal Sinuses. Edinburgh: E. & S. Livingston, 1958.
2. CR Schneiderman, RE Potter. Speech-Language Pathology. San Diego, CA: Academic Press, 2002.
3. MM Mozell, DE Hornung, PR Sheehe, DB Kurtz. What should be controlled in studies of smell? In: HL Meiselman, RS Rivlin, eds. Clinical Measurement of Taste and Smell. New York: Macmillan, 1986, pp 154–169.
4. MM Mozell, PF Kent, PW Scherer, DE Hornung, SJ Murphy. Nasal airflow. In: TV Getchell, RL Doty, LM Bartoshuk, JB Snow, Jr., eds. Smell and Taste in Health and Disease. New York: Raven Press, 1991, pp 481–492.
5. DE Hornung, MM Mozell. Smell: Human physiology. In HL Meiselman, RS Rivlin, eds. Clinical Measurement of Taste and Smell. New York: Macmillan, 1986, pp 19–38.
6. K Keyhani, PW Scherer, MM Mozell. A numerical model of nasal transport for the analysis of human olfaction. J Theor Biol 186: 279–301, 1997.
7. M Rothenberg. Incomplete velopharyngeal closure. The Breath-Stream Dynamics of Simple-Released-Plosive Production. Basel, Switzerland: S. Karger, 1968, Sect. 7.4. Available at http://www.rothenberg.org/Breath-Stream/7.htm#7.4 (Cited May 6, 2003).

8. Minnesota, University of, School of Dentistry, Speech Prosthesis Clinic. Velopharyngeal closure. Available at http://www.dentistry.umn.edu/patients/ Speech_Prosthesis_Clinic553.html (cited May 6, 2003).

9. Cincinnati Children's Hospital Medical Center, Velopharyngeal Dysfunction Clinic. Velopharyngeal Function and Velopharyngeal Dysfunction (Insufficiency). Available at http://www.cincinnatichildrens.org/svc/prog/vpi/about/ vpi.htm (cited May 6, 2003).

10. DP Kuehn. Continuous positive airway pressure (CPAP) in the treatment of hypernasality. Available at http://www.asel.udel.edu/icslp/cdrom/vol2/422/ a422 (cited August 26, 2002).

11. G Kobal, T Hummel. Olfactory evoked potentials in humans. In TV Getchell, RL Doty, LM Bartoshuk, JB Snow, Jr. eds. Smell and Taste in Health and Disease. New York: Raven Press, 1991, pp 255–275

12. AL Perlman, SL Ettema, J Barkmeier. Respiratory and acoustic signals associated with bolus passage during swallowing. Dysphagia 15: 89–94, 2000.

13. A Buettner, P Schieberle. Exhaled odorant measurement (EXOM)—a new approach to quantify the degree of in-mouth release of food aroma compounds. Lebensm Wiss Technol 33: 553–559, 2000.

14. A Buettner, A Beer, C Hannig, M Settles. Observation of the swallowing process by application of videofluoroscopy and real-time magnetic resonance imaging—consequences of retronasal aroma stimulation. Chem Senses 26: 1211–1219, 2001.

15. T Acree, H Arn. Flavornet. 1997. Available at http://www.nysaes.cornell.edu/ flavornet/chemsens.html (cited August 27, 2002).

16. JH Comroe. Human Control of Ventilation. Chicago: Yearbook Medical, 1965.

17. P Bogorodzki, R Sypniewski, A Piatkowski. Real-time respiratory signal analysis for triggered tomographic studies. Available at http://www.ire.pw.e-du.pl/zejim/binsk/spiro.pdf (cited October 25, 2002).

18. Leffingwell and Associates. Odor and Flavor Detection Thresholds in Water (in Parts per Billion). Available at http://www.leffingwell.com/odorthre.htm (cited August 27, 2002).

19. M Devos, F Patte, J Rouault, P Laffort, LJ Van Gemert. Standardized human olfactory thresholds. Oxford: IRL Press, 1990.

20. KD Deibler, EH Lavin, RST Linforth, AJ Taylor, TE Acree. Verification of a mouth simulator by in vivo measurements. J Agric Food Chem 49: 1388–1393, 2001.

21. KD Deibler. Measuring the effect of food composition on flavor release using the retronasal aroma simulator and solid phase microextraction. Ph.D. dissertation, Cornell University, Ithaca, NY, 2001.

22. V Aubry, P Schlich, S Issanchou, P Etiévant. Comparison of wine discrimination with orthonasal and retronasal profilings: Application to Burgundy Pinot Noir wines. Food Qual Preference 10: 253–259, 1999.

23. KJ Burdach, RL Doty. The effects of mouth movements, swallowing, and spitting on retronasal odor perception. Physiol Behav 41: 353–356, 1987.

24. KJ Burdach, EP Koster, JH Kroeze. Nasal, retronasal, and gustatory perception: An experimental comparison. Percept Psychophys 36: 205–208, 1984.

25. VB Duffy, WS Cain, AM Ferris. Measurement of sensitivity to olfactory flavor: Application in a study of aging and dentures. Chem Senses 24: 671–677, 1999.

26. YL Kuo, RM Pangborn, AC Noble. Temporal patterns of nasal, oral, and · retronasal perception of citral and vanillin and interaction of these odourants with selected tastants. Int J Food Sci Technol 28: 127–137, 1993.

27. JC Stevens, WS Cain. Smelling via the mouth: Effect of aging. Percept Psychophys 40: 142–146, 1986.

28. E Voirol, N Daget. Comparative study of nasal and retronasal olfactory perception. Lebensm Wiss Technol 19: 316–319, 1986.

29. J Pierce, BP Halpern. Orthonasal and retronasal odorant identification based upon vapor phase input from common substances. Chem Senses 21: 529–543, 1996.

30. The American Heritage Electronic Dictionary of the English Language, 3rd ed. Boston: Houghton Mifflin, 1992.

31. BJ Cowart, BP Halpern, EK Varga. A clinical test of retronasal olfactory function (abstract). Chem Senses 24: 608, 1999.

32. DA Wininger, BP Halpern. Retronasal and orthonasal odorant identification without sniffing (abstract). Chem Senses 24: 600, 1999.

33. BP Halpern, VG Puttanniah, M Ujihara. Retronasal and orthonasal identifica-tions of vapor-phase food-grade liquid extracts of plant materials (abstract). Chem Senses 25: 610, 2000

34. VG Puttanniah, BP Halpern. Retronasal and orthonasal identifications of odorants: Similarities and differences. (abstract). Chem Senses 26: 802, 2001.

35. BC Sun, BP Halpern. Retronasal and orthonasal odorant interactions: Masking (abstract). Chem Senses 26: 1103, 2001.

36. BC Sun, BP Halpern. Asymmetric interactions between heterogeneous retro-nasal and orthonasal odorant pairs. Chem Senses, 27: 1103, 2002. Available at http://chemse.oupjournals.org/cgi/content/full/27/7/661/DC1 (abstract).

4

Sensory Analysis and Olfactory Perception: Some Sources of Variation

T. Thomas-Danguin, C. Rouby, G. Sicard, M. Vigouroux, S. Barkat, V. Brun, and V. Farget
NSS, Université Claude Bernard, Lyon, France

F. Rousseau and J. P. Dumont
LEIMA, INRA, Nantes, France

A. Johansson, A. Bengtzon, and G. Hall
The Swedish Institute for Food and Biotechnology (SIK), Gothenburg, Sweden

W. Ormel, N. Essed, and C. De Graaf
Wageningen Agricultural University, Wageningen, The Netherlands

S. Bouzigues, S. Gourillon, L. Cunault, and S. Issanchou
UMRArômes, INRA, Dijon, France

S. Hoyer, U. Simchen, and F. Zunft
University of Potsdam, Bergholz-Rehbruecke, Germany

T. Hummel
University of Dresden Medical School, Dresden, Germany

R. Nielsen
Biotechnological Institute, Kolding, Denmark

S. Koskinen and H. Tuorila
University of Helsinki, Helsinki, Finland

I. INTRODUCTION

Olfaction is very much present and influential in our everyday life and its day-to-day quality [1]. For instance, it has been shown by Delwiche [2] that taste and smell are rated as being the most important sensations in "flavor." Thus, olfaction is the key to our relationship with food and in particular plays a major role in identifying it [3]. That is why significant changes in chemosensory perception have the capacity of interacting with aroma perception, diet selection, and to some extent nutritional status [4].

However, human olfactory sensitivity, and especially aroma perception, can be affected by a large number of factors, such as age, gender, smoking habits, disease, or injury [5–8]. Such modulation of olfactory capacity can have dramatic impact on the results of sensory evaluation studies based on olfactory tasks. This is especially important in studies using only a few panelists [gas chromatography olfactometry (GCO), quantitative flavor profiling, etc.]. It thus seems that food flavor studies could benefit from their panelists' being assessed for olfactory and other sensory performance.

In the context of a large European research project on Healthy Aging (http://healthsense.ucc.ie/), our group performed a survey of olfactory performance in the European population. Olfactory tests and questionnaires were used to assess the influence of several factors, such as age, culture, gender, illness, medication, and lifestyle, on olfactory sensitivity.

A first study was designed to assess the effectiveness of three European clinical olfactory tests for characterizing the olfactory abilities of healthy subjects qualified to take part in sensory analysis studies. On the basis of this first study, certain sources of variation in human olfactory perceptions were highlighted and their respective impacts on a number of olfactory tasks were assessed.

In a second study, a new olfactory test, the European Test of Olfactory Capabilities (ETOC), was used to assess the links between panelists' health status and their general olfactory ability.

II. STUDY 1: EFFECT OF CULTURE, GENDER, AND AGE ON OLFACTORY TASKS IN ELDERLY SUBJECTS

A. Materials and Methods

Three European olfactory tests, developed in different European countries for clinical purposes, were used [9]. The first, known as Sniffin'Sticks [10], was developed in Germany. The second test was the Scandinavian Odor

Identification Test (SOIT), which was developed in Sweden [11]; the third test was the Lyon Clinical Olfactory Test (LCOT), developed in France [12].

Three olfactory tasks are used in Sniffin'Sticks: threshold measurement, odor discrimination, and odor identification. The threshold task consists of measuring the detection threshold for n-butanol. Sixteen dilutions of n-butanol are presented to the subject in increasing order, using a three-alternative forced-choice (3-AFC) paradigm. Detection thresholds are determined by a single staircase method [13]. In the discrimination task, subjects have to discriminate 16 pairs of odors, presented as triads (two odors are identical and one is different: which one?) in a 3-AFC procedure. The identification task consists of 16 odors to be identified, following a 4-AFC procedure and using verbal alternatives. In this test, it has been recommended to use a composite score, which may be better suited for the clinical assessment of olfactory dysfunction than an isolated measure of olfactory performance [10]. However, as our goal was to evaluate the efficiency of several olfactory tasks, we did not calculate and use such a score in this study.

The Scandinavian Odor Identification Test (SOIT) relies on odor identification alone. The testing procedure is the same as the one used in the Sniffin'Sticks identification task.

The Lyon Clinical Olfactory Test is based on thresholds, suprathreshold detection, and identification tasks. The LCOT includes two threshold measurements, one for R-(+)-carvone and the other for tetrahydrothiophene (THT). The same threshold determination procedure is used for both compounds. Five concentration levels are presented in increasing order with a 4-AFC paradigm. In the suprathreshold detection task 16 odors, presented at a suprathreshold level, have to be detected by following a 4-AFC procedure. The identification task consists of 16 odors to be identified, following a 4-AFC procedure and using verbal alternatives.

A total of 121 healthy elderly subjects aged from 55 to 79 years (mean age $= 65.6 \pm 7.1$; 59.5% women) participated in the experiment. These 121 subjects were recruited in four European countries (France, Sweden, the Netherlands, and Germany). The number of subjects per country and per assessment is presented in Table 1.

All the subjects were noninstitutionalized. They were recruited in associations devoted to elderly people or by means of posters or want ads. They were not prescreened in any way, but they had to be able to travel to the laboratories by themselves and at the right time. They were not trained before testing and were rewarded either with a gift or with a voucher.

As we wanted to compare the tasks used in the three tests on the same sample of subjects, in France and Sweden, every subject performed all the tasks of the three tests. In the Netherlands only the discrimination and

Table 1 Number of Subjects per Country per Measurement[a]

Sniffin'Sticks			SOIT	Lyon Clinical Olfactory Test		
Threshold	Discrimination	Identification	Identification	Thresholds	S-T detection	Identification
F: 30 S: 25	F: 30 S: 25 D: 41	F: 30 S: 25 N: 25 D: 41	F: 30 S: 25 N: 25	F: 30 S: 25	F: 30 S: 25 N: 25	F: 30 S: 25 N: 25

[a] F, France; S, Sweden; D, Germany; N, Netherlands.

identification tasks of the three tests were performed, and in Germany, only the identification task of Sniffin'Sticks (Table 1).

All the tests were translated: that is, all the questions and answer sheets were worded in the subjects' native language.

A three-way analysis of variance (ANOVA) was performed on the test scores to assess the respective effects of country, gender, and (5-year) age group.

B. Results

1. Threshold Detection Tasks

Threshold tasks were involved in two of the tests: Sniffin'Sticks and the LCOT.

For Sniffin'Sticks, statistical analysis of our data from France and Sweden showed a significant effect of gender ($F(1, 48) = 4.5$, $p = 0.039$). Men obtained higher thresholds than women (mean difference $= 0.95$). No significant effect of country ($F(1, 48) = 0.57$, $p = 0.46$) or age group ($F(4, 48) = 0.55$, $p = 0.70$) and no significant interaction between factors were found.

The LCOT includes two threshold measurements, one for R-(+)-carvone and the other for tetrahydrothiophene (THT). For R-(+)-carvone, statistical analysis of our data from France and Sweden showed a significant gender effect ($F(1, 48) = 4.7$, $p = 0.035$). Men have higher thresholds than women (mean difference $= 0.71$). No significant effect of country ($F(1, 48) = 2.79$, $p = 0.10$) or age group ($F(4, 48) = 0.17$, $p = 0.95$) and no significant interaction between factors were found. For THT, statistical analysis of our data from France and Sweden showed no significant effect of country ($F(1, 48) = 1.47$, $p = 0.23$), age group ($F(4, 48) = 1.25$, $p = 0.30$), or gender ($F(1, 48) = 2.26$, $p = 0.14$) and no significant interaction between factors.

2. Suprathreshold Detection Task

The suprathreshold detection task is used in the LCOT. Statistical analysis of our data from France, Sweden, and the Netherlands showed a significant effect of country ($F(2, 73) = 4.62$, $p = 0.013$). The means comparison test (Scheffé; $\alpha = 0.05$) found that the Dutch (NL) sample tended to have lower scores than the Swedish (S) and French (F) ones (F − NL, mean difference $= 0.57$, $p = 0.032$; S − NL, mean difference $= 0.70$, $p = 0.009$). No significant effect of age group ($F(4, 73) = 1.75$, $p = 0.15$) or gender ($F(1, 73) = 0.85$, $p = 0.36$) and no significant interaction between factors were found.

3. Discrimination Task

This olfactory task is used in Sniffin'Sticks. Statistical analysis of our data from France, Sweden, and Germany found no significant effect of country ($F(2, 88) = 0.37$, $p = 0.69$), age group ($F(4, 88) = 0.32$, $p = 0.86$), or gender ($F(1, 88) = 0.48$, $p = 0.49$), and no significant interaction between factors. However, χ^2 analysis of the distribution of percentage discrimination error per country and odor pair revealed a significant difference for five pairs of odors (see Fig. 1).

4. Identification Task

With Sniffin'Sticks, data were collected in France, Sweden, Germany, and the Netherlands. The statistical analysis found no significant effect of country ($F(3, 113) = 1.94$, $p = 0.13$), age group ($F(4, 113) = 0.70$, $p = 0.59$), or gender ($F(1, 113) = 0.39$, $p = 0.43$), and no significant interaction between factors.

The ANOVA on the SOIT identification scores (France, Sweden, and the Netherlands) showed a highly significant effect of country ($F(3, 73) = 14.7$, $p < 0,0001$). A χ^2 analysis on percentage identification error per

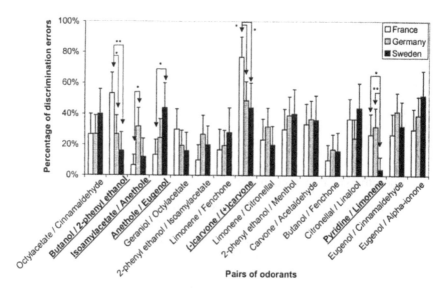

Figure 1 Percentage of discrimination errors (Sniffin'Sticks) per country and per pair of odorant. Error bars indicate the 95% confidence interval. Arrows point to the significant differences (*$p < 0.05$, **$p < 0.01$, ***$p < 0.001$).

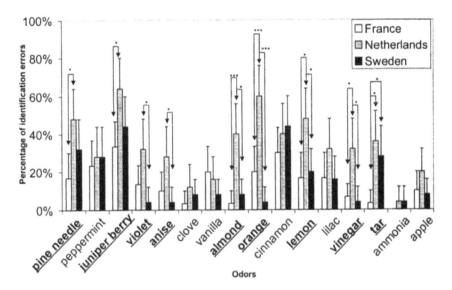

Figure 2 Percentage of identification errors (SOIT) per country and per odor. Error bars indicate the 95% confidence interval. Arrows point to the significant differences (*$p < 0.05$, **$p < 0.01$, ***$p < 0.001$). SOIT, Scandinavion Odor Identification Test.

country revealed a difference for 9 of the 16 odors (Fig. 2). The identification score ANOVA, however, found no significant effect of gender ($F(1, 73) = 0.46$; $p = 0.50$) or age group ($F(4, 73) = 1.26$; $p = 0.29$), and no significant interaction between factors.

LCOT data were collected in France, Sweden, and the Netherlands. The ANOVA found identification scores to decrease significantly with increasing age ($F(4, 73) = 3.25$, $p = 0.017$; see Fig. 3). Moreover, there was a significant effect of gender on identification ability ($F(1, 73) = 5.47$, $p = 0.022$). On average, women managed to identify one odor more than men (W: 13.0; M: 12.0). The ANOVA also found a significant effect of country on identification score ($F(2, 73) = 5.92$, $p = 0.004$). A χ^2 analysis of percentage identification error per country found a difference for 5 of the 16 odors (Fig. 4).

C. Discussion

First of all, it is to be noted that, on the basis of their scores on the three tests and norms of these tests (10–12), four subjects in Sweden and four in

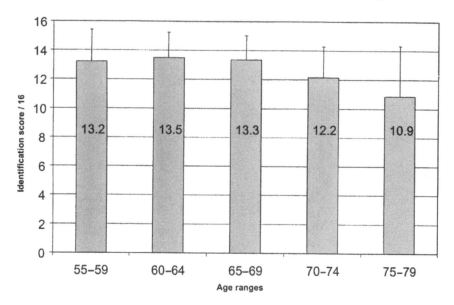

Figure 3 Change in identification score (LCOT) with age. LCOT, Lyon Clinical Olfactory Test.

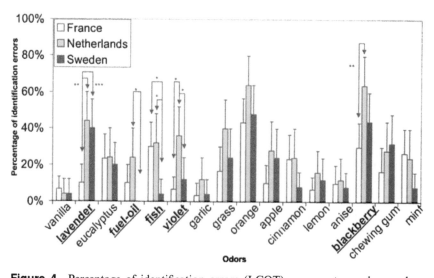

Figure 4 Percentage of identification errors (LCOT) per country and per odor. Error bars indicate the 95% confidence interval. Arrows point to the significant differences (*$p < 0.05$, **$p < 0.01$, ***$p < 0.001$). LCOT, Lyon Clinical Olfactory Test.

France were found to be anosmic or severely hyposmic. Only one subject in Sweden was suspected of a specific anosmia to R-(+)-carvone [14]. The scores of all these subjects were removed before data analysis.

In the context of our study, it was shown that one of the major factors that modulate olfactory thresholds is gender. Although threshold measurements are usually thought to be closely related to olfactory sensitivity [15–16], when only healthy elderly people are considered, the effect of age has not been shown to be significant. We did not find any effect of country on threshold tasks, which thus seem to be cross-cultural.

However, threshold measurements involve heavy and time-consuming procedures. Moreover, our results, and those of other authors [17], show a variation in performance with the odorant selected. Although some studies have found correlations between the thresholds for different odorants [18], it seems risky to extrapolate a general sensitivity level from the measurement of the threshold to only one or two odorants. This raises the question of which compound to choose, and it has been suggested that phenyl ethyl alcohol (PEA) is a good compromise candidate [19–20].

The only source of variation in suprathreshold detection scores was found to be the country. This country effect is not yet very well understood but would seem to be due to the Dutch panel as such, although not directly connected to age. In fact, even if suprathreshold detection is an easy and short procedure and seems a priori to be closely related to olfactory sensitivity, our results did not find any significant effect of healthy aging on scores. However, it is to be noted that there is a high ceiling effect on scores in this task as used in the LCOT. On average, panelists detected 15.5 odors of 16, a very high mean score that could explain the poor discrimination power of the task.

Olfactory discrimination is also an appealing procedure as it is short and easy and correlated to sensitivity measurements [15]. Moreover, this task appears to be cross-cultural as it does not rely on verbal encoding [10]. Even if our results on Sniffin'Sticks indicated no significant effect of country, gender, or age group on the global discrimination score, a variation in error rate was found from country to country and for some couples of odorants. We previously reported [9,21–22] that there is an effect of odor familiarity and verbal encoding on the discrimination task. Finally, our results indicate that the discrimination task, at least as used in Sniffin'Sticks, is not very sensitive to the decrease in olfactory performance with increasing age, as far as healthy elderly subjects, aged 55 and over, are concerned.

The identification task is the task most frequently used to assess olfactory ability. This is mainly due to the fact that identification (a) is an easy and short procedure when 4-AFC is used [23–24] and (b) is representative of olfactory capabilities, being strongly correlated with sensitivity [8,15]. It has also been previously proved that identification rests on cognitive, and especially verbal, capacity [25–27]. Our results showed that depending on the choice of odors and their intensity level, but using exactly the same procedure (4-AFC, verbal alternatives, and 16 odors to identify), country and/or gender and/or age could have an impact on test scores. Nevertheless, our data indicated that only the identification task used in the LCOT was sensitive to aging. This could be the result of a procedure that combines identification and detection and reduces the number of correct answers that can be obtained by chance to 1 of 16 [12]. Thus, on the basis of the findings of our study, the identification task appears to be the only one that could be sensitive to olfactory loss with healthy aging. However, our data also indicate that there is a variation in score with country for some odorants, and we previously reported that the identification task is influenced by the familiarity, intensity, and typicality of the odor stimulus [9,21–22].

This first study highlights several points that need to be taken into account in any sensory analysis based on olfactory tasks. Thus gender is shown to be able to influence scores in threshold and identification tasks. Cultural origin may influence scores in suprathreshold detection and identification tasks. Scores in identification tasks may also be influenced by healthy aging when corrected by detection measures. Thus it appears to be important (a) to assess and to check panelists' olfactory capabilities regularly and (b) to integrate certain parameters in sensory data processing that reflect the panelists' sensory sensitivity and could account for interindividual variance. For this purpose, we developed the European Test of Olfactory Capabilities.

III. STUDY 2: LINKS BETWEEN OLFACTION AND HEALTH

A. Materials and Methods

1. The European Test of Olfactory Capabilities

On the basis of the results obtained in Study 1, our team developed the European Test of Olfactory Capabilities [28]. This test is based on a combination of a suprathreshold detection task and an identification task. To construct the test, some of the transcultural odors previously selected in study 1 (vanilla, clove, apple, eucalyptus, cinnamon, garlic, anise, orange, lemon, and mint) were used [29]. Odor intensity was adjusted to increase the level of difficulty, especially in the detection task. Alternative odor names were also carefully selected in the identification protocol, so as to adjust the difficulty of identification [30]. The test was validated in six European countries (Germany, Denmark, France, Finland, The Netherlands, and Sweden) on more than 1100 healthy European subjects from 20 to 97 years of age (61% women; 54.6 ± 18.3 years old). We have previously reported [28] that the ETOC is a self-administered and short test (taking less than 20 min). The ETOC has been cross-culturally validated in Europe, inasmuch as the scores do not depend on the culture of the panelists (three-way ANOVA, country effect: $F(4, 985) = 0.5$, $p = 0.74$). This test is very sensitive (a) to gender differences (three-way ANOVA, gender effect: $F(1, 985) = 13.4$, $p = 0.0003$; women outperformed men) and (b) to the decline in olfactory performance over healthy aging (three-way ANOVA, age group effect: $F(13, 985) = 7.8$, $p < 0.0001$). The test has also been validated for test-retest reliability ($r_{51} = 0.90$, $n = 1.003$).

2. Olfactory Testing and Health Questionnaire

A total of 340 subjects from 20 to 97 years of age (55% women; 53.8 ± 21.1) participated in a European survey to assess links between olfactory capability and health status. They were recruited in three European countries (163 in Lyon, France, 87 in Gothenburg, Sweden, and 90 in Wageningen, Netherlands).

All the subjects were noninstitutionalized. They were recruited in associations devoted to elderly people or by means of posters or want ads. They were not prescreened in any way, but they had to be able to go to the laboratories by themselves and at the right time. They were not trained before testing and were rewarded either with a gift or with a voucher.

All the 340 subjects performed the previously described ETOC, and during the same session they filled out a 20-item questionnaire to record their health status. Both tasks were self-administrated under the control of the experimenter.

All the questions and answer sheets were worded in the subjects' native language.

3. Data Processing

On the basis of the ETOC scores, panelists were divided into two groups: (a) a Low Olfaction Group (LOG) of 88 subjects with low scores (>1 detection error and/or >4 identification errors) and a Normal Olfaction Group (NOG) comprising the remaining 252 subjects, who obtained higher scores.

In order to obtain a clear representation of the data, we first created several groups of panelists on the basis of the answers they gave to each question in the health questionnaire. Then, for each of these groups, we measured the percentages of LOG and NOG subjects. The statistical difference between each group was assessed by χ^2 tests.

B. Results

The results presented in this part are only preliminary. The data have been selected in order to illustrate some links between health problems and a decrease in olfactory capability.

First, a link was found between smoking habits and olfactory capability ($\chi^2(2) = 5.9$, $p = 0.05$). In particular, a significantly higher proportion ($\chi^2(1) = 5.3$, $p = 0.02$) of subjects with low olfactory scores (LOGs) were found in the group who answered that they had stopped smoking (LOG: 33.3%) as compared to those who answered that they were still smoking (LOG: 16.7%). In terms of the distribution of subjects with high and low olfactory performances, the difference between the group who answered that they had stopped smoking (LOG: 33.3%) and the group who answered that they had never smoked (LOG: 23.6%) just failed to _be significant at the 5% level ($\chi^2(1) = 2.9$, $p = 0.08$).

Some questions in our questionnaire addressed health problems in the mouth. Thus, a significantly higher proportion ($\chi^2(1) = 4.5$, $p = 0.03$) of LOG subjects were found in the group who answered that they had difficulty chewing (LOG: 50.0%) as compared to those who answered that they have no difficulty chewing (LOG: 24.5%). The difference became highly significant ($\chi^2(1) = 18$, $p < 0.0001$) when comparing those who answered that they always or often suffered from dry mouth (LOG: 61.5%) and those who answered that they sometimes, rarely, or never suffered from dry mouth (LOG: 22.8%). However, no statistical difference was found ($\chi^2(1) = 1.7$, $p = 0.19$) between the group who answered that they always or often felt a bad taste in the mouth (LOG: 44.4%) and the group who answered that they sometimes, rarely, or never felt a bad taste in mouth (LOG: 25.0%).

We also asked questions about 11 common health problems that could appear with aging [respiratory, cardiac, digestive, urinary, muscular, "ear, nose, or throat" (ENT), diabetes, cholesterol, thyroid, insomnia, and nervous breakdown problems]. The distribution of LOG and NOG subjects was compared between the group who reported an illness and the group who did not. No difference appeared between the two distributions for subjects who declared they had nervous breakdown, respiratory, digestive, cholesterol, thyroid, or, quite surprisingly, ENT problems (χ^2 test, $p < 0.005$; correction for multiple tests: Bonferroni adjusted $\alpha = 0.005$).

However, people suffering from the five other problems belonged significantly more often to the low olfaction group (Cardiac [$\chi^2(1) = 18$, $p < 0.0001$], diabetes [$\chi^2(1) = 13$, $p = 0.0003$], urinary [$\chi^2(1) = 14$, $p = 0.0002$], insomnia [$\chi^2(1) = 18$, $p < 0.0001$], muscular [$\chi^2(1) = 24$, $p < 0.0001$]).

One of the last questions asked for a self-rating of general health status. A significantly higher proportion of LOG subjects were found in the group who reported very poor, poor, or fair health as compared to those who reported good or very good health [$\chi^2(2) = 24.9$, $p < 0.0001$].

C. Discussion

First of all, it is to be noted that on the basis of ETOC scores, 12 subjects (8 in France, 3 in Sweden, and 1 in the Netherlands) were found to be anosmic or severely hyposmic or to have other cognitive problems that did not allow them to identify more than four odors correctly even if they detected up to 15 odors. The scores of all these subjects were removed before data analysis; the panel was thus reduced to 340 subjects.

The first results of our European survey confirmed that there were some links between olfactory capability and health status or lifestyle. First of all, on the question of the impact of smoking habits on olfactory ability, our data show that people who currently smoke and people who have never smoked have the same olfactory performance on average. Conversely, a high proportion of people who declared that they had stopped smoking also obtained lower scores on the olfactory test. The impact of smoking habits on olfactory capability today remains an open question. At the moment, we have no clear hypothesis as to what can happen when people change their smoking habits. As we have more details about smoking habits in our questionnaire, we will focus much more on these data in order to explain the effect of such a change.

Our results show that chewing problems or dry mouth problems are linked to lower olfactory capability. Dry mouth may be accompanied by dry nose, which could account for this result. At this point it seems important to mention the effect of medication, which could be responsible for dry mouth problems [31]. Thus another explanation for the link observed between olfactory performance and dry mouth problems could be that some medication has a direct impact on the olfactory function through this drying side effect.

In agreement with previous reports [4,32], our results indicated that many of the illnesses that occur with increasing age can be linked to a decrease in olfactory performance. At this level, we have to focus much more on our data in order to disentangle the direct age effect and/or the effect of the illness itself and/or the effect of the medication prescribed. The mechanisms by which drugs alter the chemical senses are not well understood [33]. Drugs may alter receptor cell turnover or interfere with transduction mechanisms. They may also alter neurotransmitters in the central nervous system that process chemosensory information. Nevertheless, it is important to note that there was a higher proportion of subjects with olfactory deficits among those who declared that they suffered from at least one disease. It would also be interesting to know the impact of the illness or the medication used in relation to the decrease in olfactory sensitivity or in cognitive performance.

. The last result presented concerns the link between the decrease in olfactory performance and general self-rated health status. It appears that the more people think they are in good health, the higher their olfactory performance, and vice versa. To our mind, this is one of the major proofs of the impact of olfactory perception, especially in the context of food flavor, on people's wellness and willingness to keep healthy. Conversely, a deficit in olfactory capability could contribute to a decrease in the perception of food flavor and thus to a decrease in appetite for food, in turn contributing to the low nutritional status frequently associated with the health problems of the elderly.

IV. CONCLUSIONS

Challenges in the characterization of flavor compounds rest, at least in part, on olfactory and/or gustatory assessment. However, in sensory analysis studies, the measuring instrument is "the human element" [34–35]. That is the reason why it appears very important to know, as well as possible, the capabilities of this "measuring instrument."

We have seen that age, culture, and gender modulate the three main abilities of the olfactory function, which are stimulus detection, discrimination, and identification. These factors, and especially age, can also affect other olfactory tasks such as intensity or pleasantness ratings [6]. It has further been shown that general health status and illness (medication), as well as mouth problems and smoking habits, are linked to olfactory abilities. All these factors have to be kept in mind when one uses olfactory tasks and recruits subjects for sensory panels.

One way to "standardize" the sensory abilities of panelists is to use olfactory tests. Some of them, such as the ETOC, have been designed for sensory evaluation purposes in order to provide sensory and cognitive clues to olfactory capability. The test scores could thus usefully be included in the processing of sensory data. The assessment of panelists for olfactory and for other sensory sensitivity may, in part, account for the difference between flavor perception and flavor release, which remains a major challenge in the characterization of flavor compounds.

ACKNOWLEDGMENTS

This work was carried out with financial support from the European Commission Quality of Life Fifth Framework Programme QLK1-CT 1999-00010, HealthSense project: "Healthy ageing: How changes in sensory physiology, sensory psychology and socio-cognitive factors influence food choice."

The authors wish to thank Claire Chabanet (UMRArômes, INRA, Dijon, France) for statistical advice.

REFERENCES

1. DA Deems, RL Doty, GR Settle, V Moore-Gillon, P Shaman, AF Mester, CP Kimmelman, VJ Brightman, JB Snow. Smell and taste disorders: A study of 750 patients from the University of Pennsylvania Smell and Taste Center. Arch Otolaryngol Head Neck Surg 117:519–528, 1991.
2. JF Delwiche. Attributes believed to impact flavor: An opinion survey. J Sens Stud. In press.
3. C Murphy. Cognitive and chemosensory influences on age-related changes in the ability to identify blended foods. J Gerontol 40:47–52, 1985.
4. SS Schiffman. Taste and smell in disease. New Engl J Med 308:1275–1279, 1337–1343, 1983.
5. RL Doty. Olfaction. Annu Rev Psychol 52:423–452, 2001.
6. CJ Wysocki, AN Gilbert. National geographic smell survey: Effects of age are heterogeneous. Ann NY Acad Sci 561:12–28, 1989.
7. WS Cain. Testing olfaction in a clinical setting. Ear Nose Throat J 68:322–328, 1989.
8. WS Cain, JF Gent, RB Goodspeed, G Leonard. Evaluation of olfactory dysfunction in the Connecticut Chemosensory Clinical Research Center. Laryngoscope 98:83–88, 1988.
9. T Thomas-Danguin, C Rouby, G Sicard, M Vigouroux, V Farget, A Johansson, A Bengtzon, G Hall, W Ormel, C De Graaf, S Tricot, JP Dumont. Assessing olfactory performances: How much odour identification and discrimination are influenced by culture, typicality and intensity.
10. T Hummel, B Sekinger, SR Wolf, E Pauli, G Kobal. "Sniffin'Sticks": Olfactory performance assessed by the combined testing odor identification, odor discrimination and olfactory threshold. Chem Senses 22:39–52, 1997.
11. S Nordin, A Brämerson, E Lidén, M Bende. The Scandinavian Odor-Identification Test: Development, reliability, validity and normative data. Acta Otolaryngol (Stockh) 118:226–234, 1998.
12. C Rouby, G Sicard, M Vigouroux, T Jiang, I Gallice, M Demolis, T Thomas-Danguin, M Bensafi. The Lyon Olfactory Clinical Test: Development and normative data.

13. RL Doty. Olfactory system. In: TV Getchell, RL Doty, LM Bartoshuk, JB Snow, eds. Smell and Taste in Health and Disease. New York: Raven Press, 1991, pp 1803.

14. JE Amoore. Specific anosmia and the concept of primary odours. Chem Senses Flav 2:267–281, 1977.

15. RL Doty, R Smith, DA McKeown, J Raj. Tests of human olfactory function: Principal components analysis suggests that most measure a common source of variance. Percect Psychophys 56:701–707, 1994.

16. WS Cain, MD Rabin. Comparability of two tests of olfactory functioning. Chem Senses 14:479–485, 1989.

17. DB Kurtz, TL White, PR Sheehe, DE Hornung, PF Kent. Odorant confusion matrix: The influence of patient history on patterns of odorant identification and misidentification in hyposmia. Physiol Behav 72:595–602, 2001.

18. M Yoshida. Correlation analysis of detection threshold data for "standard test" odors. Bull Faculty Sci Eng Chuo Univ 27:343–353, 1984.

19. RL Doty, TP Gregor, RG Settle. Influence of intertrial interval and sniff-bottle volume on phenyl ethyl alcohol odor detection thresholds. Chem Senses 11:259–264, 1986.

20. RL Doty, WE Brugger, PC Jurs, MA Orndorff, PJ Snyder, LD Lowry. Intranasal trigeminal stimulation from odorous volatiles: Psychometric responses from anosmic and normal humans. Physiol Behav 20:175–185, 1978.

21. T Thomas-Danguin, C Rouby, G Sicard, M Vigouroux, A Johansson, A Bengtzon, G Hall, W Ormel. Odour identification and discrimination: Influence of culture and typicality on performance (abstract). Chem Senses 26:1062, 2001.

22. T Thomas-Danguin, C Rouby, G Sicard, M Vigouroux, A Johansson, A Bengtzon, G Hall, W Ormel, S Tricot, JP Dumont. Testing olfactory performances: Influence of cultural and psychophysical factors on olfactory performance (abstract). Proceedings of the 4th Pangborn Sensory Science Symposium, Dijon, 2001, pp 59.

23. D Dubois, C Rouby. Names and categories for odors: The "Veridical Label." In C Rouby, B Schaal, D Dubois, R Gervais, A Holley, eds. Olfaction and Cognition. New York: Cambridge University Press, 2002, pp 47–66.

24. WS Cain, RJ Krause. Olfactory testing: Rules for odor identification. Neurol Res 1:1–9, 1979.

25. M Larsson, D Finkerl, NL Pedersen. Odor identification: Influences of age, gender, cognition and personality. J Gerontol 55B:304–310, 2000.

26. J Corwin, M Serby, P Larson, D Kelkstein. Olfactory identification in aging and Alzheimer's disease: Task-specific effects. J Clin Exp Neuropsychol 12:1–8, 1990.

27. T Schemper, S Voss, WS Cain. Odor identification in young and elderly persons: Sensory and cognitive limitations. J Gerontol 36:446–452, 1981.

28. T Thomas-Danguin, C Rouby, G Sicard, M Vigouroux, V Farget, A Johanson, A Bengtzon, G Hall, W Ormel, C De Graaf, F Rousseau JP Dumont. Development of the ETOC: A European Test of Olfactory Capabilities. Rhinology. In press.

29. M Vigouroux, T Thomas-Danguin, C Rouby, G Sicard, V Farget, W Ormel, C De Graaf, A Johansson, A Bengtzon, G Hall, T Hummel. Testing olfactory performances in four European countries: Influence of culture (abstract). Chem Senses. In press.

30. C Rouby, D Dubois. Odor discrimination, recognition and semantic categorization. Chem Senses 20:78–79, 1995.

31. LM Sreebny, SS Schwartz. A reference guide to drugs and dry mouth, 2nd ed. Gerodontology 14:33–47, 1997.

32. SS Schiffman. Taste and smell losses in normal aging and diseases. J Am Med Assoc 278:1257–1362, 1997.

33. SS Schiffman. Changes in taste and smell: Drug interactions and food preferences. Nutr Rev 52:S11–S14, 1994.

34. M Meilgaard, GV Civille, BT Carr. Sensory Evaluation Technique. Boca Raton, FL: CRC Press, 1999.

35. EP Köster. Sensory evaluation in natural environment. In P Shreier, ed. Flavour 81. Berlin, New York: Walter de Gruyter, 1981, pp 93–100.

5

Similarity and Diversity in Flavor Perception

Terry E. Acree, Kathryn D. Deibler, and Katherine M. Kittel
New York State Agricultural Experiment Station, Cornell University, Geneva, New York, U.S.A.

I. INTRODUCTION

Specific anosmias are likely due to differences in the expression of specific olfactory receptor protein genes. Given the number of human olfactory receptor genes (~ 40) and their distribution across most human chromosomes (all except C 20 and Y), it would not be surprising to find differences in the responses of most individuals to a well designed chemical probe presented to subjects using gas chromatography olfactometry (GCO). If such a probe were combined with a reproducible and efficient GCO protocol, subjects could be "phenotyped" on the basis of their peripheral response to odorants. Such data would greatly assist an investigation of functional diversity in human olfaction and its impact on perception. In the future, when methods are developed that identify an individuals olfactory receptor gene composition, techniques that measure the phenotypic expression of these genes will be needed to identify their associated function. The data needed to assign function to specific olfactory genes could be provided by GCO.

A. Challenges in Odor Measurement

In 1963, Rothe and Thomas proposed the Odor Unit (concentration/ threshold) to quantify the potency of odorants in a flavor system, i.e., beverages, foods, fragrances, and other natural products [1]. This relationship has been used to distinguish constituents that may contribute to aroma (potent odorants) from the vast majority of volatiles in natural products that are at concentrations below their odor threshold [2–7]. Although odorant concentration can be measured with good precision (<10%) odorant thresholds are extremely variable (>100%). The large "error" in threshold measurements is a combination of the variance between and within subjects [8,9]. The between subject variance is often partitioned into three factors: general anosmia, specific anosmia, and temporal effects. The variation in thresholds due to general anosmia, i.e., an equal difference in response to all odorants, has been circumvented by selection of people based on their olfactory responses to a standard set of compounds [10–12]. General anosmia can result from a number of factors; the most common is due to differences in nasal airflow caused by genetic and congenital factors, or disease and injury. The use of standard odorants such as n-butanol to select subjects with "little general anosmia" is common in the assessment of air quality by olfactometry [13]. However, it is possible to use standards to correct for general anosmia in olfactometry data, as has been demonstrated for GCO data [12].

The numerator in the Odor Unit is the odorant concentration, which can be measured very precisely. Techniques such as isotope dilution analysis, external standards, and ultraviolet spectroscopy have been used to measure odorants at their functional concentrations in food systems [2,14–19]. Furthermore, improvements in chemical sampling have increased sensitivity and yielded data more relevant to the human eating experience. Among these, mouth simulators and in vivo techniques [20] have produced more theoretically satisfying results. In vivo measurements of exhaled nasal gas using mass spectrometry directly measure the actual concentration of a compound during eating [21,22] with great precision but at present with limited sensitivity. Mouth simulators produce a pattern of volatility similar to in vivo retronasal concentrations with great sensitivity [19,23–27]. However, the measured threshold is the limiting factor to both precision and accuracy of Odor Units.

Many methods exist for determining odorant detection thresholds, the denominator of the Odor Unit, yet the variability of most of these measurements remains greater than 100% [11,9,28,29]. Among the techniques used to deliver odorants to panelists are olfactometers, food models, GCO, and water [30–35]. The enormous range of thresholds reported for a single compound is partially due to the different delivery methods used and partially due to the variation inherent in the human olfactory system itself. Using GCO, Marin and associates [11] demonstrated that variation between people was greater than variation due to either the age or the sex of the subject and was significantly different from the within subject variation. In the process of developing a more accurate GCO protocol to compare individual thresholds across a large [~30] set of odorants simultaneously, Friedrich et al. observed that all of their test subjects [~8] showed large differences in their thresholds of 2 to 5 of the 30 compounds tested [12]. These studies imply that the functional difference in the olfactory system of individuals may be responsible for the large variation observed in human olfactory thresholds and Odor Units. An understanding of these differences will be helpful in developing a method to standardize human threshold data, especially when associated with genetic and psychophysical data [36–38]. Taken together, this information substantiates the idea that specific anosmia is not a disorder but an expression of normal human diversity and that GCO could be useful in the study of olfactory phenomics. Specific anosmia is the difference in the potency responses to just a few odorants by individuals while they exhibit the same potency response to all other odorants. This is probably not an "error" but reflects the extensive differences in the genetically determined potency responses of individuals. In this way specific anosmia, originally defined as a genetic defect, is a phenotypic expression of normal human diversity. Ideally, a method that removes the effects of general anosmia from threshold measurements, while averaging out the effects of specific anosmia, would yield more precise and accurate populations thresholds. However, a more important question is the impact of specific anosmia on the perception and behavior of individuals. To study the "phenomics of anosmia," tools that yield precise measures of odorant potency in individuals will be needed. Phenomics is a study of relationships between the multitude of olfactory phenotypes, defined by the odor potency and perceptual character of individual odorants, and the resulting perceptions that govern behavior in individual humans. By using GCO the potency and character of individual odorants can be measured for individuals, corrected for general anosmia, to yield data suited for the study of olfactory phenomics.

The diversity of human olfactory acuity, as defined by specific anosmia, may be caused by the diversity of the olfactory receptor protein repertoire in humans. Physiologically, humans have a limited number of clearly defined olfactory receptor types that modulate odor perception in response to a small recurrent set of odorants distributed throughout the natural world. For example, the Flavornet lists about 500 chemicals reported to be above their threshold in natural products [39,40]. Natural products contain a subset of functional odorants, above their threshold, that invoke perceptions that humans experience as odor. The same odorants are found in many natural products but in different relative concentrations. Tested individually, these odorants produce precepts that have been organized into psychophysical aroma classes and genera [41]. Although they must reflect the human olfactory receptors expressed in the nose, there is no clear and predictable relationship between aroma class and the odorant composition that causes the perception. Illuminating these relationships is one of the most pressing challenges in flavor and sensory science today.

II. OLFACTORY PERCEPTION

It is important to realize that olfactory precepts are formed from potency (quantitative) and character (qualitative) information. Most psychophysical tests collect only potency data. Although the perception of an odorant is derived from the response of specific receptors, there is not a one-to-one correspondence between odorant or precept and receptor. Multiple odorant types can form a single precept and multiple precepts can be formed by a single odorant: i.e., the olfactory receptor is broadly tuned. There are estimated to be 400 full-length odorant receptor genes in the human genome [38,42]. These genes code for odorant receptor proteins that mediate the initial event in olfaction: binding of odorants. Olfactory sensory neurons expressing the same receptor protein are believed to converge in the brain at a single location in the olfactory bulb, called *a glomerulus* [43]. Each glomerulus then transmits the signal farther into the brain for higher processing. When a mixture of chemicals binds to olfactory receptors, signals first are sent to the glomerulus then travel on to the brain, producing precepts analogous to an "image" in a movie (Fig. 1A). These "images" may be unique to a particular chemical or be due to the synthesis or overlaying of "images" produced by several chemicals. Individual "images" register in milliseconds then combine over the course of seconds to form an olfactory experience—an olfactory "movie" (Fig. 1B–D). These signals are sent for further processing in the brain, where they are interpreted by memory, emotion, cognition, and other higher brain functions, which may further contribute to variation in perception of odors. Ultimately this leads to specific behavior, such as purchasing a food.

III. THE GENOMICS OF OLFACTION

Determining the functional identity of the olfactory receptor genome is a massive undertaking given the large number of genes in the olfactory genome [The largest gene class in human deoxyribonucleic acid (DNA)] and the complexity of olfaction reception processes. Determining the molecular receptive range of an olfactory receptor protein through structure-activity

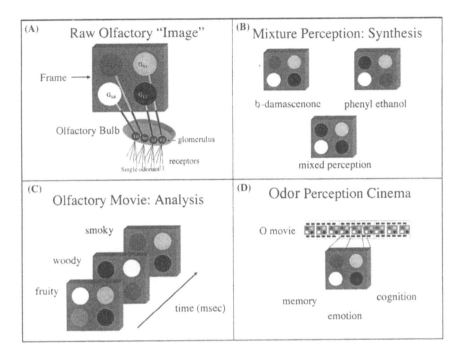

Figure 1 Drawings simplifying the structure of the olfactory process by using a motion picture analogy. A) Raw olfactory "image" created by the translation of information from odorant receptors to the brain. B) Synthesis of perceptions from mixed stimuli. C) Time-dependent processing of multiple precepts creates a movie scene, processed by other brain functions. D) The movie is edited by the brain functions to produce the final cinematographic experience: flavor.

relationships between a receptor and its possible ligands is a lengthy and difficult process, yet has been successful for at least one rat odor receptor protein [44,45]. Other experiments utilizing a "rational drug design method" to characterize receptors require an expression system for the olfactory receptor protein, which has yet to be developed. When it is possible to sequence the olfactory genome of many people, it will be a simple matter to use GCO data to associate chemosensory function with specific receptor proteins and their genes. This noninvasive approach gives direct human correlations, without using modeling estimation or approximation from other species. Using a human-based approach, GCO could connect the function of human olfactory receptor proteins directly to the chemicals they bind, through the study of specific anosmias. Once all olfactory receptor gene expressions can be measured easily, sensory panels may be normalized on the basis of the participant's olfactory genome. This, however, is a very distant dream.

IV. CONCLUSIONS

Recent developments in flavor science have seen an increase in the accuracy and precision of chemical measurements, at the same time that the complexity of flavor perceptions and their variation within and between people remain a confounding problem. Future studies of the olfactory genome should give insight into the reason for this variation across the population because understanding the genomes' relationship to flavor chemistry is the ultimate goal. The use of gas chromatography olfactometry (GCO) to correlate variation in the human olfactory response in both quantitative and qualitative terms can clarify the relationship between patterns of odorant mixtures and the perceptions they invoke. Ultimately, quantifying the diversity of the olfactory genome and associating it with the complexities of flavor perception will provide tools for more precise sensory studies. This will yield methods to interpret flavor chemistry data, explaining individual attitudes toward different foods.

REFERENCES

1. M Rothe, B Thomas. Aroma of bread: Evaluation of chemical taste analyses with the aid of threshold value. Z Lebensm Untersuch Forsch 119:302, 1963.
2. W Grosch, A Sen, H Guth, G Zeiler-Hilgart. Quantification of aroma compounds using a stable isotope dilution assay. In Y Bessiere, AF Thomas, eds, Flavour Science Technology, 6th ed. Weurman Symposium. Chichester, England: John Wiley & Sons, 1990, pp 191–194.
3. W Grosch. Detection of potent odorants in foods by aroma extract dilution analysis. Trends Food Sci Technol 4:68–73, 1993.
4. P Schieberle. New developments in methods for analysis of volatile flavor compounds and their precursors. In: AG Gaonkar, ed. Elsevier Science, 1995, pp 403–431.
5. V Audouin, F Bonnet, ZM Vickers, GA Reineccius. Limitations in the use of odor activity values to determine important odorants in foods. In: JV Leland, P Schieberle, A Buettner, TE Acree, ed. Gas Chromatography-Olfactometry. Washington, DC: American Chemical Society, 2001, pp 156–171.
6. BS Mistry, T Reineccius, LK Olson. Gas chromatography olfactometry for the determination of key odorants in foods. In: R Marsile, ed. Techniques for Analyzing Food Aroma. New York: Marcel Dekker, 1997, pp 265–292.

7. P Munch, T Hofmann, P Schieberle. Comparison of key odorants generated by thermal treatment of commercial and self-prepared yeast extractions: Influence of the amino acid composition on odorant formation. J Agric Food Chem 45:1338–1344, 1997.
8. SS Stevens, WS Cain, RJ Burke. Variability of olfaction threshold. Chem Senses 13:643–653, 1988.
9. HT Lawless, CJ Corrigan, M Johnston. Variation in odor thresholds for l-carvone and cineole and correlations with suprathreshold intensity ratings. Chem Senses 20:9–17, 1995.
10. RV Golovyna, LA Semina, VN Yakovleva, NG Enikeeva. Standardized method for evaluating the ability of tasters to differentiate odors. Pishch Prom-st 47–49, 1984.
11. AB Marin, TE Acree, J Barnard. Variation in odor detection thresholds determined by charm analysis. Chem Senses 13:435–444, 1988.
12. JE Friedrich, TE Acree, EH Lavin. Selecting standards for gas chromatography-olfactometry. In: JV Leland, P Schieberle, A Buettner, TE Acree, ed. Gas Chromatography-Olfactometry. Washington, DC: ACS, 2001, pp 148–155.
13. TJ Schulz, AP van Harreveld. International moves towards standardization of odour measurement using olfactometry. Water Sci Techol 34 (3–4):541–547, 1995.
14. Z Dobkowski. Stable isotope dilution analysis. I. General discussion. Chem Anal (Warsaw) 10:295–300, 1965.
15. Z Dobkowski. Stable isotope dilution analysis. II. Application for determination of elements and chemical compounds. Chem Anal (Warsaw) 10:519–530, 1965.
16. V Ferreira, R Lopez, A Escudero, JF Cacho. Quantitative determination of trace and ultratrace flavor active compounds in red wines through gas chromatographic-ion trap mass spectrometric analysis of microextracts. J Chromatogr A 806:349–354, 1998.
17. M Preininger, F Ullrich. Trace compound analysis for off-flavor characterization of micromilled milk powder. In: JV Leland, P Schieberle, A Buettner, TE Acree, ed. Gas Chromatography-Olfactometry. Washington, DC: ACS, 2001, pp 46–61.
18. M Rychlik, JO Bosset. Flavour and off-flavour compounds of Swiss Gruyere cheese: Identification of key odorants by quantitative instrumental and sensory studies. Int Dairy J 11:903–910, 2001.
19. KD Deibler, EH Lavin, RST Linforth, AJ Taylor, TE Acree. Verification of a mouth simulator by in vivo measurements. J Agric Food Chem 49:1388–1393, 2001.
20. DD Roberts, AJ Taylor. Flavor Release. Washington, DC: ACS, 2000.
21. A Boelrijk, K Well, P van Mil, K de Kruif, G Smit. Measuring flavour release in real time and in vivo. In: S Benita, P Couvreur, ed. International Symposium on Controlled Release of Bioactive Materials. Paris, France: 2000, pp 1354–1355.

22. AJ Taylor, RST Linforth. Techniques for measuring volatile release in vivo during consumption of foods. In: DD Roberts, AJ Taylor, ed. Flavor Release. Washington, DC: American Chemical Society, 2000, pp 8–21.
23. P Overbosch, WGM Achterof, PGM Haring. Flavor release in the mouth. Food Rev Int 7:137–184, 1991.
24. DD Roberts, TE Acree. Simulation of retronasal aroma using a modified headspace technique: Investigating the effects of saliva, temperature, shearing, and oil on flavor release. J Agric Food Chem 43:2179–2186, 1995.
25. JS Elmore, KR Langley. Novel vessel for the measurement of dynamic flavor release in real time from liquid foods. J Agric Food Chem 44:3560–3563, 1996.
26. JP Roozen, A Legger-Huysman. Models for flavor release in the mouth and its relation to new food product development. In: ed. Chemistry of Novel Foods. 1997, pp
27. MS Brauss, B Balders, RST Linforth, S Avison, AJ Taylor. Fat content, baking time, hydration and temperature affect flavour release from biscuits in model-mouth and real systems. Flavour Fragrance J 14:351–357, 1999.
28. D Mannebeck, H Mannebeck. Quality and reproducibility of olfactometric measurements. Gefahrstoffe-Reinhaltung der Luft 62:135–140, 2002.
29. N Abbott, P Etievant, S Issanchou, D Langlois. Critical evaluation of two commonly used techniques for the treatment of data from extract dilution sniffing analysis. J Agric Food Chem 41:1698–1703, 1993.
30. C Huyer. Olfactology of aniline and its homologs. In: ed. Utrecht, 1917, pp
31. A Chaplet. Olfactometers and olfactometry. Parfumerie Moderne 30:3–13, 1936.
32. A Dravnieks. Measuring industrial odors. Chem Eng 81:91–95, 1974.
33. HT Lawless. Theoretical note: Tests of synergy in sweetener mixtures. Chem Senses 23:447–451, 1998.
34. PKC Ong, TE Acree. Similarities in the aroma chemistry of Gewürztraminer variety wines and lychee (*Litchi chinesis* Sonn.) fruit. J Agric Food Chem 47:665–670, 1999.
35. H Gygax, H Koch. The measurement of odors. Chimia 55:401–405, 2001.
36. P Mombaerts. Seven-transmembrane proteins as odorant and chemosensory receptors. Science 286:707–711, 1999.
37. S Firestein. How the olfactory system makes sense of scents. Nature 413:211–218, 2001.
38. S Zozulya, F Echeverri, T Nguyen. The human olfactory receptor repertoire. GenomeBiologycom 2:RESEARCH0018, 2001.
39. H Arn, TE Acree. Flavornet: A database of aroma compounds based on odor potency in natural products. In: ET Contis, CT Ho, CJ Mussinan, TH Parliament, F Shahidi, AM Spanier, ed. Food Flavors: Formulation, Analysis and Packaging Influences. Elsevier, 1998
40. Flavornet: http://www.nysaes.cornell.edu/flavornet
41. GV Civille, BG Lyon. Lexicon for Aroma and Flavor Sensory Evaluation. ASTM, DS 66, 1996.

42. G Glusman, I Yanai, I Rubin, D Lancet. The complete human olfactory subgenome. Genome Res 11:685–702, 2001.
43. P Mombaerts. Molecular biology of odorant receptors in vertebrates. Annu Rev Neurosci 22:487–509, 1999.
44. H Zhoa, L Ivic, JM Otaki, M Hashimoto, K Mikoshiba, S Firestein. Functional expression of mammalian odorant receptor. Science 279:237–242, 1998.
45. RC Araneda, AD Kini, S Firestein. The molecular receptive range of an odorant receptor. Nat Neurosci 3:1248–1255, 2000.

6

Implications of Recent Research on Olfaction for the Neural Basis of Flavor in Humans: Challenges and Opportunities

Gordon M. Shepherd
Yale University School of Medicine, New Haven, Connecticut, U.S.A.

I. INTRODUCTION

Studies of the brain mechanisms underlying the perception of flavor face several daunting challenges. We use the term *flavor* to characterize the perception of foods and liquids before, during, and after ingestion. To begin with, it is a multisensory perception, combining input from several sensory systems: the taste and olfactory sensory systems, the trigeminal somatosensory system for the tactile feel of food in the mouth, and the proprioceptive system for the feedback from the muscles involved in mastication. From this perspective, flavor is an example of a sensory modality arising out of the coordinated actions of several sensory systems.

An understanding of the neural mechanisms underlying flavor must therefore begin by separating and identifying the contributions of each component system. Of these, olfaction plays a major role in most kinds of flavors of ingested foods and beverages. Because of this, an understanding of the brain mechanisms in smell must be central to an understanding of the neural mechanisms in flavor. I therefore focus on smell while recognizing the important role played by the other systems in producing the combined perception.

A second difficulty in studying the brain mechanisms of flavor is that the flavor modality is large and ill defined. One can appreciate this by considering other sensory systems. Visual perception is characterized by the specific attributes of a visual stimulus—wavelength, luminance, shape, orientation, motion. Similarly, hearing is characterized by the specific qualities of sound frequency, loudness, and frequency modulation. Touch is characterized by place, quality, and intensity. In all these cases what we may call the "sensory space" is clearly limited and clearly defined.

By contrast, flavor space is difficult to define. Taste, of course, has its traditional four qualities—sweet, salt, sour, and bitter—and possibly a fifth, umami. However, smell seems to have a virtually limitless "odor space." It is common to say that a human can identify up to 10,000 different odors, but that number as far as I can determine is only a speculation. Since odors are the main component of most flavors, "flavor space" is correspondingly difficult to define. This underlines the reason why no description of brain mechanisms in flavor can arise until there is a better understanding of the complex roles of the mechanisms in smell.

A final difficulty in studying flavor for the purposes of the present discussion is that the different experimental approaches are compartmentalized. Food scientists carry out highly sophisticated state-of-the-art psychophysical tests of foods and beverages, along with chemical analyses of the flavor attributes of different foodstuffs. This represents a tremendous accumulated knowledge of the organic chemistry of flavor compounds and their relation to perception that neuroscientists would find of great value if they knew about it. On the other hand, there is intense activity in neuroscience directed toward the molecular, cellular, and systems basis of olfactory processing. Unfortunately, these studies are largely unknown to food scientists yet could provide valuable insights and tools to guide and extend future research.

This chapter is intended to convey the significance of some of the recent advances in the neurobiology of olfaction for the food sciences. I begin by briefly describing key areas of recent progress in studies of the neural basis of smell in laboratory animals. It is now possible to organize this description in terms of the sequence of neural mechanisms that are involved in the early stages of odor processing by the peripheral olfactory system. These steps begin with differential activation by different odor molecules of a large family of olfactory receptors expressed in the olfactory sensory cells in the nose. Activated receptor cells project to the glomerular layer of the olfactory bulb, which forms "odor images" of the smell stimuli in the form of glomerular activity patterns. These activity patterns are then processed by the neural circuits in the olfactory bulb for output to the olfactory cortex and ultimately the neocortex as the basis for the perception of smell.

These studies provide new challenges to food science, in the form of new opportunities in analyzing the role of smell in flavor and contributing to understanding of the neural basis of smell. Some areas that would have the potential for new collaborations between neuroscience and food science are discussed. Pursuing these collaborations should have the added benefit of helping to test the traditional belief that humans have a poor sense of smell. In this connection, I discuss some intriguing recent studies of behavioral tests of smell perception that suggest that humans and subhuman primates have much stronger senses of smell when compared with other mammals than has been generally recognized. This implies a more important role for smell and flavor in human behavior and in human evolution than heretofore recognized.

II. CHEMICAL COMPOSITION OF FOODS AND FOOD PRODUCTS: FLAVOR OBJECTS AND THE NATURE OF FLAVOR SPACE

The neural basis of the sense of smell begins with the action of an odor molecule on receptors on the sensory neurons in the nose. An important first step in the analysis is to characterize the odor molecules that make up "odor space."

A valuable method for this was introduced in psychophysical studies many years ago by the use of homologous series of aliphatic and other simple types of compounds, such as alcohols, acids, and aldehydes [1]. Recent experiments on the physiological responses of different cells to different odors have built on this approach. These studies have revealed much more structure of odor space as represented in the brain than had been seen previously [2]. Thus, it was found that if the odor of a given member of a homologous series is able to activate a particular cell, the cell's responsiveness is characteristically limited to that member and the ones immediately on each side with longer and shorter carbon lengths. In virtually no cases are responses seen to compounds with much longer or shorter carbon lengths. The odors to which a cell is responsive are referred to as its *molecular receptive range* [2], by analogy with the *spatial receptive field* that characterizes cells in the visual and somatosensory system.

This is an extremely useful step in revealing order within odor space and how it is represented within neural space. However, it must be kept in mind that natural stimuli usually occur in the form of complex combinations of related stimuli, such as visual objects, auditory phonemes, and tactile textures. In the olfactory system one can refer to these as *odor objects* [3]; in invertebrates they are referred to as *odor blends*.

Studies of these odor objects need to be built on a rational exploration of chemical structure similar to that for the individual compounds. Neuroscientists therefore need to have the best possible information about the salient odor compounds in different foodstuffs. An example is the Flavornet website [4], where current knowledge about the major flavor components of different foodstuffs is systematically organized and may be easily accessed (see also Ref. [5]).

III. STRUCTURE-ACTIVITY RELATIONS BETWEEN ODORS AND RECEPTORS: TRANSDUCING ODOR SPACE INTO BRAIN SPACE

What is the information carried in an odor molecule that determines its molecular receptive range? The classical method for answering this question for other types of receptors is to express it in a convenient type of cell and test with different ligands. This is now routine for such receptors as the beta-adrenergic receptor, which shows a narrow preference for its natural agonist, epinephrine. This approach has been difficult with olfactory receptors, for two main reasons: expressing the olfactory receptors in heterologous systems has turned out to be extremely difficult, and the ligand preferences are broad and initially unknown.

The best evidence is from two types of approach: overexpressing a given receptor by viral transfer into olfactory receptor cells [6] and carrying out polymerase chain reaction (PCR) on single olfactory receptor cells after determining their molecular receptive range [7]. These results have supported previous classical studies of receptor cell responses in showing that the odors are transduced by a combinatorial code: a given receptor interacts with multiple odors, and a given odor interacts with multiple receptors. These and other studies have shown that olfactory receptors are more broadly tuned than other members of the G-protein-coupled receptor (GPCR) family. It is believed that this overlapping combinatorial code enables the system to encode a virtually limitless odor space.

In order to understand the processing mechanisms in a sensory system, it is necessary to identify the fundamental units of information, the sensory determinants or "primitives," conveyed in the sensory stimulus. The best example is the visual system, where the sensory primitives are simple spots and lines [8]. The application of this approach to the olfactory system has been discussed at some length [2,3,9]. In the olfactory system, these fundamental units obviously are carried in the odor molecules.

Primitives in an odor molecule have been identified by modeling the interactions between olfactory receptors and odor molecules. This has provided consistent evidence that the olfactory receptors form a binding pocket within the plane of the membrane, similar to that of other GPCRs, and that specific attributes of the odor molecules interact within this pocket with specific amino acid residues at several sites facing the pocket [10–12]. The attributes include the functional group at a terminal carbon, such as acids and aldehydes; functional groups within a carbon chain, such as double bonds; carbon chain length; functional groups attached to a carbon chain, such as a phenol group; steric shape; and chirality. Note that carbon length and functional group were already recognized as fundamental determinants of odor perception in the pioneering psychophysical studies of aliphatic homologous series [1] and in physiological responses of olfactory cells to homologous series [2]. These modeling studies await testing on expressed receptors.

Together these attributes can be considered as the fundamental information-carrying units of an odor molecule, similar to the determinants or pharmacophores of a membrane signaling molecule. They are also similar to (though much smaller than) the epitopes of an antigen, and by analogy have been referred to as *odotopes* [2,3]. They function as the primitives of the perception of smell. An odor molecule is thus encoded by its combination of primitives (determinants). An odor object is encoded as the collective profile, the envelope so to speak, of the determinants of its constituent odors.

IV. COMPARING THE ODOR SPACES OF HUMANS AND OTHER SPECIES

How does human odor space compare with that of other primates and other mammals? One way to characterize this space is by the odor molecules themselves, but as already noted it is impossible to catalogue all the potential members. Another way is by characterizing the olfactory receptors.

Potential insight arises from the recent genome projects, which have shown that the human has a total of some 950 olfactory receptor genes, of which some 350 are functional and some 600 are pseudogenes, that is, genes that have mutations introduced into them that render them nonfunctional [13,14]. By comparison, the mouse has some 1400 olfactory receptor genes, of which some 1200 are functional and only approximately 200 are nonfunctional ([15,16]; this information is available publicly at Ref. [17]). This molecular evidence therefore appears to support the popular view that human olfaction is much less sensitive than that of most other mammals. Is extrapolation directly from gene data to psychophysical data in this manner justified? We return to this question later.

In the primate series, a study has shown a progressive reduction in functional olfactory genes from old world primates with a high proportion of functional genes through the apes, which have a progressively reduced proportion, to the human [18]. However, there are inconsistencies. Lemurs, heavily dependent on smell, have a lower proportion of functional genes than several old world primates, which appear less dependent. The simplest conclusion nonetheless is that the decline in numbers and functional olfactory genes through most of the primate series refects a reduction in the abilities and importance of smell, ending with humans at the bottom. We consider the psychophysical evidence for this idea later.

V. ODOR IMAGES IN MICE AND HUMANS: MAPPING ODOR SPACE WITHIN BRAIN SPACE

Sensory transduction involves the transfer of information carried in the stimulus to information carried in the receptor cells. For smell, one can refer to this as *transfer* from odor space to neural space.

After the initial transduction of odor space into neural space by the olfactory receptors, a second messenger cascade amplifies the response in each cell, leading to activation of a cyclic nucleotide gated (CNG) ionic channel. The resulting ion flows depolarize the membrane, generating the receptor potential, which spreads to the site of action potential initiation [19].

The responses of the cells in the olfactory epithelium can be recorded by patch electrodes or by imaging of the Ca^{2+} that enters through the CNG channels. The latter method has enabled one to remove swatches of olfactory epithelium from different zones within the nasal cavity and visualize the responses of up to several hundred receptor cells at a time [20]. These experiments are providing the first glimpses of the spatial organization of the cells responding to given odors. With weak odor concentrations only one or a few cells within a swatch are activated; with increasing concentrations more cells are activated. The cells form a functional mosaic in the olfactory epithelium, with cells responding to different odors intermingled. But this mosaic is not the basis for odor perception; that step involves the projection of the receptor cells to the olfactory bulb.

The action potentials generated in the receptor cells propagate through the axons to reach the glomeruli of the olfactory bulb. Within the glomeruli the axons make synapses onto the dendrites of the bulbar neurons. In the 1970s the 2-deoxyglucose (2DG) activity mapping method showed that stimulation with a given odor activates glomeruli in a specific pattern, which is generally reproducible between animals [21,22]. The weakest odor concentrations activate only single or a few glomeruli; with stronger concentrations nearby glomeruli also become activated. The 2DG method shows the global pattern throughout the entire olfactory bulb. Maps for many different odor molecules are now available [23]. These results have recently begun to be extended by high-resolution functional magnetic resonance imaging (fMRI) [24], which enables the global activity patterns in response to different odors at different concentrations to be recorded in the same animal [23]. Parallel studies with optical methods have provided complementary data from the dorsal surface of the olfactory bulb (see [25–27]).

These functional studies have been strongly supported by molecular studies in which the individual receptor types are labeled with green fluorescent protein (GFP) and their projections to the olfactory glomeruli are mapped [28]. This has revealed a common connection rule: the subset of cells expressing a given receptor typically project their axons to one of two glomeruli, on the medial and lateral aspects of the olfactory bulb. A map of the projections of olfactory sensory neuron subsets expressing different odor receptors thus underlies the activity maps. This relationship is dynamic because it reflects the moment-to-moment functional state of the system.

Taking the mapping results together, there is a strong consensus emerging that odor space is mapped into two-dimensional odor maps in the glomerular layer. Each glomerulus thus constitutes a combined molecular, anatomical, and functional module in processing the information. There is evidence for multiple mechanisms for signal-to-noise enhancement within each glomerulus. The odor map serves as an "odor image" [3] in the sense of a two-dimensional representation of odor space, analogous to a visual image in the retina or the body image in the somatosensory pathway. This image is processed by the synaptic circuits in the olfactory bulb and projected onto the olfactory cortex, to serve eventually as the basis for the perception of smell.

As already indicated, this consensus is based primarily at this stage on the analysis of responses to single odors. A challenge for the future will be to extend these studies to odor mixtures as odor objects, and then to flavor.

VI.   BEHAVIORAL TESTING AND PSYCHOPHYSICS: HOW MACROSMATIC ARE HUMANS?

The foregoing constitutes a brief summary of the main areas of agreement in current neuroscience studies of the neural basis of olfactory processing.

It is obviously critical to test these proposed mechanisms with psychophysical experiments. Of immediate interest is whether the lower number of functional olfactory receptor genes in humans than in rodents and subhuman primates is paralleled by lower psychophysical performance in odor detection and discrimination.

A beginning has been made in testing this hypothesis by studies in 2000 that compared the detection thresholds of primates and other mammals with those previously reported for humans [29]. A surprising finding has been that humans compare very well. For example, when tested with a series of aliphatic alcohols, humans were less sensitive than dogs but more sensitive than rats for shorter-chain alcohols and were roughly equivalent to both for the longer-chain compounds. From these results, Laska has suggested that it is time to reconsider the traditional view that the sense of smell is weak in humans and consider the possibility that, rather than being microsmats (having a weak sense of smell), humans in fact are macrosmats (having a strong sense of smell). Perhaps, one might speculate, humans are supersmellers.

The implications of this mismatch are potentially profound and raise critical challenges for both neuroscience and food science. It seems to me an exciting prospect: that smells, and by extension flavors, are not such minor factors in human life as we traditionally assume. Many workers in the food sciences as well as in enology and the fragrance industry already know this. However, there has not yet been any follow-up testing of this proposal. It is, I think, important to pursue these implications, for both neuroscience and the food sciences. I would like to end by mentioning three aspects that I think warrant attention and further investigation.

VII.   IMPLICATIONS FOR RELATING GENES TO BEHAVIOR

The comparisons of psychophysical tests in different species [29] are in direct conflict with the implications of the decline of functional olfactory receptor genes from rodents through the primate series to humans. It appears that we have a lot more to learn about the relation between counting receptor genes and testing for detection and discrimination of odors. This implies that there is not a one-to-one relation between genes and behavior.

This result has in fact already been known, from the finding that up to 80% of the olfactory bulb can be removed without affecting psychophysical tests for odor detection and discrimination [30]. It has been pointed out that this could reflect the great deal of redundancy built into the specificity of this system [31]. It could also reflect the inability of the behavioral methods, however sophisticated, to test adequately for the ways that the system is employed in natural behavior in the wild. The olfactory system may be useful for exploring in greater detail this relation between genes and behavior, as a possible model for this relation in other parts of the brain.

VIII. IMPLICATIONS FOR FOOD SCIENCE

It seems to me that these findings have important implications for food science. It means that one may be dealing with a sense for smell and flavor that is relatively more powerful than is generally assumed. The standards developed by the food industry for testing thresholds and discriminability therefore have a significance not only for testing humans, but also for comparing those abilities with other supersmellers, such as rodents and carnivores. Careful testing should be able to reveal more precisely the capabilities of each species, so that precise comparisons can be made with the olfactory genomes. In discussions of this question with several colleagues (W Cain, S Youngentob, B Slotnick, D Laing, R Doty, P Laffort, personal communications) it has been emphasized that it will be important to carry out psychophysical tests on different species by the same laboratory with the same methods. From this perspective, psychophysical testing of human smell by food scientists will be of great interest, especially if it can be extended to subhuman primates and even other mammals for comparison by the same laboratory. Such studies should provide direct testing of the postulate that humans are macrosmats [29].

An intriguing question raised by the genomic data is whether there are areas of the mouse genome not covered by the fewer human genes, suggesting that these would reflect corresponding areas of odor space from which we are excluded, or whether the loss in the human is random. The initial evidence suggests that the fewer functional genes in the human are distibuted relatively randomly throughout the mouse genome. This suggests that humans are not excluded from large areas of rodent olfactory experience [15,16]. However, there appear to be smaller areas that may represent losses of narrower types of odor experience in humans when compared with rodents. These questions require careful testing.

IX. IMPLICATIONS FOR HUMAN EVOLUTION

Comparing the senses of flavor and smell in different species also has immediate implications for the roles of these senses in human evolution. The traditional view is that primate evolution involved movement of the eyes to the .front of the face for more complex stereovision, which required reduction of the size of the snout and with it reduction in the size and complexity of the nasal cavity, directly paralleled by a reduction in the sense of smell [32]. The decrease in the olfactory receptor gene repertoire appears consistent with this view. This implies that the smaller nose and nasal cavity of the human are inferior for mediating the sense of smell. However, this implication to my knowledge is another untested aspect of the traditional view. A more interesting hypothesis is that the evolution of the human nose has involved adaptations that enabled it to retain its capacities, or perhaps even gave it advantages, in sensing both inspired and retropharyngeal odor molecules as humans and their societies evolved. In addition, humans have the possibility of enhancing their perceptions of smell and flavor through more complex neural processing at higher brain levels.

If the sense of smell remained strong during human evolution, that suggests that smells and flavors have played larger roles in human societies than heretofore appreciated. These roles could have included preferences for specific smells and flavors involved in migrations of populations and conflicts between people with different preferences. That such preferences and differences are present between populations today is not the least reason for neuroscientists and food scientists to be interested in studying them.

ACKNOWLEDGMENT

Our work has been supported by the National Institutes of Health (NIDCD), the Human Brain Project, and MURI.

REFERENCES

1. P Laffort. A linear relationship between olfactory effectiveness and identified molecular characteristics, extended to fifty pure substances. In Olfaction and Taste III (ed. C Pfaffman). New York: Rockefeller, 1969, pp 150–157.
2. K Mori, Y Yoshihara. Molecular recognition and olfactory processing in the mammalian olfactory system. Prog Neurobiol 45:585–619, 1995.

3.   GM Shepherd. Computational structure of the olfactory system. In Olfaction: A Model for Computational Neuroscience (ed. H Eichenbaum and J Davis). Cambridge, Mass.: MIT Press, 1991, pp 3–42.

4.   http://www.nysaes.cornell./flavornet/

5.   http://www.leffingwell.com/

6.   H Zhao, L Ivic, JM Otaki, M Hashimoto, K Mikoshiba, S Firestein. Functional expression of a mammalian odorant receptor. Science 279:237–242, 1998.

7.   B Malnic, J Hirono, T Sato, LB Buck. Combinatorial receptor codes for odors. Cell 96:713–723, 1999.

8.   D Marr. Vision. San Francisco: Freeman, 1983.

9.   JE Amoore, G Palmieri, E Wanke. Molecular shape and odour: Pattern analysis of PAPA. Nature 216:1084–1087, 1967.

10. Y Pilpel, D Lancet. The variable and conserved interfaces of modeled olfactory receptor proteins. Protein Sci 8:969–977, 1999

11. MS Singer. Analysis of the molecular basis for octanal interactions in the expressed rat olfactory receptor. Chem Senses 25:155–165, 2000.

12. WB Floriano, N Vaidehi, WA Goddard III, MS Singer, GM Shepherd. Molecular mechanisms underlying differential odor responses of a mouse olfactory receptor. Proc Natl Acad Sci U S A 97:10712–10716, 2000.

13. T Fuchs, G Glusman, S Horn-Saban, D Lancet, Y Pilpel. The human olfactory subgenome: From sequence to structure and evolution. Hum Genet 108:1–13, 2001.

14. S Zozulya, F Echeverri, T Nguyen. The human olfactory receptor repertoire. Genome Biol 2, 2001.

15. X Zhang, S Firestein. The olfactory receptor gene superfamily of the mouse. Nat Neurosci 5:124–133, 2002.

16. JM Young, C Friedman, EM Williams, JA Ross, L Tonnes-Priddy, BJ Trask. Different evolutionary processes shaped the mouse and human olfactory receptor gene families. Hum Mol Genet 11:535–546, 2002.

17. http://www.senselab.edu/odormapdb

18. S Rouquier, A Blancher, D Giorgi. The olfactory receptor gene repertoire in primates and mouse: Evidence for reduction of the functional fraction in primates. Proc Natl Acad Sci U S A 97:2870–2874, 2000.

19. F Zufall, SD Munger. From odor and pheromone transduction to the organization of the sense of smell. Trends Neurosci 24:191–193, 2001.

20. M Ma, GM Shepherd. Functional mosaic organization of mouse olfactory receptor neurons. Proc Natl Acad Sci U S A 97:12869–12874, 2000.

21. WB Stewart, JS Kauer, GM Shepherd. Functional organization of rat olfactory bulb analysed by the 2-deoxyglucose method. J Comp Neurol 185:715–734, 1979.

22. http://www.senselab.med.yale.edu/ordb

23. http://www.leonlab.bio.uci.edu/

24. X Yang, R Renken, F Hyder, M Siddeek, CA Greer, GM Shepherd, RG Shulman. Dynamic mapping at the laminar level of odor-elicited responses in rat olfactory bulb by functional MRI. Proc Natl Acad Sci USA 95:7715–7720, 1998.

25. AR Cinelli, JS Kauer. Voltage-sensitive dyes and functional activity in the olfactory pathway. Annu Rev Neurosci 15:321–351, 1992.
26. L Belluscio, LC Katz. Symmetry, stereotypy, and topography of odorant representations in mouse olfactory bulbs. J Neurosci 21:2113–2122, 2001.
27. H Nagao, M Yamaguchi, Y Takahash, K Mori. Grouping and representation of odorant receptors in domains of the olfactory bulb sensory map. Microsc Res Tech 58:168–175, 2002.
28. P Mombaerts, F Wang, C Dulac, SK Chao, A Nemes, M Mendelsohn, J Edmondson, R Axel. Visualizing an olfactory sensory map. Cell 87:675–686, 1996.
29. M Laska, A Seibt, A Weber. "Microsmatic" primates revisited: Olfactory sensitivity in the squirrel monkey. Chem Senses 25:47–53, 2000.
30. XC Lu, BM Slotnick. Olfaction in rats with extensive lesions of the olfactory bulbs: Implications for odor coding. Neuroscience 84:849–866, 1998.
31. GM Shepherd. Specificity and redundancy in the organization of the olfactory system. Microsc Res Tech 24:106–112, 1993.
32. S Jones, R Martin, D Pilbeam. The Cambridge Encyclopaedia of of Human Evolution. New York: Cambridge University Press, 1992.

7

Effect of Texture Perception on the Sensory Assessment of Flavor Intensity

K. G. C. Weel, A. E. M. Boelrijk, A. C. Alting, and G. Smit
NIZO Food Research, Ede, The Netherlands

J. J. Burger
Quest International, Naarden, The Netherlands

H. Gruppen and A. G. J. Voragen
Wageningen University, Wageningen, The Netherlands

I. INTRODUCTION

Sensory properties such as taste, aroma, texture, and appearance are important determinants of the consumer's acceptance and liking of food products. These properties are assessed by looking at, sniffing at, or tasting the product. As the eyes judge the appearance, the taste buds in the tongue perceive the taste, the olfactory region in the nose "smells" the aroma, and the tongue and jaw muscles register the mouth feel (or perceived texture). Interactions on a psychological level among these senses are well known, for instance, the effect of color on aroma [1,2] and the interactions between taste and aroma [3–5]. An interesting example of interaction between taste and aroma was found for chewing gum. The temporal perception of mint was shown to be determined by sucrose release, rather than by menthone release, suggesting perceptual interaction between volatiles and nonvolatiles [6].

Interactions between texture and aroma have been described, as well. For liquid systems containing hydrocolloids, it has been shown that an increased viscosity due to an increase in hydrocolloid concentration results in a reduction of the aroma perception [7–9]. For solid systems, gels of protein and carbohydrate origin have been used as a model system. In general, gel systems show the same trend as liquid systems. An increase in gelling agent concentration, and thus an increase in gel hardness, causes a decrease in the sensory rating of the aroma perception [10–12]. Time-intensity (TI) methodology has been used for gels, as well, to study the effect of gel structure on aroma intensity [13–16]. In most of these studies, an increase in gel hardness was achieved by an increase in gelling agent concentration. Gel structure was shown to influence the maximal perceived *aroma* intensity (Imax). These studies, however, focused on the perception of aroma compounds, not on their release.

In recent years, several methods have become available to measure aroma release in real time in the nosespace of test persons during eating [17–19]. In the present study, we used the MS-Nose^{TM}, developed by Taylor and coworkers [18,19]. The system allows sensitive and fast monitoring of the in vivo aroma release.

The release of various aroma compounds from gelatin gels has been measured with this instrument [20,21]. In these studies, in vivo aroma release measurements were combined with TI recordings. Tmax (time to maximal intensity) values were compared for in vivo aroma release and sensory time intensity for menthol and dimethylpyrazine. When the maximum was reached quickly, the sensory perception was found to lag behind the aroma release. When the maximum was reached after a longer period, the Tmax of perception preceded the Tmax of aroma release, as a result of a sensory adaptation effect [20]. Baek and associates [21] described a significant decrease in Imax and an increase in Tmax for TI as the gelatin concentration increased. This was found for gels flavored with furfuryl acetate. Imax values of time intensity and in vivo aroma release were not correlated. However, the rate of volatile release seemed to correlate well with the sensory data [21].

In most studies, an increase in concentration of gelling agent was used to increase the hardness of the gels. We have developed a gel system based on whey protein, which allowed different rheological properties at equal protein concentration [22]. By variation in protein concentration during heating it was possible to obtain rheological variation between the gels. No addition of salts or sugars, which might possibly influence the aroma perception, was necessary for this purpose. An additional advantage of this gelation system is that the level of protein-related off-flavors is minimized as a result of mild heating conditions. Furthermore, the flavors used are not exposed to heat, as gelation was performed at ambient temperature by mild acidification, which prevents degradation of aroma compounds during gel formation.

The purpose of our study was to determine whether sensory aroma perception is influenced by textural properties, and whether this is through an effect in aroma release or a psychophysical effect. Release and perception of these flavor compounds from five different whey protein gels were studied by in vivo aroma release and by time intensity.

II. MATERIALS AND METHODS

A. Materials Used

Whey protein isolate was purchased from Davisco Foods International Inc. (Le Sueur, Minnesota). Glucono-δ-lactone (GDL) was purchased from Sigma Chemical Co. (St. Louis, Missouri) and ethanol (>99.9%) from J. T. Baker (Deventer, The Netherlands). Quest International (Naarden, The Netherlands) kindly provided diacetyl and ethylbutyrate.

B. Preparation of the Gels

Solutions of whey protein isolate (4%, 7.5%, and 11% w/w in demineralized water) were heated at 68.5°C for 3 hr. After cooling to ambient temperature, parts of the protein solutions of 7.5% and 11% were diluted to final concentrations of 4% by addition of demineralized water. Ethylbutyrate or diacetyl was added to all solutions to final concentrations of 150 ppm; GDL was added to the solutions with final protein concentrations of 4%, 7.5%, and 11% (w/w), to a final concentration of 0.32%, 0.52%, and 0.86% (w/w), respectively. After addition of GDL the pH decreased slowly to a final value of 5.0, toward the pI of the protein (pH 5.1). Table 1 lists the protein concentrations of the five gels used during heating and gelation.

Additionally, gels with increasing concentrations of diacetyl and ethylbutyrate were prepared, according to the procedure described. The protein concentrations of the solutions used for these gels during heating and gelation were 11% and 4%, respectively. Diacetyl and ethylbutyrate were both present in each gel at concentrations of 100, 125, 150, 175, and 200 ppm.

C. Characterization of Gel Structure

Gel hardness was determined by a texture analyzer (type TA-XT2, Stable Micro Systems Ltd., Godalming, UK). Gel hardness was expressed as the stress (pascals) at the maximal peak of the force-time curve [23].

Water holding capacity (WHC) was determined by mincing 40 g of gel with a plunger in 60 crushing movements and putting it into a centrifuge

Table 1 Protein Concentration During Heating and Gelation Steps of the Gel Formation[a]

Gel number	Protein concentration	
	Heating step, percentage	Gelation step, percentage
1	4	4
2	7.5	4
3	11	4
4	7.5	7.5
5	11	11

[a] Percentage weight/weight.

tube. The gel was centrifuged for 30 min at 160 g. After centrifugation the serum was removed and weighted. The *WHC* (percentage) of the gel was defined as the mass fraction of the water retained from the total amount of water present in the gel [24,25]. Imaging was performed by using a confocal scanning laser microscope (CSLM) (Leica), type TCS-SP, configured with an inverted microscope, and an ArKr laser for single-photon excitation. The protein gels were stained by applying 2 μL of an aqueous solution of 0.05% (w/w) Rhodamine B to 200 μL gel. The 568-nm laser line was used for excitation, inducing a fluorescent emission of Rhodamine B, detected between 600 and 700 nm.

D. Headspace Gas Chromatography

With the headspace gas chromatography (HSGC) technique, each gel was directly prepared in a 10-mL headspace vial (3 mL of gel per vial) according to the procedure described (5, 10, or 25 ppm of diacetyl and ethylbutyrate). The headspace concentrations were allowed to reach equilibrium during overnight storage at ambient temperature. To analyze the headspace aroma concentration 1.0 mL of headspace air was injected splitless on the column after 20-min incubation at 35°C. A GC-8000[top] gas chromatograph (CE Instruments, Milan, Italy) was equipped with a CP-SIL 5 CB low-bleed column (30 m × 0.25 mm, film thickness, 1.0 μm, Chrompack, Middelburg, The Netherlands) and a flame ionization detector (FID). The oven temperature was initially 40°C for 5 min, then increased by 15°C/min to 150°C and was kept at 150°C for 5 min. Inlet and detector temperatures were 250°C and 225°C, respectively. Gas flow rates were as follows:

hydrogen, 35 mL/min; air 350 mL/min, makeup nitrogen, 30 mL/min. The headspace concentrations were expressed as peak areas in arbitrary units.

E. In Vivo Aroma Release Measurements and Time-Intensity Recordings

Ten panelists were familiarized with the aroma of diacetyl and ethylbutyrate and trained to produce TI curves, their in vivo aroma release was measured simultaneously by the MS Nose, during five training sessions in which three samples were judged by each panelist. The structurally different gels flavored with either diacetyl or ethylbutyrate were presented in triplicate in random order to each panelist as cylinder-shaped samples of 2 mL. The samples were presented in five sessions of three gels to prevent sensory fatigue. Before every session a nonflavored gel with medium hardness (no. 3) was presented as a blank, followed by a flavored gel with medium hardness (no. 3) as aroma reference. The panelists were instructed to chew regularly (independently of the gel hardness) for 30-sec without swallowing, then to swallow the entire bolus, and, after that, to continue chewing for 60 sec.

In vivo aroma release was monitored by on-line sampling of a part of the exhaled air (75 mL/min, through a capillary tube, 0.53-mm internal diameter, heated to 100°C) into the MS Nose, an atmospheric pressure chemical ionization gas phase analyzer (APCI-GPA) attached to a VG Quattro II mass spectrometer (Micromass UK Ltd., Manchester, UK). The compounds were ionized by a 3.0-kV discharge. Source and probe temperatures were 80°C. Diacetyl and ethylbutyrate were analyzed in selected ion mode (0.2-sec dwell on each ion) and cone voltages of 19 V and 20 V, respectively. Acetone was measured as an indicator of the panelists' breathing pattern. TI curves were recorded by FIZZ software (Biosystemes, Couternon, France) over a 1.5-min period. Panelists were instructed to rate the perceived diacetyl or ethylbutyrate *aroma* on a scale from 0 to 10. The maximal intensity of the reference gel was agreed by the panelists to have an aroma intensity of 5 on the 0–10 scale. The panelists were instructed not to let textural effects influence their perception of the aroma intensity.

III. RESULTS AND DISCUSSION

A. In Vivo Aroma Release

The panel assessed the in vivo aroma release from gels with increasing concentrations of diacetyl and ethylbutyrate (100–200 ppm). The averaged curves are shown in Fig. 1. A clear linear increase in Imax is graphically found for gels with increasing aroma concentrations. The regression

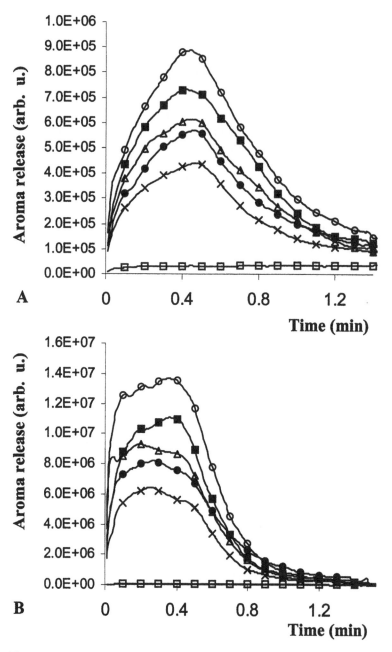

Figure 1 Averaged release profiles for gels flavored with 0 (□), 100 (×), 125 (•), 150 (Δ), 175 (■), and 200 (o) ppm of diacetyl (A) and ethylbutyrate (B).

coefficients of the linear regression between *aroma* concentration and Imax of all individual curves were 0.99 and 0.97 for diacetyl and ethylbutyrate, respectively. This shows that for this technique and gel system a 10-person panel is sensitive enough to detect (relatively large) differences in aroma concentration.

The temporal aspect of the in vivo release of diacetyl and ethylbutyrate can be visualized when the curves are normalized to 100% of Imax, as displayed in Fig. 2. The release of ethylbutyrate is faster than the release of diacetyl, both in increasing and in decreasing phase. This difference is probably associated with the difference in hydrophobicity between the two compounds. Diacetyl, a hydrophilic compound, partitions into the aqueous film coating of mouth and throat more easily than the hydrophobic ethylbutyrate, causing a slower increase in release and a prolonged release of diacetyl. A quick release seems to be correlated with high hydrophobicity, as was also concluded by Taylor and coworkers, who followed the release from mixed-phase gels of aroma compounds with different physicochemical properties [26].

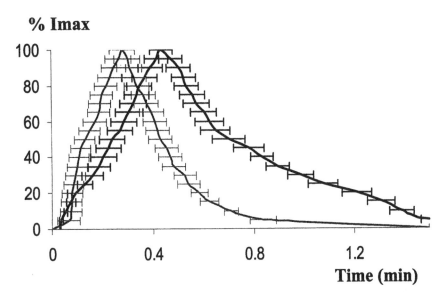

Figure 2 Averaged normalized release curves for ethylbutyrate (gray) and diacetyl (black). The maximal release is equal to 100%.

B. Physical Characterization of the Gels

Figure 3 shows CSLM images of the microstructure of the five different gels used. Some physical data (WHC and gel hardness) have been listed as well. It is clearly seen in Fig. 3 that the openness of the gels decreases from gel 1 to gel 5. This is in good agreement with the results of the hardness and WHC measurements (Fig. 3). A gel with a more open structure is generally softer than a gel with a more compact structure [27]. Furthermore, a relatively weak gel with an open structure has a lower WHC [28].

The molecular interactions of diacetyl and ethylbutyrate with whey protein gels were studied by static headspace measurements. This was done in order to check whether the differences in aroma release and perception of the studied gels could be explained directly by physical gel parameters, without the need to take final protein concentration and *aroma*-protein interactions into account. There were no differences in static headspace

Gel	Hardness (kPa)	WHC (g/g)	CSLM image
1	6	0.71	2 µm
2	18	0.74	
3	15	0.76	
4	50	0.84	
5	92	0.86	

Figure 3 Physical properties and CSLM images of the whey protein gels. (From Ref. [30]. Reprinted with permission from Weel et al, J. Agric. Food Chem. 50:5149-5155. Copyright 2002 American Chemical Society.)

concentrations observed between the gels (Fig. 4). If the molecular binding of diacetyl or ethylbutyrate with whey protein was significant, a lower static headspace concentration would be expected for gels 4 and 5 compared to gels 1, 2, and 3, as the final protein concentration of gels 4 and 5 is higher than those of gels 1, 2, and 3. This is not the case.

C. Influence of Gel Structure on Aroma Release and Perception

Figure 5 shows the averaged curves of time-intensity and in vivo aroma release of the five gels, eaten by 10 panelists in triplicate for diacetyl (A) and ethylbutyrate (B). Each curve in both figure parts is the average of 30 single curves. The perceived intensity of diacetyl and ethylbutyrate decreases with an increase in gel hardness (from gel 1 to gel 5) averaged over the 10-person panel. The averaged in vivo aroma release curves, however, show no clear difference between the gels.

It could be argued that the absence of an effect of gel structure on in vivo aroma release is caused by a lack of sensitivity of the current setup of the in vivo aroma release measurements, using 10 panelists who assessed each gel in three replicates. However, in Fig. 1 it was shown that an obvious change in aroma concentration in the gels results in a clear change in in vivo aroma release, using the same experimental setup. If clear differences between the variously structured gels in in vivo aroma release occur, our experimental setup registers them.

The results have been analyzed statistically. Each TI and in vivo aroma release curve was summarized by a Tmax and an Imax value. For every individual panelist linear regression has been applied to the dependent variables Imax and Tmax of both TI and in vivo aroma release. Gel hardness and WHC were used as predictors. TI-Imax is the only dependent variable for which a considerable number (6 of 10) of significant negative slopes was found, for both ethylbutyrate and diacetyl. For the other tested dependent variables, only a few significant slopes were observed. This confirms the conclusions drawn from the graphical presentation in Fig. 4, namely, that only the TI-Imax is significantly correlated with the gel structure.

No significant difference was found between the standard error values of data of panelists who did have a significant correlation and those who did not. This shows that the absence of a significant negative correlation is not caused by nonconsistent panelist performance.

A decrease in aroma perception with increasing gel hardness has been reported [11–15]. These studies focused on aroma perception; the in vivo aroma release itself was not measured. On the basis of our results we suggest

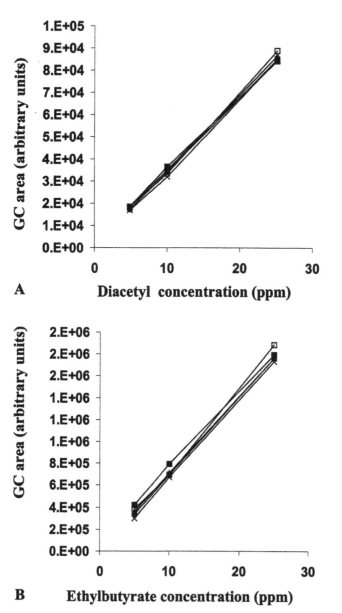

Figure 4 Equilibrium HSGC concentration of diacetyl (A) and ethylbutyrate (B) above whey protein gels 1 (●), 2 (□), 3 (Δ), 4 (×), and 5 (■). (From Ref. [30]. Reprinted with permission from Weel et al, J. Agric. Food Chem. 50:5149–5155. Copyright 2002 American Chemical Society.)

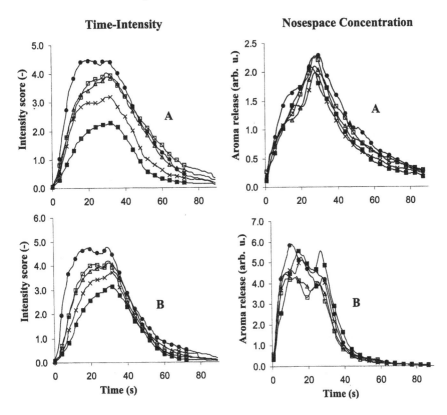

Figure 5 Averaged time-intensity recordings and averaged release profiles for gels 1 (●), 2 (□), 3 (Δ), 4 (×), and 5 (■), flavored with diacetyl (A) and ethylbutyrate (B). (From Ref. [30]. Reprinted with permission from Weel et al, J. Agric. Food Chem. 50:5149–5155. Copyright 2002 American Chemical Society.)

that the release of aroma compounds from gels might not necessarily decrease with firmer gels, but that the aroma perception could be influenced by the textural properties of the gels in a psychophysical way. Psychophysics has been described as the study of the relationship of what exists in the real world (stimulus or stimuli) and the human experience of these events [29]. Interactions between perception of aroma and texture might behave in a similar way to interactions of perception of color, taste, and aroma.

Although the panelists were explicitly instructed not to let textural effects influence their perception of the aroma intensity, an effect of texture on aroma perception was found for a considerable number of panelists. The aroma perception of these panelists is subconsciously influenced by gel texture, whereas the panelists without a significant correlation between TI-Imax and gel structure are able to separate aroma perception from texture perception. In the situation of a normal consumer, whose perception occurs spontaneously, psychophysical interactions of the senses are likely to occur. This emphasizes the potential importance of texture to aroma perception.

D. Conclusions

The main conclusion is that for the whey protein gel system used the sensory assessment of aroma intensity is influenced by a change in gel structure. The change in aroma perception could not be attributed to a change in in vivo aroma release. Therefore, it was concluded that the perception of aroma intensity could be influenced by textural parameters in a psychophysical way.

ACKNOWLEDGMENT

The work presented in this chapter has been funded by Imperial Chemical Industries PLC (project SRF 2015).

REFERENCES

1. CN DuBose, AV Cardello, O Maller. Effects of colorants and flavorants on identification, perceived flavor intensity, and hedonic quality of fruit-flavored beverages and cake. J Food Sci 34:1393–1399, 1415, 1980.
2. J Johnson, FM Clydesdale. Perceived sweetness and redness in colored sucrose solutions. J Food Sci 47:747–752, 1982.
3. E Von Sydow, H Moskowitz, H Jacobs, H Meiselman. Odor-taste interactions in fruit juices. Lebensmittel Wissenschaft Technol 7:18–24, 1974.
4. RA Frank, J Byram. Taste-smell interactions are tastant and odorant dependent. Chem Senses 13:445–455, 1988.
5. DF Nahon, JP Roozen, C DeGraaf. Sweetness flavour interactions in soft drinks. Food Chem 56:283–289, 1996.
6. JM Davidson, RST Linforth, TA Hollowood, AJ Taylor. Effect of sucrose on the perceived flavor intensity of chewing gum. J Agric Food Chem 47:4336–4340, 1999.
7. RM Pangborn, AS Szczesniak. Effect of hydrocolloids and viscosity on flavor and odor intensities of aromatic flavor compounds. J Texture Stud 4:467–482, 1974.

8. RM Pangborn, ZM Gibbs, C Tassan. Effect of hydrocolloids on apparent viscosity and sensory properties of selected beverages. J Texture Stud 9:415–436, 1978.

9. ZV Baines, ER Morris. Flavour/taste perception in thickened systems: The effect of guar gum above and below c*. Food Hydrocolloids 1:197–205, 1987.

10. R Clark. Sensory-texture profile analysis correlation in model gels. In: R Chandrasekaran, ed. Frontiers in Carbohydrate Research. New York: Elsevier Applied Science, 1992, pp 85–89.

11. I Jaime, DJ Mela, N Bratchell. A study of texture-flavor interactions using free-choice profiling. J Sens Stud 8:177–188, 1993.

12. J Carr, D Baloga, JX Guinard, L Lawter, C Marty, C Squire. The effect of gelling agent type and concentration on flavor release in model systems. In: RJ McGorrin, JV Leland, ed. Flavor-Food Interactions. Washington, DC: American Chemical Society, 1996, pp 98–108.

13. JX Guinard, C Marty. Time-intensity measurement of flavor release from a model gel system: Effect of gelling agent type and concentration. J Food Sci 60:727–730, 1995.

14. J Bakker, WE Brown, BP Hills, N Boudaud, C Wilson, M Harrison. Effect of the food matrix on flavour release and perception. In: AJ Taylor, DS Mottram, ed. Flavour Science: Recent Developments. Cambridge: Royal Society of Chemistry, 1996, pp 369–374.

15. CE Wilson, WE Brown. Influence of food matrix structure and oral breakdown during mastication on temporal perception of flavor. J Sensory Stud 21:69–86, 1997.

16. EA Gwartney. Texture and Flavor Release. PhD dissertation, North Carolina State University, Raleigh, 1999.

17. W Lindinger, A Hansel, A Jordan. On-line monitoring of volatile organic compounds at pptv levels by means of proton-transfer-reaction mass spectrometry (PTR-MS): Medical applications, food control and environmental research. Int J Mass Spectrom Ion Processes 173:191–241, 1998.

18. AJ Taylor, RST Linforth, BA Harvey, A Blake. Atmospheric pressure chemical ionisation mass spectrometry for in vivo analysis of volatile flavour release. Food Chem 71:327–338, 2000.

19. AJ Taylor, RST Linforth. Flavour release in the mouth. Trends Food Sci Technol 7:444–448, 1996.

20. RST Linforth, I Baek, AJ Taylor. Simultaneous instrumental and sensory analysis of volatile release from gelatine and pectin/gelatine gels. Food Chem 65:77–83, 1999.

21. I Baek, RST Linforth, A Blake, AJ Taylor. Sensory perception is related to the rate of change of volatile concentration in-nose during eating of model gels. Chem Senses 24:155–160, 1999.

22. AC Alting, RJ Hamer, CG de Kruif, RW Visschers. Formation of disulfide bonds in acid-induced gels of preheated whey protein isolate. J Agric Food Chem 48:5001–5007, 2000.

23. MC Bourne. Texture profile analysis. Food Technol 3:62–66, 72, 1978.

24. VR Harwalkar, M Kalab. Susceptibility of yoghurt to syneresis: Comparison of centrifugation and drainage methods. Milchwissenschaft 38:517–522, 1983.

25. PN Kocher, EA Foegeding. Microcentrifuge-based method for measuring waterholding of protein gels. J Food Sci 58:1040–1046, 1993.

26. AJ Taylor, S Besnard, M Puaud, RST Linforth. In vivo measurement of flavour release from mixed phase gels. Biomol Eng 17:143–150, 2001.

27. MME van Marle. Structure and Rheological Properties of Yoghurt Gels and Stirred Yoghurts. PhD dissertation, Twente University, Enschede, The Netherlands, 1998.

28. M Verheul, SPFM Roefs. Structure of whey protein gels, studied by permeability, scanning electron microscopy and rheology. Food Hydrocolloids 12:17–24, 1998.

29. H Meiselman. Human taste perception. CRC Crit Rev Food Technol 3:89–119, 1972.

30. KGC Weel et al. Flavour release and perception of flavoured whey protein gels: perception is determined by texture rather than by release. J Agric Food Chem 50:5149–5155, 2002.

8

Difficulty in Measuring What Matters: Context Effects

Bonnie M. King, Paul Arents, and C. A. A. Duineveld
Quest International, Naarden, The Netherlands

I. INTRODUCTION

The lack of a simple, direct link between stimulus and response in human evaluation of flavor has frustrated many product developers and marketing managers who want to predict customer purchase from ingredient formulae. Psychophysical models have been proposed to explain how stimulus energy is transformed to a subjective experience and how that experience is then translated into an observed response. It is also known that humans react differently to the same stimulus under different circumstances. For example, response tasks such as evaluating flavor recognition, flavor intensity, and flavor appreciation are dependent on stimulus context. Poulton [1] has shown that context is responsible for many biases in judging sensory magnitude. Lawless et al. [2–4] have asked whether context effects can be eliminated or minimized by training. One could imagine that sensory scientists involved in industrial product testing at least have come to grips with the problem since context effects pose a major obstacle to measuring what really matters for product development. Another hurdle in the acquisition of meaningful data for product development is the assumption of a direct correspondence between instrumental measurements and human evaluation. Fortunately there are occasions when physicochemical data have been modeled successfully to explain some sensory experience. On the other hand, there is still great frustration at having to admit that a parameter as measurable as flavor release in the MS-Nose (Micromass,

Manchester, UK) experiment [5] does not necessarily correlate with flavor perception.

This chapter examines context effects with respect to intensity measurements of some aroma volatiles and one tastant, sucrose. It shows that the difficulties in obtaining unbiased data can be dealt with by a judicious choice of experimental procedures, followed by the proper data treatment and interpretation.

II. MATERIALS AND METHODS

A. Panelists

The Sensory Research Panel (SRP) is a paid professional panel consisting of a pool of 31 women, aged 42–64, who work 2-hr sessions 4 days in the week for Quest-Naarden. Panelists' experience ranged from 2 to 17 years at the time the experiments were conducted. SP1 and SP2 are both trained industrial sensory panels with, at the time the experiments were conducted, 5–7 years of experience doing descriptive analysis. These two panels also operate with pools of paid panelists working 2-hr sessions twice a week. SP1 consists of 16 panelists (4 men, 12 women) aged 25–50. SP2 has 23 panelists (5 men, 18 women) aged 35–70. The MS-Nose panel ($N = 3$) used in Experiment 1 consists of male staff (25–36 years old) from NIZO Food Research in Ede, The Netherlands.

B. Experiment 1: Comparison of Flavor Perception and Flavor Release

Experiment 1 compared perceived intensities for a middle concentration (M) of a flavor volatile preceded by either a high (H) concentration or a low (L) concentration of the same flavor. The NIZO MS-Nose panelists made intensity judgments, without prior training, of *flavor strength* on a 10-point scale for each sample immediately before the MS-Nose experiments for which they were trained. Aqueous concentrations (grams per liter) of ethyl butyrate (0.002, 0.0063, 0.020), amyl acetate (0.001, 0.0045, 0.020), linalool (0.003, 0.007, 0.015), and methyl cinnamate (0.00032, 0.001, 0.0032) were made from 1% ethanol solutions; for *cis*-3-hexenol (0.005, 0.016, 0.050) the stock solution was 10% in ethanol. For the MS-Nose measurements, 15 mL of solution was taken into the mouth and held for 3 seconds without movement while the panelist exhaled nasally into the sampling unit, thus providing an acetone signal. The panelist inhaled for the next 3 seconds, making one chewing movement per second, and then swallowed the sample.

Exhalation continued for another 3 seconds to produce the in vivo aroma release peak.

The MS-Nose consists of an atmospheric pressure chemical ionization gas phase analyzer (APCI-GPA) attached to a VG Quattro II MS mass spectrometer [5]; the cone voltage was 20 V. The ions monitored (MH^+, $M - H_2O + H^+$) were 58 (acetone in the panelists' breath), 71 (amyl acetate fragment), 83 (cis-3-hexenol), 117 (ethyl butyrate), 131 (amyl acetate), 137 (linalool), and 163 (methyl cinnamate).

Two replications of MS-Nose evaluations with intensity scaling for each flavor solution were conducted in the same session, using the measuring sequence *water, 4 flavor solutions (H-M-L-M); water, 4 flavor solutions (L-M-H-M); water.* Linalool, cis-3-hexenol, and methyl cinnamate were measured on one day; ethyl butyrate and amyl acetate were measured on another day a month later. Ion chromatograms were transformed to release curves in Excel and integrated. The parameter $^{10}log(Area)$ of the in vivo aroma release peak was used for all analyses in this chapter, although Imax and Tmax were also obtained from the release curves. The means and standard deviations for interaction terms (flavor component*presentation order) from an unbalanced analysis of variance (ANOVA) were determined for both the MS-Nose data and the panelists' intensity scores.

GenStat (2000, release 4.2, fifth edition), was used to perform all calculations and statistical analyses discussed in this chapter.

C. Experiment 2: Contrasts Established Before or After a Target Stimulus

The SRP ($N = 25$) evaluated all of the solutions described in Experiment 1 by using the audio method for intensity profiling [6]. Two procedures, defined by Lawless et al. [3] as either simple contrast or reversed-pair contrast, were followed, with the exception that only one combination of procedure and condition occurred in any given session. Thus, for the contrast procedure either condition HM or condition LM was used; for the reversed-pair procedure, either condition MH or ML. Samples (15 mL) were served under red lighting in plastic cups coded with three-digit numbers. Panelists emptied the entire contents of each sample into their mouths and swallowed the solutions. The descriptors for which retronasal flavor intensity was measured were *candy* (amyl acetate), *green (cis-3-hexenol)*, *fruity* (ethyl butyrate), *linalool* (linalool), and *cinnamon/metallic* (methyl cinnamate).

Data for the entire experiment were scaled per panelist/procedure as follows:

$$X'_{ijkl} = \frac{X_{ijkl} - \bar{X}_{i...}}{\bar{S}_{i...}} \cdot I \sqrt{\prod_{v=1}^{I} \bar{S}_{v...} + \frac{1}{I} \sum_{v=1}^{I} \bar{X}_{Iv...}}$$

where $i = 1,\ldots, I$ stands for the panelists, $j = 1,\ldots, J_i$ for the products for the ith panelist, $k = 1,\ldots, K_{ij}$ for the replications for the ith panelist and her jth product, $l = 1,\ldots, L$ for the descriptors; $\bar{X}_{i...}$ and $\bar{S}_{i...}$ are the mean and standard deviation, respectively, over all product, replication, descriptor combinations for the ith panelist.

Panel means of intensity per flavor/concentration were obtained from the scaled data by fitting variance components using restricted maximum likelihood (REML). Panelists were considered a random effect.

D. Experiment 3: Comparison of Panels/Scaling Techniques

Nine concentrations of aqueous sucrose (Kristal suiker extra fijn, Suiker Unie, Dinteloord, The Netherlands) were evaluated for sweetness by swallowing. All members of the three industrial panels participated in this experiment. SRP used the audio profiling procedure as in Experiment 2. SP1 and SP2 used line scales of, respectively, 10 cm or 15 cm. Contrast, frequency, and range effects were examined for each panel by presenting various selections of the sucrose concentrations (50 mL) served in plastic cups coded with three-digit numbers.

The contrast effect for 70 g/L sucrose was studied by placing it in the middle of a full range (with the following concentrations in grams per liter: 17.1, 34.2, 137, 274), at the high end of a range of low concentrations in grams per liter (17.1, 27.4, 34.2, 43.8) or at the low end of a range of high concentrations in grams per liter (110, 137, 174, 274). The three panels evaluated one range per session in the order full, high, low, and replicated this procedure four times. Data were treated as in Experiment 2, using a REML model having the combination ContextConcentration as fixed term. Random terms were (session) panelist and (session) ContextConcentration. Means for the 70-g/L sample were compared by using Fisher's least significant difference (LSD). Frequency distributions for data combined over the three concentration ranges were studied per panel by dividing the scale range used into 10 equal portions. For SRP 1000 evaluations were thus distributed over the range of 160 to 879 Hz. SP1 (578 evaluations) and SP2 (515 evaluations) both used their maximal available range of, respectively, 10 cm or 15 cm.

Frequency and range effects were further investigated by presenting six samples in a positively skewed context (3×12 g/L, 2×38 g/L, 1×120 g/L) or in a negatively skewed context (1×12 g/L, 2×38 g/L, 3×120 g/L). The number of evaluations analyzed for each panel was SRP (540), SP1 (144), SP2 (186). The effect of not spacing samples equally was studied with SRP and SP2 by presenting the following concentrations in grams per liter: 17.1, 27.4, 43.8, 174, 274. There were 255 evaluations for equal spacing and 175 evaluations for unequal spacing for the SRP. The corresponding numbers of evaluations for SP2 were 175 and 155.

E. Experiment 4: Procedures That Might Influence Context Effects by Emphasizing the Previous Sample

Experiment 3 was repeated with two additional measuring procedures: R2P (relative-to-previous sample) and R2R (relative-to-reference). Only members of the SRP participated in this experiment. Evaluation of the sucrose solutions took place under red lighting. The R2P procedure for scaling with the audio method has been described in detail [7]. Instead of hearing a fixed start tone (500 Hz for normal profiling) before each measurement, the previous response tone was audible through the headphones. Panelists changed the frequency of this tone to correspond to the perceived sweetness intensity of the target sample. The sample presentation for this method was monadic.

When scoring R2R, panelists were served a pair of samples in each round: one reference sample (70 g/L sucrose), which was tasted first, and one target sample. The intensity of the reference sample was arbitrarily set equal to the start tone (500 Hz). Panelists changed the frequency of this tone to correspond to the perceived sweetness intensity of the target sample.

Data for the profiling procedure consisted of 1000 evaluations distributed over the low, high, and full ranges as 380, 365, and 255 observations, respectively. Scaling and REML analysis were as described for Experiment 3. Data for the R2P procedure consisted of 540 evaluations: 170 low, 150 high, 220 full range. Scaling and REML analysis were as for profiling. The R2R data (93 low, 87 high, and 136 full range for a total of

307 evaluations) were scaled in a slightly different manner, as follows:

$$X^*_{ijkl} = \frac{X_{ijkl} - 500}{U_{i\ldots}} \cdot I \sqrt{\prod_{v=1}^{I} U_{v\ldots}} + 500 \quad \text{with}$$

$$U_{i\ldots} = \frac{1}{L \sum_{\tau=1}^{J_i} K_{i\tau}} \sum_{l=1}^{L} \sum_{j=1}^{J_i} \sum_{k=1}^{K_{ij}} (X_{ijkl} - 500)^2$$

where $i = 1, \ldots, I$ stands for the panelists, $j = 1, \ldots, J_i$ for the products for the ith panelist, $k = 1, \ldots, K_{ij}$ for the replications for the ith panelist and her jth product, $l = 1, \ldots, L$ for the descriptors.

The number of evaluations in the experiments examining unequal spacing were profiling (175), R2P (145), and R2R (145). For the skewed distributions, 540 evaluations by profiling were compared to 438 evaluations by R2P. Skewed distributions were not examined by R2R.

III. RESULTS AND DISCUSSION

The discrepancy between flavor perception and flavor release as shown in Fig. 1 is the result of a classical contrast effect. These results indicate that there was no significant difference in *nosespace concentration* of volatiles for the middle concentration due to the sample presentation order (Fig. 1a). On the other hand, the *flavor intensity* of the middle concentration preceded by the lower concentration was always rated higher than when it was preceded by the higher concentration (Fig. 1b), although statistical significance at the 5% level was achieved only for ethyl butyrate. This lack of statistical significance reflects the limited data ($N = 3$) and panelist scoring inconsistencies due to lack of training. Training and familiarity with the substances evaluated do not eliminate this contrast effect, however, as can be seen in Fig. 2. All the contrast tests with the SRP showed a significant difference, as did the reversed-pair tests for *candy* and *green*.

These effects confirm those for a basic taste evaluated by the same procedure: Lawless et al. [3] reported a contrast effect for sweetness judgments of 10% aqueous sucrose when preceded by either 5% or 20% aqueous sucrose. This effect was present for measurements made by a 15-point category scale, a magnitude estimation scale (10% sucrose assigned a rating of 10), and two versions of the labeled magnitude scale proposed by Green et al. [8]: strongest imaginable sensation of sweetness or strongest imaginable oral sensation excluding pain. The effects all persisted in the

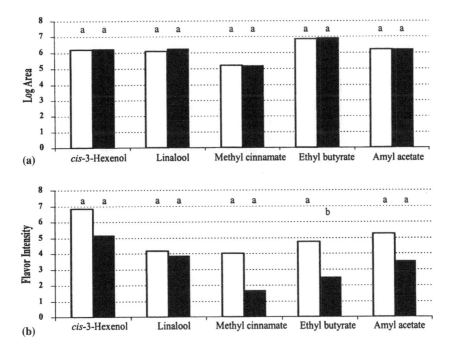

Figure 1 A comparison of flavor release from MS-Nose measurements (a) and flavor perception (b) of the middle concentrations of five flavor compounds used in Experiment 1. Light bars represent the sequence LM; dark bars represent the sequence HM. Per flavor compound, sequences with the same letter (a, b) are not significantly different at the 5% level according to Fisher's LSD. HM, LM, LSD,

corresponding reversed-pair tests with these methods. Thus it does not seem possible to remove such a contrast effect by either choice of scale or choice of procedure, such as the use of a reference standard [4].

It is interesting to see whether the two-sample presentation used in Experiments 1 and 2 is more conducive to creating contrast effects than procedures using a larger number of samples, such as Experiment 3. The sequential monadic presentation is generally used for descriptive analysis by profiling. Product development often requires measurement of more samples than can be evaluated in one session. Moreover, for certain tests (e.g., shelf life) evaluations need to be compared over time. Experimental designs that propose serving orders fully connected over the sessions should be used. The serving design may, on the other hand, be governed by the product design, which is based on experimental variables such as those relating to ingredients or production. In that case the number of repetitions

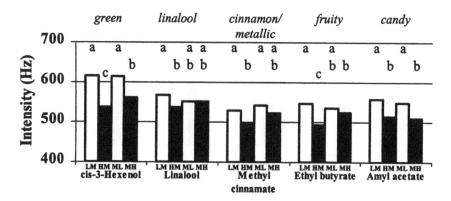

Figure 2 Mean SRP scores for the middle concentration of the five flavor compounds used in Experiment 2. Sequence of sample presentation is indicated by LM or HM for the simple contrast tests and ML or MH for the reversed-pair tests. Per flavor compound, sequences with the same letter (a, b, c) are not significantly different at the 5% level according to Fisher's LSD.

should be adequate to prevent introducing potential contrast effects. Nonetheless, problems can arise when limited sample availability prohibits some preliminary testing to determine the sensory range, as with market products for which the ingredients are not known a priori. Contrast effects have been reported when measuring sweetness in different aqueous sucrose concentrations served as a sequence [9,10]. As in the two-sample presentation, solutions immediately preceded by a higher sucrose concentration were judged significantly less sweet than the same samples preceded by a lower concentration.

Lawless and Heymann [2] ask "whether it is possible to 'inoculate' a trained panelist against the day-to-day context of the experimental session by sufficient training." Industrial panelists are trained in test methodology and given substantial practice in order to increase sensory perception and discrimination. In a task such as intensity rating, in which various scales and procedures are possible, the effects of methodology and practice are confounded for these panelists. In the current study of contrast effects and biases, Experiment 3, the three panels were matched so that the scaling methodology could be compared. Figure 3 shows at least one significantly different evaluation of sweetness among the context-dependent 70-g/L samples for all three panels. However, the contrast effect is least evident for data from the SRP. The audio method used by the SRP can be considered a matching procedure, and Stillman [11] showed less susceptibility to range

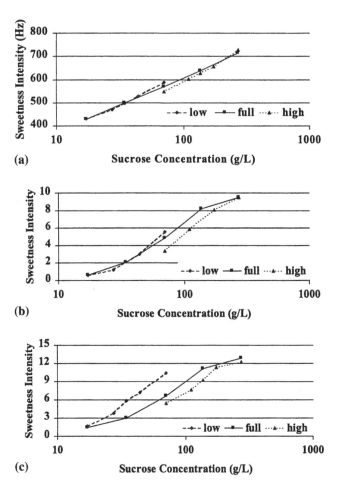

Figure 3 Sweetness intensity ratings for the three ranges of sucrose solutions used in Experiment 3. SRP/audio scaling (a), SP1/10-cm line scale (b), SP2/15-cm line scale (c). SRP, SP1, and SP2 refer to three industrial sensory panels.

bias with a matching procedure (in his case, variable lines) than with a 13-point category scale.

The sucrose solutions evaluated in Experiment 3 represent the full range of sweetness likely to be encountered in any panel's experience, with the high and low ranges postulated as "extremes" in order to study the contrast effect. Frequency distributions for the data from Fig. 3 are given in Fig. 4. The data for SRP are normally distributed. Data for each sucrose concentration show pretty much the same distribution, albeit around a

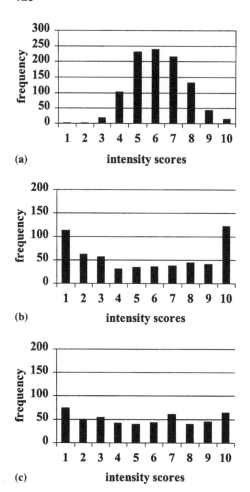

Figure 4 Frequency distributions for data from Fig 3. SRP/audio scaling (a), SP1/ 10-cm line scale (b), SP2/15-cm line scale (c). SRP, SP1, and SP2 refer to three industrial sensory panels.

different mean. Data from SP1, on the other hand, show large end effects or bunching in categories 1 and 10, as was also the case for the individual distributions of the highest and lowest concentrations. This scale did not extend far enough in either direction for panelists to accommodate the extreme solutions that were outside the range of sweetness they usually encountered. Data from SP2 reflect range mapping to a much greater extent than was seen with data from the other two panels. These data are almost evenly distributed over the 10 categories, albeit with slight bunching at the ends.

If one wishes to compare sweetness not only within a product type but also across product types, the scale for sweetness must provide sufficient differentiation within general categories (e.g., bread or tomato sweetness as opposed to sweetness in candies, ice cream, and soft drinks) without resorting to range mapping, i.e., using the full scale for every set of samples presented. Moreover, if one wishes to measure several other attributes besides sweetness in all these products, it is convenient to have the same scale for each attribute so that the flavor profile can be constructed: i.e., sweetness can be compared to the intensity of vanilla flavor, green notes, etc. Although we have treated scale equivalence testing in a separate paper [12], it is mentioned here as another challenge to the characterization of flavor compounds by sensory profiling.

All three panels evaluated the samples in the negatively skewed presentation as slightly less sweet than those in the positive distribution, as is shown in Fig. 5. This difference was significant for the middle concentration (38 g/L sucrose) for panels SRP and SP2, conforming to the curvature suggested by the Parducci range-frequency theory [13,14]. For panel SP1 only the lowest concentration showed a significant difference in sweetness. Thus the scaling method/training had little influence on reducing context effects. One could argue that six samples are too few to establish a skewed context, but the goal of this experiment was to examine skewness within the confines of a product development panel's typical experience. Riskey et al. [9] used 21 samples in their skewed presentations (seven concentrations of sucrose), which were evaluated for sweetness on 9-point category scales. Their data clearly indicated that the samples in the negatively skewed presentation were evaluated as less sweet. Lee et al. [15] used a 15-point category scale to evaluate 20 samples (five concentrations) of either aqueous NaCl or sucrose in orange juice. They were unable to remove the context effect by treating their data with a signal detection analysis.

The unequal spacing of samples had no effect for the SRP but did perturb the intensity function for SP2 (Fig. 6).

Experiment 4, which used only the SRP to examine different procedures for evaluating a series of concentrations, showed no decrease in contrast effects by introducing a reference. It made little difference whether the reference was the preceding sample or a fixed concentration, as shown in Fig. 7. Significant differences between identical concentrations presented in different contexts (i.e., the high as opposed to the full range) were evident for both R2P and R2R. When the samples were presented in a skewed distribution, the trend was the same for R2P and the profiling data

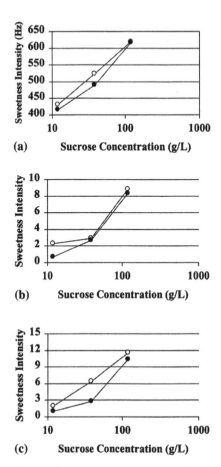

Figure 5 Sweetness intensity ratings for the six sucrose solutions presented in positively skewed context (open circles) or negatively skewed context (closed circles) according to Experiment 3. SRP audio scaling (a), SP1/10-cm line scale (b), SP2/15-cm line scale (c). SRP, SP1, and SP2 refer to three industrial sensory panels.

(Fig. 5), although the difference for the middle concentration was not significant for R2P. Unequal spacing of samples had no effect in any of the three methods.

One way of preventing context effects, albeit not a very practical solution for industrial sensory analysis, would be to evaluate only one sample per session. When all the data from SRP profiling in Experiments 3 and 4 were analyzed using only the first serving round, as opposed to all rounds, the intensity functions did not change. We see the explanation for

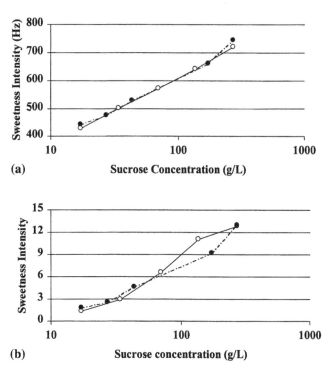

Figure 6 Sweetness intensity ratings for the equally spaced (open circles) and unequally spaced (closed circles) sucrose solutions according to Experiment 3. SRP/audio scaling (a), SP2/15-cm line scale (b). SRP, Sensory Research Panel.

this fact as also the reason why we encounter little difficulty with context effects under our normal profiling procedures, i.e., monadic presentations of four stimuli per product type per session and use of connected experimental designs to balance repetitions and sessions. Experience indicates that under these conditions, the model used by the SRP to evaluate all descriptor/product combinations is most likely a comparison to "life experience." The SRP indicates different ranges of the audio scale to correspond to intensities in what amounts to an "absolute," as opposed to a product-specific or attribute-specific fashion.

This model for SRP intensity evaluation corresponds to the weaker definition of an absolute model as given by Koo et al. [16]:

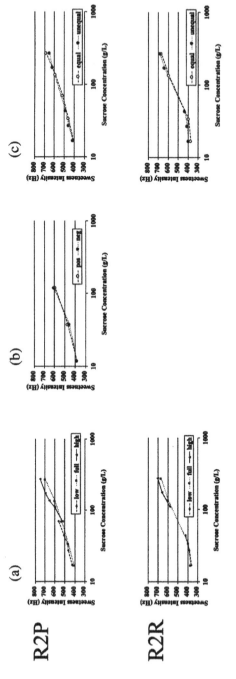

Figure 7 SRP comparison of sweetness intensity ratings for sucrose solutions evaluated relative-to-previous sample (R2P) and relative to a fixed reference (R2R) according to Experiment 4; the three ranges (a), the skewed context for R2P (b), and the equal/unequal spacing context (c). SRP, Sensory Research Panel.

[J]udges have a set of exemplars stored in memory, yet each judge's exemplars and the numbers [frequencies in our measurement] associated with them are not necessarily the same. Thus, each judge would be internally consistent in their assigning of [tones] but there would not necessarily be agreement among judges on which [tones] should be assigned.

This individual (but very consistent) SRP scaling behavior has developed solely on the basis of a panelist's experimentation with the audio method, her exposure to a wide variety of stimuli, and her constant practice. It is not the result of anchoring or calibration with references. Reference substances are used qualitatively, however, to define descriptors.

Lawless et al. [3] have suggested that contrast effects occur at the level of response output transformation. Although some experimental procedures and types of scales can reduce context effects, we do not have any reason to believe that such effects can be eliminated completely. Given proper awareness of these effects, the sensory scientist should be able to draw proper conclusions from experiments in which context effects could be influential.

ACKNOWLEDGMENTS

Maurits Burgering, Eduard Derks, Diane Honn, Nathalie Moreau, Susanne Schroff, Seeta Soekhai, Gerard Vermeer, Koen Weel, and 70 panelists contributed to the data collection for this chapter.

REFERENCES

1. EC Poulton. Models for biases in judging sensory magnitude. Psychol Bull 86(4):777–803, 1979.
2. HT Lawless, H Heymann. Sensory Evaluation of Food. New York: Chapman & Hall, International Thomson Publishing, 1998, pp 301–340.
3. HT Lawless, J Horne, W Speirs. Contrast and range effects for category, magnitude and labeled magnitude scales. Chem Senses 25:85–92, 2000.
4. J Diamond, HT Lawless. Context effects and reference standards with magnitude estimation and the labeled magnitude scale. J Sens Stud 16:1–10, 2001.
5. AJ Taylor, RST Linforth, BA Harvey, A Blake. Atmospheric pressure chemical ionisation mass spectrometry for in vivo analysis of volatile flavour release. Food Chem 71:327–338, 2000.
6. BM King. Sensory profiling of vanilla ice cream: flavor and base interactions. Lebensm.-Wiss. u.-Technol. 27:450–456, 1994.

7. BM King. Odor intensity measured by an audio method. J Food Sci 51:1340–1344, 1986.
8. BG Green, GS Shaffer, MM Gilmore. Derivation and evaluation of a semantic scale of oral sensation magnitude with apparent ratio properties. Chem Senses 18:683–702, 1993.
9. DR Riskey, A Parducci, GK Beauchamp. Effects of context in judgements of sweetness and pleasantness. Percept Psychophys 26:171–176, 1979.
10. HNJ Schifferstein, JER Frijters. Contextual and sequential effects on judgments of sweetness intensity. Percept Psychophys 52:243–255, 1992.
11. JA Stillman. Context effects in judging taste intensity: A comparison of variable line and category rating methods. Percept Psychophys 54(4):477–484, 1993.
12. CAA Duineveld, P Arents, BM King. Determination of equivalence between intensity scales by means of paired comparison. Food Quality Preference 14(5/6):419–424, 2003.
13. A Parducci. A range-frequency compromise in judgment. Psychol Monogr 77(2, Whole No. 565), 1963.
14. A Parducci, DH Wedell. The category effect with rating scales: Number of categories, number of stimuli, and method of presentation. J Exp Psychol Hum Percept Perf 12(4):496–516, 1986.
15. H-S Lee, K-O Kim, M O'Mahony. How do the signal detection indices react to frequency context bias for intensity scaling? J Sens Stud 16:33–52, 2001.
16. T-Y Koo, K-O Kim, M O'Mahony. Effects of forgetting on performance on various intensity scaling protocols: Magnitude estimation and labeled magnitude scale (Green scale). J Sens Stud 17:177–192, 2002.

9

Measuring the Sensory Impact of Flavor Mixtures Using Controlled Delivery

David J. Cook, Jim M. Davidson, Rob S. T. Linforth, and Andrew J. Taylor
The University of Nottingham, Loughborough, Leicestershire, England

I. INTRODUCTION

Research into the sensory perception of food flavor has generally been conducted for each of the senses (taste, aroma, texture, or color) in isolation or at best in simple mixtures, e.g., looking at the influence of one component on the perception of another. Although it is commendable to develop a fundamental understanding of the senses in this manner, we may draw inaccurate conclusions if we attempt to extrapolate findings from simple model systems to the complex world of real foodstuffs. For example, Stevens [1] showed that taste components in a mixture may be detected at a fraction of their sensory threshold concentrations as measured individually. This effect is thought to result from so-called generous integration at a neural level. Drewnowski [2] highlighted one implication of this in relation to caffeine, which exerts a considerable influence on the flavor experience of foods, despite typically being present at near or below threshold values. Experience from research into other sensory modalities has shown that it is simplistic to investigate the senses individually and that such an approach fails to recognize the multisensory nature of our interaction with the world around us. Cross-modal integration, the neurological combination of signals from different sensory systems [3], can mean that subjective experience within one sensory modality is affected by the stimulation of another. Examples may be drawn from each of the senses, for example, the McGurk effect (visions influencing what we hear [4]) or, in the field of flavor research, the effect of color on perceived flavor. DuBose and associates [5] showed that the color of a beverage could influence judgments of its flavor. For example, changing the color of a cherry flavored beverage from red to green caused a significant number of panelists to believe that the drink was lime-flavored.

Taste-aroma interactions [6] are a further example of cross-modal perception. Dalton and colleagues [7] provided evidence of the interdependency of the gustatory and olfactory systems, by demonstrating integration of subthreshold taste and smell. In this experiment concurrent presentation of a subthreshold taste stimulus (saccharin) was shown to lower the detection threshold for benzaldehyde aroma (presented orthonasally). It was concluded that cross-modal summation occurred, as central neural integration of taste and smell generated a representation of flavor perception. Such interactions are thought to occur only when the taste and aroma concerned are congruent with one another: i.e., they are perceptually linked because they have been experienced together in foods [8]. This suggests an experiential learning angle to perception, which could not be elucidated by studying the senses in isolation and which helps to account for cultural differences in flavor perception and preference.

In addition to mixture effects, there is a temporal element to flavor perception, which is difficult to simulate in sensory testing. Grab and Gfeller [9] correlated sensory perception with breath by breath analysis of aroma release using atmospheric pressure ionization-mass spectrometry (API-MS) to show how flavor character changed as a strawberry was eaten. The sensory impressions moved from fruity to fruity-estery to fermented, then green and metallic/woody, as the matrix of the strawberry was broken down, hydrated, and salivated. This could be correlated with the changing volatile profile in-nose over time. Clearly it would be of interest to the flavor scientist to see whether recreating this order of release artificially in a food product would lead to a more convincing perception of fresh strawberry flavor.

The authors have devised a system that facilitates studies on the perception of complex flavor mixtures and allows controlled delivery in the temporal dimension. The idea behind the invention is simple. Multiple streams of flavor stimuli in solution are pumped, mixed, and delivered in-mouth through polytetrafluoroethene (PTFE) tubing. The mixing proportions and individual and overall flow rates can be programmed in real time.

A panelist records changes in perception of a specified sensory attribute over time using a time-intensity (TI) technique.

As a research tool, the multiple-channel flavor delivery system (McFlads) offers many opportunities for study. Presentation of flavor stimuli in solution is, of course, a simplification of many eating experiences (because of the absence of food matrix and the mastication required to break it down). However, this simplification facilitates the study of fundamental interactions among the senses, without variations caused by mastication. McFlads allows the delivery of multiple stimuli in-mouth in a temporally controlled fashion and results in retronasal aroma delivery, which is representative of human ingestion. Taste, aroma, and tactile stimuli (e.g., viscous liquids) can be presented in this fashion.

A prototype system using computer-controlled syringe pumps to deliver up to four streams of stimuli simultaneously was developed. This prototype has been used to investigate the perception of flavor mixtures over time. Results are presented here for experiments looking at (a) adaptation to flavors delivered in this fashion and (b) the nature of taste-aroma interactions in a strawberry flavored solution that comprised a commercial strawberry flavor, sucrose, and organic acids in solution.

II. MATERIALS AND METHODS

A. Determining the Degree of Adaptation to Compounds in Solution

I. Materials

The six solutions tested were sucrose (20 and 100 g/L; British Sugar, Peterborough, UK), citric acid (0.5 and 2.0 g/L; Lancaster Synthesis, Morecambe, UK), and isoamyl acetate (30 and 300 ppm; Firmenich SA, Geneva, Switzerland).

2. McFlads Configuration

The adaptation experiments utilized the most basic pumping setup, using just one syringe pump (Harvard Apparatus, South Natick, MA) with no changes in stimulus type or concentration. Solutions were pumped via plastic tubing (PFA, Swagelok Co., Solon, Ohio) to a mouthpiece for delivery to the panelist (100-μL Gilson pipette tip fitted over the tubing and renewed for each panelist).

3. Sensory Analysis

Fourteen members of an experienced paid panel were used in the trial (aged from 32 to 61 years; 11 females and 3 males), and each appraised all of the six test solutions. The panel had been trained in the techniques of magnitude estimation and time-intensity analysis and had demonstrated ability in these techniques over a period of several years. A familiarization session was held before the trial, to ensure that the panel were comfortable with appraising samples delivered in-mouth by McFlads.

Flavorings were delivered in-mouth at a constant rate (5 mL/min) for 10 min. The panelist was seated in a booth designed for sensory evaluation and was isolated from the pumping equipment. The panelist end of the tubing was introduced via a serving hatch and a clamp stand used to hold the mouthpiece at a suitable height.

4. Time-Intensity Technique

In the case of the sucrose solution, panelists were asked to rate changes in perceived sweetness over time; for the citric acid they were asked to rate changes in acidity and for the isoamyl acetate, to rate changes in perceived fruity/banana aroma. Panelists recorded these changes in intensity by marking a piece of graph paper (A4 size) at specific intervals with a pen. Time was marked on the x axis and the panelist was asked to plot perceived intensity on the y axis after 15 and 30 sec and subsequently at 30-sec intervals. The y-axis range was 0–200 and the initial intensity (whether sweet, acid, or fruit) was defined as 100. The panelists then had to rate the intensity relative to this value over the 10-min period. No specific instructions were given regarding the timing of mouth movements and swallowing, although panelists were asked not to suck on the mouthpiece as this would affect stimulus delivery.

B. Perceptual Interactions of Strawberry Aroma with Sugar and Acids in Solution

1. Materials

A balanced strawberry flavor was blended from 150 ppm strawberry aroma (Fraise 504877T, Firmenich SA, Geneva, Switzerland), 60 g/L sucrose, 1.0 g/L citric acid, and 0.2 g/L malic acid.

2. McFlads Configuration

Two syringe pumps, configured as shown in Fig. 1, were used to pump and mix solutions of strawberry aroma, sucrose, acids (citric and malic

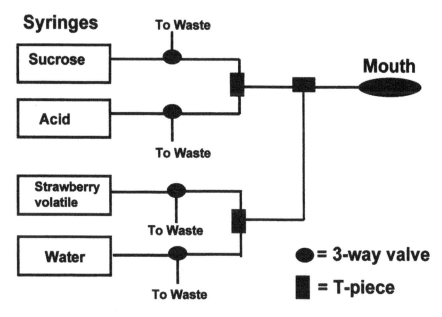

Figure 1   The multiple-channel flavor delivery system (McFlads) pumping setup for the strawberry flavor experiment.

combined), and water, such that the initial concentrations of each delivered in-mouth were as specified for the balanced strawberry flavor. The overall flow rate was 10 mL/min.

Initially, panelists received all three flavorings (aroma, acid, and sucrose). Then, after 30 sec of each run, the composition of the flavor was changed by completely removing either one, two, or all three of the components (seven different possible runs). Whenever solutions were removed, a compensating flow of water was introduced, so that the concentration of the remaining components and overall flow rate remained constant. After 3 min of run time, changes made at 30 sec were reversed, and the flavor mixture restored to the initial condition.

3. Sensory Analysis

A minimum of three and a maximum of six panelists were used for each run. Not all panelists carried out each run; however, because each stream of flavoring was removed singly, in pairs, and in triplicate, there was considerable replication built into what was effectively a factorially designed experiment.

4. Time-Intensity Technique

Panelists were asked to record changes in perceived strawberry flavor (*not* sweetness or acidity) by turning a dial. The dial was connected to an analog-to-digital converter (ADC-16 High Resolution Data Logger, Pico Technology Ltd., Cambridge, UK). The digitized output was captured by a data acquisition software package (Picolog, Pico Technology Ltd.). Turning the dial moved a cursor over a graph on a computer screen. Panelists could see the screen and were asked to move the cursor up and down over time to represent changes in strawberry flavor perception. The cursor started in the middle of the computer screen, representing an arbitrary score of 100, which was specified as the initial intensity of the strawberry flavor. The top and bottom of the screen represented 200 and 0 strawberry flavor intensity, respectively.

C. Analytical Checks on the Pumping Methodology

To verify the appropriate delivery of flavorings using McFlads, separate runs of the strawberry flavor experiment were performed with accompanying analysis of the volatile and nonvolatile flavor components over the time course.

1. Analysis of Volatile Release

In-nose release of strawberry aroma was measured by using the MS-Nose (Micromass, Manchester, UK). A pumping regimen, in which the strawberry aroma was removed after 30 sec and replaced after 3 min, was delivered to each of seven panelists, while conducting simultaneous nose-space sampling to the MS-Nose. The ion with m/z 117 (ethyl butyrate) was used to follow strawberry aroma release. The in-nose concentration of ethyl butyrate was calculated by comparison with the signal for a calibrant of known concentration [10].

2. Analysis of Nonvolatile Release

In-mouth release of sugars and acids was measured using a cotton bud swabbing technique [11,12] with subsequent liquid chromatography-mass spectrometry (LC-MS) analysis (LCZ, Micromass, Manchester, UK). Measurements were taken at 30-sec intervals, by five panelists, during a run in which both nonvolatile components were switched off. Sucrose was monitored at m/z 341, citric acid at m/z 191, and malic acid at m/z 133).

III. RESULTS AND DISCUSSION

A. Determining the Degree of Adaptation to Compounds in Solution

Adaptation can be defined as a reduction in sensitivity following stimulation and is common to all senses. It is thought that adaptation is due to processes occurring at the receptor level. For example, in response to a steady-state stimulus, the neural discharge of the chorda tympani (the main nerve involved in gustatory perception) progressively declines until zero is reached [13]. The extent of adaptation to taste and aroma stimuli depends not only on the particular compound but also on the strength of stimulus [14–16], the duration of stimulation [17], the method of stimulation [15,18], and prior beliefs about the stimulus [19].

The degree of adaptation to the flavorings used in our study was surprisingly low over a 10-min period (Table 1). When results from all 14 panelists were averaged together, the mean time-intensity curves showed only small reductions in perceived intensity throughout the experiment (Fig. 2). The greatest mean decreases in perception were found with 0.05 g/L citric acid and 30 ppm isoamyl acetate (Table 1), although perception at the end of 10 min was still between 80% and 90% of that initially. The most likely explanation for the observed low adaptation rates is that mouth movements and swallowing, associated with consuming the flow of flavoring, may reduce the tendency for adaptation. Theunissen and Kroeze [20] asked panelists to hold a sucrose solution in-mouth for 25 sec and judge the sweetness every 5 sec. They found that when mouth movements were made, there was much less adaptation, although the rate of such movements

Table 1 The Average Change in the Perceived Intensities of Three Flavor Stimuli Delivered Continuously by McFlads over a 10-Minute Period[a]

Solution	Average intensity after 10 min[b]	Change in intensity, %
300 ppm isoamyl acetate	97.5	−2.5
30 ppm isoamyl acetate	90	−10
2 g/L citric acid	95	−5
0.5 g/L citric acid	84	−16
100 g/L sucrose	100	0
20 g/L sucrose	92.5	−7.5

[a] Data are the mean of 14 panelists. McFlads, multiple-channel flavor delivery system.
[b] Perceived intensity at 10 min has been expressed as a percentage of initial intensity (100).

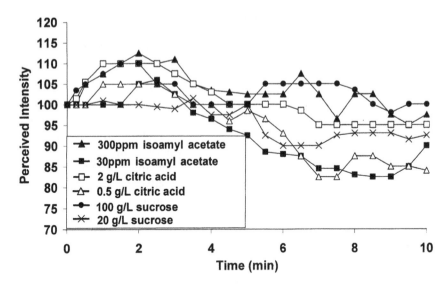

Figure 2 Changes in the perception of constant levels of taste and aroma compounds pumped into the mouth for 10 min. (Curves are the average of 14 panelists' time-intensity assessments.)

did not influence the degree of adaptation. It was suggested that when mouth movements are made, stimulus material moves through the mouth and thus different receptor cells are stimulated, whereas without mouth movements, the same receptor cells are continuously stimulated. Those cells that are intermittently stimulated have time to recover from adaptation, and this leads to a lower total degree of adaptation. Additionally, when the tongue is moved, there are changes in other sensory stimuli (e.g., pressure, or the tactile stimulus locating the bolus in-mouth), which may counteract adaptation to gustatory stimuli.

Adaptation to aroma delivered in solution (isoamyl acetate) was also low (e.g., 90% of initial intensity after 10 min). In the case of aroma transport to the olfactory receptors, mouth movements and swallowing actions are the main determinants of retronasal delivery. The timing of such events would be expected to produce pulsed, rather than continuous, stimulation, which probably accounts for the low levels of adaptation to retronasal aromas. This results in a pattern of stimulation that is more representative of the consumption of real foods than in studies involving orthonasal aroma delivery, in which greater degrees of adaptation (e.g., to 30% to 40% of initial intensity over a 10-min period) have been reported [19,21]. The degree of adaptation to flavors during normal food consumption appears to be relatively low, in accordance with our results. In one study, it was shown that the degree of adaptation to sweetness while eating yogurt was significantly less than the adaptation to sweetness stimulated by placing sucrose-impregnated filter paper on the tongue [15].

In general, the higher stimulus concentration of each flavoring exhibited slightly less adaptation (Table 1 and Fig. 2). There are conflicting reports in the literature as to how concentration affects the rate of adaptation. For instance, some studies have shown that the degree of adaptation increases (i.e., there is a faster decline in intensity over a fixed period) with the concentration of the odorant or tastant [20–23]; others have shown the opposite effect [14,15,18,24]. The reason published results differ so widely is that different workers use different methods to investigate adaptation. For example Bujas and coworkers [13] showed that adaptation to solutions of sucrose, citric acid, sodium chloride, and quinine was dependent upon the area of the tongue that was stimulated (less adaptation for smaller area of stimulation). Meiselman and Buffington [18] showed that stimulation of the taste buds using filter paper soaked in sodium chloride produced a faster rate of adaptation to sodium chloride than when using a flow method. However, both methods lowered the perceived saltiness to practically zero after 180 sec. Theunissen and coworkers [15] likewise reported that sucrose-soaked filter paper in contact with the tongue resulted in more adaptation than sucrose delivered by a flow method.

B. Perceptual Interactions of Strawberry Aroma with Sugar and Acids in Solution

The removal of any of the flavorings (acids, sucrose, or strawberry aroma) during the pumping experiment resulted in reduced sensory perception of strawberry flavor (Fig. 3). By far the greatest impact on overall strawberry flavor was due to sucrose. In any run in which sucrose was removed from the mixture (runs 2, 4, 6, and 7 in Fig. 3) there was a large and sudden decrease in strawberry perception, practically to zero. This effect was reversed when the flow of sucrose was switched on again after 3 min and perceived intensity returned to its original level of around 100.

When the acid and strawberry aroma were turned off either separately (runs 1 and 3) or together (run 5) they had less of an effect on the perceived strawberry intensity than sucrose. Stopping the flow of acids caused the perceived strawberry intensity to fall to just above 50%, whereas removal of the strawberry aroma caused the intensity to decline to 60% of initial perception. A minimum of 60% was also reached when both the acid and the volatile were turned off together. These three curves were considerably different from the rest. In runs in which the aroma was removed, but sucrose

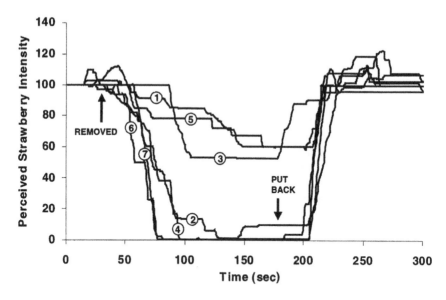

Figure 3 Mean time-intensity curves for strawberry flavor perception, resulting from pumping and mixing solutions of citric/malic acids, sucrose, and a strawberry aroma. Initially all three components (acids, sucrose, strawberry aroma) were present. After 30 seconds either one, two, or all three of the solutions were replaced by an equivalent flow of water (see key). After 180 sec, the mixture was restored to its initial state. 1, Aroma off; 2, sucrose off; 3, acid off; 4, aroma and sucrose off; 5, aroma and acid off; 6, acid and sucrose off; 7, all off.

remained [1] and [5], the rate of decrease in perceived strawberry intensity was slower than in any run in which sucrose flow was stopped.

When acid alone was removed from the solution (run 3) there was a "delayed reaction" in the decrease in strawberry perception. However, when the acid was turned back on, strawberry perception increased rapidly (more quickly than any other curve). From this it would seem that acidity has a greater impact on strawberry perception when it is suddenly added to a mixture in real time than when it is removed.

The analytical checks on delivery of flavorings to the tongue (Fig. 4) and nosespace (Fig. 5), respectively, confirmed that McFlads was producing the desired changes in stimulus. After the flow of each stream was stopped at 30 sec, there was a drop in stimulus concentration as flavoring that had already been delivered to the mouth was dissipated from the system (persistence). For instance, with sucrose, there was still approximately 2 g/100 g saliva present after 60 sec, and concentrations only fell close to zero

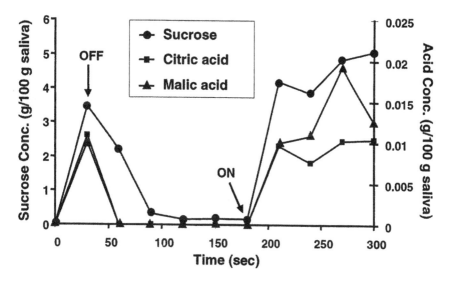

Figure 4 Analytical check on nonvolatile release in-mouth during the strawberry flavor experiment (using a tongue-swabbing/LC-MS technique).

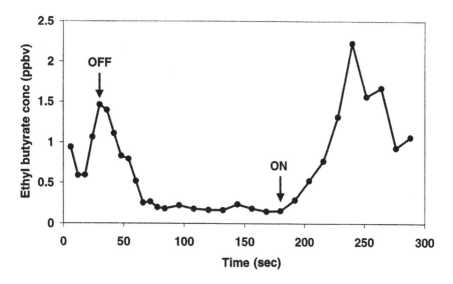

Figure 5 Analytical check on the delivery of ethyl butyrate (m/z 117) in-nose during the strawberry flavor experiment (using the MS-Nose).

thereafter. With reference to Fig. 3, it can be seen that these analytical data correspond closely with perceptual data, since the steep decrease in perceived strawberry flavor following the removal of sucrose (runs 2, 4, 6, and 7) occurred after 60 sec.

A further consideration when interpreting the sensory data is the potential for "halo dumping" effects [25]. It is conceivable that panelists might have reported reduced perception of strawberry flavor because of an awareness that the sweetness intensity of the solution was dropping. This theory suggests that because the response of panelists is limited to scoring one sensory dimension (strawberry), when they are aware of an obvious change in another dimension (sweetness), they "dump" the perceived change into the parameter they are scoring. Thus the observed influence of sweetness on strawberry perception could be a result of the sensory methodology. Several factors suggest that this is not the cause. First, panelists in our experiment were instructed to score strawberry flavor and *not* to respond to changes solely in sweetness or acidity. Second, the effects observed are extremely clear-cut, as strawberry perception practically disappeared on any run when sucrose was removed. If there was not a perceptual interaction between sweetness and strawberry flavor and strawberry perception was governed solely by the aroma, then there would clearly be a greater residual intensity in the absence of sucrose. Third, there is now sufficient literature concerning the multisensory nature of perception and specifically taste-aroma interactions to support our conclusions. The elegant experiment of Dalton and associates [7] has, as discussed earlier, demonstrated the perceptual linkage between taste and aroma. Davidson and colleagues [12] showed how the loss of sucrose from a chewing gum matrix caused a significant decrease in the perceived mintiness of the gum. Breath by breath analysis, using the MS-Nose, showed that the mint aroma compounds did not decline during the chewing period, and that the perceptual interaction between sweetness and the mint aroma appeared to be the sole cause of this decline in mint intensity. Bonnans and Noble [26] reported that when a fruity odorant was tasted in solution, its intensity was enhanced by both sweetness and sourness. Valdés and coworkers [27,28] also found that when sucrose was added to a raspberry odorant or apricot nectar in solution, panelists ascribed more fruitiness to the sweeter samples when they were tasted. In such experiments it can be difficult to judge whether observed increases in aroma perception are simply due to physicochemical effects on aroma partition caused by addition of the nonvolatile. However, von Sydow and associates [29] showed that increasing the sugar concentration of blueberry and cranberry fruit juices increased their fruitiness (evaluated by sipping), even though no difference in aroma was perceived by sniffing alone.

The phenomenon of halo dumping may in itself represent how difficult it is for humans to deconstruct multisensory percepts. A McFlads experiment to test our hypothesis would involve taking a taste and aroma that are not perceptually linked and seeing whether, in this case, time-intensity perception of the aroma were influenced by switching tastant flow off and on. Preliminary results (not shown) using sucrose and lime flavor (a noncongruent pairing) show clearly that lime perception does *not* follow sucrose concentration in-mouth during such experiments.

IV. CONCLUSIONS

McFlads allows in-mouth delivery of complex flavor mixtures, with controlled delivery in the temporal dimension. A simple experiment looking at adaptation to flavors delivered in this fashion over a 10-min period showed surprisingly little adaptation. This was probably due to the effects of mouth movements and swallowing actions in minimizing adaptation. Adaptation to aromas delivered retronasally was significantly lower than that reported in experiments using orthonasal presentation of aromas [16,30].

Studies on pumped mixtures of sucrose, citric/malic acids, and strawberry aroma gave evidence of the perceptual interactions among these three components in determining overall perceived strawberry flavor. Sucrose was the key driver of strawberry flavor perception and caused the most drastic and rapid reductions in perceived intensity when it was removed from the mixture.

Aside from the applications highlighted in this chapter, a variety of research applications can be envisaged for McFlads. With minor developments in software and sensory methodology, the system could be developed into a rapid tool for sensory threshold determination. Controlled delivery of mixed stimuli could prove useful in magnetic resonance imaging (MRI) studies of brain function, provided that pumping equipment could be operated satisfactorily within the associated magnetic field. The system might even be adapted to investigate the effects of mastication on flavor perception, if, for example, a nonflavored chewing gum base were chewed at the same time as delivery of the solution ("chewing solutions").

Finally, McFlads may prove a useful tool for the commercial flavorist. Real-time mixing of flavor streams could be used to approach the preferred blend of taste, aroma, and viscous components in flavor development. Equally, the system could be programmed for batch preparation of flavor mixtures for appraisal.

ACKNOWLEDGMENT

The authors wish to thank Masterfoods UK (a division of Mars Inc.) for their support of this work.

REFERENCES

1. JC Stevens, Physiol. Behav. 1997, 62, 1137–1143.
2. A Drewnowski, Nutr. Rev. 2001, 59, 163–169.
3. GA Calvert, MJ Brammer, SD Iversen, Trends Cognit. Sci. 1998, 2, 247–253.
4. H McGurk, J MacDonald, Nature 1976, 264, 746–748.
5. CN DuBose, AV Cardello, OJ Maller, Food. Sci. 1980, 45, 1393–1399.
6. AC Noble, Trends Food Sci. Technol. 1996, 7, 439–444.
7. P Dalton, N Doolittle, H Nagata, PAS Breslin, Nat. Neurosci. 2000, 3, 431–432.
8. HNJ Schifferstein, PWJ Verlegh, Acta Psychol. 1996, 94, 87–105.
9. W Grab, H Gfeller, In Flavour Release; DD Roberts, AJ Taylor, eds.; ACS Symposium Series: Washington DC, 2000; Vol. 763, pp 33–43.
10. AJ Taylor, RST Linforth, BA Harvey, A Blake, Food Chem. 2000, 71, 327–338.
11. JM Davidson, RST Linforth, AJ Taylor, In Flavor Release; DD Roberts, AJ Taylor, eds.; ACS Symposium Series: Washington DC, 1999; Vol. 763, pp 99–111.
12. JM Davidson, TA Hollowood, RST Linforth, AJJ Taylor, Agric. Food Chem. 1999, 47, 4336–4340.
13. Z Bujas, D Ajdukovic, S Szabo, D Mayer, M Vodanovic, Physiol. Behav. 1995, 57, 875–880.
14. HT Lawless, EZ Skinner, Percept. Psychophys. 1979, 25, 180–184.
15. MJM Theunissen, JHA Kroeze, HNJ Schifferstein, Percept. Psychophys. 2000, 62, 607–614.
16. WS Cain, Percept. Psychophys. 1970, 7, 271–275.
17. BP Halpern, ST Kelling, HL Meiselman, Physiol. Behav. 1986, 36, 925–928.
18. HL Meiselman, C Buffington, Chem. Senses 1980, 5, 273–277.
19. P Dalton, Chem. Senses 1996, 21, 447–458.
20. MJM Theunissen, JHA Kroeze, Chem. Senses 1996, 21, 545–551.
21. EP Koster, RA Wijk, In The Human Sense of Smell; DG Laing, RL Doty, W Breipohl, eds.; New York: Springer-Verlag, 1991, chapter 10.
22. P Dalton, Chem. Senses 2000, 25, 487–492.
23. S Szabo, Z Bujas, D Ajdukovic, D Mayer, M Vodanovic, Percept. Psychophys. 1997, 59, 180–186.
24. JF Gent, DH McBurney, Percept. Psychophys. 1978, 23, 171–175.
25. CC Clark, HT Lawless, Chem. Senses 1994, 19, 583–594.
26. S Bonnans, AC Noble, Chem. Senses 1993, 18, 273–283.
27. RM Valdés, EH Hinreiner, MJ Simone, Food Technol. 1956, 10, 282–285.

28. RM Valdés, MJ Simone, EH Hinreiner, Food Technol. 1956, 10, 387–390.
29. E von Sydow, H Moskowitz, H Jacobs, H Meiselman, Lebensm.-Wiss. Technol. 1974, 7, 18–24.
30. H Stone, GT Pryer, G Steinmetz, Percept. Psychophys. 1972, 12, 501–504.

10

Nosespace Analysis with Proton-Transfer-Reaction Mass Spectrometry: Intra- and Interpersonal Variability

D. D. Roberts, P. Pollien, and C. Yeretzian
Nestlé Research Center, Lausanne, Switzerland

C. Lindinger
University Innsbruck, Innsbruck, Austria

I. INTRODUCTION

The aroma of foods during eating is related to volatile organic compounds (VOCs) that reach the olfactory epithelium in the upper part of the nose. When food is eaten or a beverage consumed, VOCs are released in the mouth (oral cavity). During mouth movements (mastication), and in particular during swallowing, volatile compounds are transported from the oral cavity to the pharynx. At the subsequent exhalation, volatiles are swept by the airflow from the lungs and are exhaled through the nose. During their transport from the pharynx through the nasal cavity, aroma compounds pass along the olfactory epithelium and can trigger a sensory perception. The critical point in aroma perception during eating is the retronasal transfer of volatile aroma compounds from the oral to the nasal cavity, via the pharynx.

The significance of retronasal aroma stems from the fact that it reflects the aroma perceived by a consumer during eating and drinking. One method to explore retronasal aroma analytically is nosespace analysis. Over the last few years, two analytical approaches, which are capable of monitoring the aroma exhaled through the nose on-line and with high speed and sensitivity, have been developed. These are atmospheric-pressure chemical ionization mass spectrometry (MS) (APCI-MS) [1–3] and proton-transfer-reaction mass spectrometry (PTR-MS) [4–6].

Here we have applied PTR-MS to the nosespace analysis of a series of 12 liquid samples. These included one water sample and three milk samples with different fat content (0.033%, 2.7%, and 3.8% fat), containing five aroma compounds at three different concentrations. The nosespace from all samples was analyzed by five different panelists (two women and three men) in five repetitions. After a short discussion of recent technical developments of our nosespace setup, the discussion focuses on the inter- and the intrapersonal variability of the nosespace intensity for five-compounds—β-damascenone, hexanal, ethylbutyrate, benzaldehyde, and 2,3-butanedione. Further interpretation of the results is found in Ref. 7.

II. MATERIALS AND METHODS

A. Samples

Four matrices were studied: water, whole milk (3.8% fat), semiskim milk (2.7% fat), and skim milk (0.033% fat). Bottled water (Vittel) and commercial shelf-stable ultrahigh-temperature milk were purchased locally. Five aroma compounds were used at the following concentrations: β-damascenone (3.125, 12.5, and 50 mg/L), hexanal (1, 4, and 16 mg/L), ethylbutyrate (1, 4, and 16 mg/L), benzaldehyde (1, 4, and 16 mg/L), and 2,3-butanedione (10, 40, and 160 mg/L) (Firmenich SA, Genève, Switzerland, for β-damascenone and Fluka Chemie GmbH, Buchs, Switzerland, for the others). The three different concentrations that were chosen differed by a factor of 4: a low-, a middle-, and a high-concentration series. The range of concentrations was chosen so that they would give a detectable signal by nosespace analysis and an acceptable range sensorially and be soluble at the working concentration. The five compounds were analyzed together for the low-, middle-, and high-concentration series. Compounds were dissolved in 500 mL of milk or water with manual shaking using a high-precision Microman Pipette (Gilson Inc., Middleton, WI, USA) directly at the concentration indicated. Samples were analyzed after 1 hr on the day that they were prepared at room temperature.

B. Nosespace Analysis

The design of the experiment has been led by the desire to make the sampling of the breath air and the simultaneous eating as comfortable as possible. Three elements seemed important to us. *First*, the panelist shall be able to breathe freely without feeling uncomfortable or perturbed by the nosepiece. *Second*, the assessor should have some freedom of movement. *Third*, the panelist shall be able to eat and drink under conditions that are close to a natural eating situation, simultaneously with sampling the breath air.

The solution we propose is illustrated in Fig. 1. The air exhaled through the nose is sampled via two glass tubings fitted comfortably into the nostrils. Care is taken that the tube endings touching the nostrils are smooth and properly dimensioned. Each panelist has an individual nosepiece. The separation and diameters of the tubings are tailor-made and allow the panelist to breathe freely, while eating and drinking. The whole nosepiece is fixed on laboratory eyeglasses, which are connected to the PTR-MS via flexible and heated tubing. The panelist can move the upper body and the head and can sit comfortably during experiments. Finally, the two tubings, sampling the breath air, are curved upward to leave space in front of the mouth open. Overall, the design depicted in Fig. 1 allows the panelist to perform the experiments under conditions that mimic as much as possible a natural consumption situation. A more complete discussion of the novel nosespace setup is reported in a forthcoming publication [7].

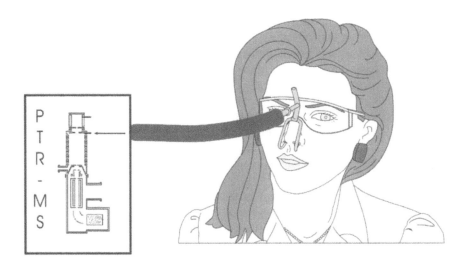

Figure 1 Schematic of the nosepiece for sampling breath-by-breath the air exhaled through the nose during eating and drinking. The nosepiece is connected to a PTR-MS, for on-line monitoring of the concentration of volatile aroma compounds in the breath air.

The air exhaled through the nose is collected and combined into one larger tube of 7-mm inner diameter, which is open to the laboratory. A person with regular breathing exhales approximately 4–5 L of air per minute. In our experiments, the majority of the breath air is released into the laboratory air. Only 80 mL/min of the breath air is drawn up for analysis into a heated stainless steel tubing of 0.53-mm inner diameter. The tube is inactivated with an inner quartz coating (Silcosteel® tube from RESTEK, Bad Homburg, Germany). The 80 mL/min is split into two fractions: 14 mL/min is introduced for analysis into the drift tube of the PTR-MS, and the remainder is released through a flow controller and membrane pump into the laboratory air. All tubings are heated to 70°C to prevent condensation.

The PTR-MS technique has been extensively discussed in a series of review papers [4–6,8–13]. Briefly, it combines a soft, sensitive, and efficient mode of chemical ionization, adapted to the analysis of trace VOCs, with a quadrupole mass filter. The gas to be analyzed is continuously introduced into the chemical ionization cell (drift tube) and ionized by proton transfer from H_3O^+. The protonated VOCs are extracted by a small electrical field from the drift tube and mass analyzed by a quadrupole mass spectrometer. The specific aspect of the chemical ionization scheme in PTR-MS is that the generation of the primary H_3O^+ ions, and the chemical ionization process, $VOC + H_3O^+ \rightarrow VOCH^+ + H_2O$, are spatially and temporally separated and can therefore be individually optimized.

PTR-MS is capable of monitoring the full breath-by-breath dynamic of a series of aroma compounds simultaneously and at high sensitivity. Technically, a time resolution of about 0.2 sec can be achieved in our setup [14]. In actual experiments, the time resolution is limited by the number of compounds that are simultaneously recorded and by their respective concentrations in the breath air (which determine the required dwell time in order to have a signal-to-noise ratio larger than 3). In most practical situations, a time resolution of 1–2 sec is sufficient. Here we have opted for a time resolution of 0.5 sec, corresponding to 100 msec dwell time per compound.

Five panelists participated in the study. They were introduced to the nosespace technique through previous studies or through training sessions beforehand. The panelists' breathing was regulated by following a light on/ light off timer, with a new breath every 11 sec (5 sec inspiration, 6 sec expiration). A standardized method for tasting was used in which 10-mL samples were swallowed about every 2 min. Just before exhalation the samples were consumed and were swallowed immediately. A period of 2 min was chosen between samples because after this time, aroma compound persistence in the breath returned to the baseline level. Although the sample order was randomized, five replicates of each sample were analyzed one after the other. The protonated mass ions were monitored simultaneously with the following m/z: 2,3-butanedione (87), benzaldehyde (107), β-damascenone (191), hexanal (83), and ethylbutyrate (117). Each of the five compounds was monitored at the mass of the protonated parent compounds. Blank milk analyses for each milk type without aroma compounds were also run to verify that the ions followed were not present at substantial quantities in the milk.

III. RESULTS AND DISCUSSION

In this chapter, we first describe in greater detail the nosespace setup developed at the Nestlé Research Center. We then discuss the variability among panelists as well as the in-mouth release variability for repetitions for the same person. Further interpretation of the effect of milk type and fat on nosespace release and the correlation with sensory intensity quantitation is reported in a separate publication (7).

Fig. 2 shows some example data from the five replicates of one panelist drinking milk spiked with five aroma compounds. Each large peak corresponds to the first exhalation of a sample being consumed, just after it was swallowed. This peak height is the value used for the nosespace release level. Besides the five aroma compounds, we show the trace of the water cluster, which measures the humidity of the breath air and hence reflects the breathing rhythm regularity of the panelist.

Between the samples, the amount released during breathing drops almost to the baseline level. One can also see in this graph the variation that occurs even when the same person consumes the same sample. Whereas most analytical gas chromatography mass spectrometry (GCMS) headspace sampling devices have a coefficient of variation (CV) of between 5% and 20%, this study showed an average CV of 33% for nosespace analysis. As humans who are performing according to a strict protocol still can show variations from one swallow to another and breathing differences, this larger CV seems reasonable. For the variation from the five replicates, Table 1 shows that the order of CV is 2,3-butanedione/β-damascenone < benzaldehyde < ethylbutyrate < hexanal. In comparing these values with their volatility in water (Table 2), we see a trend that the more volatile compounds in water are also more variable in their release. Table 1 also shows the error that is derived directly from the PTRMS measurements. The value varies between 6% and 22%, depending on the compound. Thus, the difference between the average 33% value for nosespace and these values from the measuring device would be the variation due to the human.

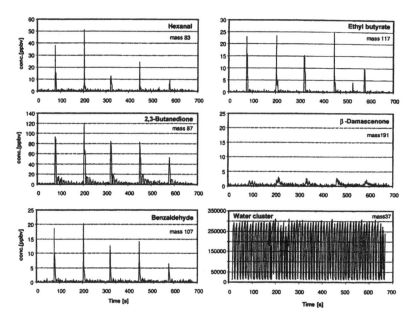

Figure 2 Example of nosespace analysis output showing five replicates of liquid sample consumption, with 2 min breathing between each sample. This example was with Female A, skim milk, at the moderate flavor concentrations showing release of aroma compounds and the water cluster $(H_2O\text{-}H_3O)^+$ formed from the addition of the H_3O^+ to water that is naturally present in the breath.

Monitoring one ion at a time can increase the sensitivity of the technique. In this experiment, the sensitivity was not optimized, as all five ions were monitored simultaneously. In any case, the compounds in water were released at levels well above the background noise levels. 2,3-Butanedione at 10 ppm showed an average signal-to-noise ratio of 18:1. Benzaldehyde at 1 ppm showed an average signal-to-noise ratio of 5:1. Ethylbutyrate at 1 ppm showed an average signal-to-noise ratio of 19:1. Hexanal at 1 ppm showed an average signal-to-noise ratio of 7:1. β-damascenone at 3.1 ppm showed an average signal-to-noise ratio of 6:1. These were the lowest concentrations that were tested. Obviously, at higher concentrations in water, the signal-to-noise ratio was even higher. However, in high-fat milk, the signal-to-noise ratio dropped, falling below the limit only for β-damascenone at 3.1 ppm. With atmospheric pressure chemical ionization, a signal-to-noise ratio lower limit was set at 3:1 (2), which is also the lower limit we used.

Table 1 Variation Among Panelists and Among Compounds for Their Release of Flavor Compounds in Nosespace Analysis[a]

Compound	Male A Conc	CV[e]	Male B Conc	CV	Male C Conc	CV	Female A Conc	CV	Female B Conc	CV	Error of PTR-MS[g]	CV of panelist means[c]	CV of panelist means—1[f]	Sum of Ranks CV[d]
2,3-Butanedione	158	27	180	29	187	25	139	24	125	20	6	17	13	117
Benzaldehyde	33	23	20	32	20	22	21	41	19	28	15	26	5	133
Ethylbutyrate	41	40	60	56	102	41	38	39	51	29	8	44	21	179
Hexanal	56	44	38	58	75	43	35	48	39	30	9	34	23	204
β-Damascenone	2.5	21	3.4	23	4.2	23	2.5	25	2.9	34	22	24	17	117
Sum of ranks concentration[b]	15		14		7		19		19					

[a] Concentration is in ppb-volume: average of all samples and at all concentrations with replicates (60 total in each compound). PTR-MS, proton-transfer-reaction mass-spectrometry; CV, coefficient of variation.

[b] The panelists were ranked for their release (1 = highest, 5 = lowest) of each compound and the ranks for all compounds were added together (to indicate which panelists released more than others).

[c] The five panelists' mean values were used and a CV calculated from them (to describe degree of differences between panelists).

[d] The compounds were ranked for their CV (1 = lowest, 5 = highest) for each of the 60 samples, and the ranks were added together (to indicate the variability of each compound).

[e] The CV values of the five replicates.

[f] The four panelists' mean values were used and a CV calculated from them (to indicate degree of difference between panelists). The highest panelist was not included.

[g] The error is based on a calculation (counts $^{-0.5}$). Thus, compounds with greater nosespace levels have a lower error. The value given here is the average for all concentrations, except for β-damascenone, for which it is the average of the two highest concentrations.

Table 2   Physical Chemical Data for Compounds Studied

Compound	Air-water partition coefficient at 30°C[a]	Compound lipophilicity, k_w value[b]
2,3-Butanedione	0.0011	−0.3
Benzaldehyde	0.0019	1.02
Ethylbutyrate	0.052 (at 37°C)	1.44
Hexanal	0.011	2.13
β-Damascenone	0.0037	2.79

[a] Measured by using a static headspace cell (according to Ref. 18).
[b] Measured using the HPLC method (see Ref. 19). Higher values indicate higher lipophilicity.

For the nosespace analysis, the basic hypothesis is that the release increases in a linear way as a function of compound concentration. For a simple analysis of headspace, this hypothesis was confirmed on xylene, 2-butanone, 2,3-butanedione, and propanol over an extended concentration range (10^{-4} ppm to 10 ppm in water) [15]. These four compounds were chosen because of the great range of volatility they cover. Nosespace analysis represents a more complicated system as the volatile compounds travel through the mouth, nose, and lung channels and interact with the human mucous membranes in the process. Thus, the linearity of values was not a given and was important to check. Fig. 3 shows that, indeed, a close to linear relationship was observed for benzaldehyde and 2,3-butanedione for all matrices. The more lipophilic compounds, β-damascenone, hexanal, and ethylbutyrate, appear to have linear relationships for water and skim milk but show some deviations from linearity with the high-fat milks. Carvone in gels was also shown to exhibit a linear relationship between concentration and in-nose release [16].

Table 1 shows the average data from the 12 samples repeated five times each. Although all panelists drank the same sample volume, there are rather large ranges in the amount released, depending on the person. From one person to another, the release can often double in value. On average, the two females released less than the three males. Male C released the most for all but one compound. Males A and B showed intermediate release levels. There was some variation depending on the compound but this was generally the order of personal release amounts. The concentration of flavor compounds that reach the olfactory receptors can vary between people as a result of physiological and eating differences. It is, thus, not surprising that people have different flavor perceptions. Greater differentiation among

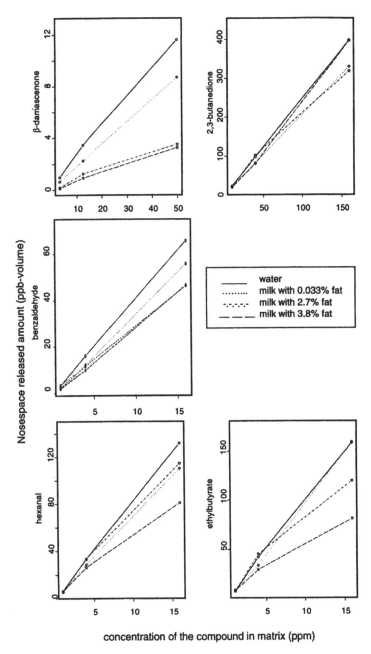

Figure 3 Correlation graphs between compound concentration in the sample and amount released in nosespace.

people is additionally caused by differences in types and numbers of olfactory receptors.

The compounds also do not behave similarly in their variation between people and their variation from the five replicates. In looking at the variation between people, the panelists' mean values were used and a CV was calculated from this. This was done for all panelists and also for all panelists minus the highest-release panelists. Ethylbutyrate and hexanal have a larger variation and 2,3-butanedione and benzaldehyde have a lower variation. In another study, ethylbutyrate was also found to have high persistence variation among panelists [17], although the authors acknowledge difficulty in monitoring its rapid change in volatile concentration. The compounds exhibiting a larger variation among people are those with a higher air-water partition coefficient, and the compounds exhibiting a lower variation among people are those with lower air-water partition coefficients. Indeed, the compounds with higher volatility in water seem to be more affected by the swallowing and physiological differences between people.

IV. CONCLUSIONS

Compared to traditional headspace analysis, the nosespace setup combined with on-line PTRMS has the advantage that it provides a closer and dynamic view of the aroma as it is perceived during eating or drinking. This is not only an analytical method but one that combines humans with the analytical method. The introduction of humans increased the coefficient of variation (CV) of the method from about 12% to 33%. Performing five replicates of each sample, thus, still allows significant differences in samples to be determined. Interestingly, compounds of higher water volatility showed a greater CV and also greater differences between people. This nosespace method has the potential to explore what is actually released in the mouth during consumption and what reaches the nasal cavity to be smelled. Finally, it reveals natural interpersonal differences in aroma stimulation and may help in better understanding individual differences in aroma preferences.

ACKNOWLEDGMENT

The authors acknowledge Nicolas Antille for his statistical advice and analysis of the results.

REFERENCES

1. R Linforth, KE Ingham, AJ Taylor. Time Course Profiling of Volatile Release from Foods During the Eating Process. AJ Taylor, DS Mottram, eds. Cambridge: The Royal Society of Chemistry, 1996; pp 361–368.
2. AJ Taylor. Volatile flavor release from foods During eating. CRC Crit Rev Food Sci Nutr 36:765–784, 1996.
3. R Linforth, F Martin, M Carey, J Davidson, AJ Taylor. Retronasal transport of aroma compounds. J Agric Food Chem 50:1111–1117, 2002.
4. C Yeretzian, A Jordan, H Brevard, W Lindinger. Time-resolved headspace analysis by proton-transfer-reaction mass-spectrometry. In: DD Roberts, AJ Taylor, eds. Flavour Release. ACS Symposium series 763: Washington, DC; 2000; pp 58–72.
5. M Graus, C Yeretzian, A Jordan, W Lindinger. In-Mouth Coffee Aroma: Breath-by-Breath Analysis of Nose-Space While Drinking Coffee. In: ASIC-19eme Colloque Scientifique International sur le Café. http://www.asic-cafe.org, 2002.
6. LB Fay, C Yeretzian, I Blank. Novel mass spectrometry methods in flavour analysis. Chimia 55:429–434, 2001.
7. DD Roberts, P Pollien, N Antille, C Lindinger, C Yeretzion. Comparison of nosespace, headspace, and sensory intensity ratings for the evaluation of flavor absorption by fat. J Agric Food Chain. In press.
8. C Yeretzian, A Jordan, W Lindinger. Analysing the headspace of coffee by proton-transfer-reaction mass-spectrometry. Int J Mass Spectrom. 223/224:115–139, 2003.
9. C Yeretzian, A Jordan, H Brevard, W Lindinger. Time-resolved headspace analysis by proton-transfer-reaction mass-spectrometry. In: DD Roberts, AJ Taylor, eds. Flavour Release. ACS Symposium series 763: Washington, DC, 2000; pp 112–123.
10. A Hansel, A Jordan, R Holzinger, P Prazeller, W Vogel, W Lindinger. Proton transfer reaction mass spectrometry: On-line trace gas analysis at the ppb level. Int J Mass Spectrom Ion Phys 149/150:609–619, 1995.
11. W Lindinger, A Hansel. Analysis of trace gases at ppb levels by proton transfer reaction mass spectrometry (PTR-MS). Plasma Sources Sci Technol 6:111–117, 1997.
12. W Lindinger, A Hansel, A Jordan. On-line monitoring of volatile organic compounds at pptv levels by means of proton-transfer-reaction mass spectrometry (PTR-MS): Medical applications, food control and environmental research. Int J Mass Spectrom 173:191–241, 1998.
13. W. Lindinger, A Hansel, A Jordan. Proton-transfer-reaction mass spectrometry (PTR-MS): On-line monitoring of volatile organic compounds at pptv levels. Chem Soc Rev 27:347–354, 1998.
14. T Karl, A Guenther, A Jordan, R Fall, W Lindinge. Eddy covariance measurement of biogenic oxygenated VOC emissions from hay harvesting. Atmospher Environ 35:491–494, 2000.

15. P Pollien, A Jordan, W Lindinger, C Yeretzian. Liquid-air partitioning of volatile compounds in coffee: dynamic measurements using proton-transfer-reaction mass spectrometry. Int J Mass Spectrom. In press.

16. TA Hollowood, RST Linford, AJ Taylor. The relationship between carvone release and the perception of mintyness in gelatine gels. In: DD Roberts, AJ Taylor, eds. Flavour Release. ACS Symposium series 763: Washington, DC, 2000; pp 370–380.

17. R Linforth, AJ Taylor. Persistence of volatile compounds in the breath after their consumption in aqueous solutions. J Agric Food Chem 48:5419–5423, 2000.

18. A Chaintreau, A Grade, R Munoz-Box. Determination of partition coefficients and quantitation of headspace volatile compounds. Anal Chem 67:3300–3304, 1995.

19. N El Tayar, H Van de Waterbeemd, B Testa. Lipophilicity measurements of protonated basic compounds by reversed-phase high-performance liquid chromatography. II. Procedure for the determination of a lipophilic index measured by reversed-phase high-performance liquid chromatography. Chromatogr 320:305–312, 1985.

11

Correlation Between Sensory Time-Intensity and Solid-Phase Microextraction Analysis of Fruity Flavor in Model Food Emulsions

M. Fabre and E. Guichard
INRA-UMRA, Dijon, France

V. Aubry
Nestlé PTC, Beauvais, France

A. Hugi
Nestlé Research Center, Lausanne, Switzerland

I. INTRODUCTION

Interactions between flavor compounds and a variety of nonflavor matrix components, for example, protein and fat, influence flavor perception in foods.

Interactions of flavor compounds with proteins are known to have a strong influence on flavor release from model foods [1]. Proteins often cause a decrease in the volatility of flavor compounds. It is well known that proteins interact with volatiles both reversibly [2–3] and irreversibly [4–5].

Flavor release depends also on oil content, which affects the partition of aroma compounds during the different emulsion phases (lipid, aqueous, and vapor) [6]. In fact, lipids absorb and solubilize lipophilic flavor compounds and reduce their vapor pressures [7–8], as explained by mathematical models [9], headspace analysis [10], and sensory analysis [11–12].

Food flavor, taste, and texture sensations are perceived over time during consumption, and intensity of flavor perception can change over time. The temperature of the product, after time, equilibrates with the mouth temperature. Physical manipulations such as tongue movements, mastication, and salivary dilution affect the product and its sensory characteristics over time. The classic methods of descriptive sensory analysis do not take into account this temporal dimension. For that purpose, dynamic methods such as time-intensity analysis were developed [13].

In the present chapter, we have thus chosen to study the influence of the nature of fat and protein on the aroma release by solid-phase microextraction measurements and on the flavor perception by time-intensity sensory analysis.

II. PROCEDURE

A. Preparation of the Model Food Emulsions

Emulsions were composed of 3% proteins (milk proteins in powdered form), 9% fat, and 0.5% emulsifier. The model food emulsions were flavored with an aroma mixture at 0.04%. The composition of food model emulsions is indicated in Table 1. Because of the confidential nature of this project, a more detailed description of the fats used cannot be made available.

The flavor mixture was composed of flavor compounds from different chemical classes dissolved in propylene glycol; these compounds were obtained from flavor suppliers (Food Ingredients Specialities, [FIS], York, England). They appear by concentration order in the mixture in Table 2. Final concentrations of flavor compounds varied between 0.16 and 10 ppm.

Table 1   Composition of the Six Food Model Emulsions

	Ingredients		
		Commercial fats	
Emulsions	Commercial protein powders	Name	Solid fat content, percentage; 25°C
1	B: Milk proteins 1	X: Vegetable oil 1	49.97
2	B: Milk proteins 1	Y: Vegetable oil 2	30.41
3	B: Milk proteins 1	Z: Animal fat	7.53
1	L: Milk proteins 2	X: Vegetable oil 1	49.97
2	L: Milk proteins 2	Y: Vegetable oil 2	30.41
3	L: Milk proteins 2	Z: Animal fat	7.53

Table 2 Aroma Compounds by Concentration Order in the Mixture: Odor Threshold and Odor Description

Aroma compound	Odor threshold in water, mg/kg (literature cited)	Odor description
Ethyl butanoate	0.001 [14]	Fruity
Ethyl hexanoate	0.0003 [14]	Fruity
Mesifuran	0.01 [14]	Burnt sugar, caramelized
Methyl hexanoate	0.087 [14]	Fruity
Hexenol	0.5 [15]	Green
Diacetyl	0.003 [16]	Buttery
Linalol	0.006 [15]	Flower
γ-Octalactone	0.0017 [17]	Sweet, flower

They represented some of the major chemical functions in food flavor compounds (ketones, esters, alcohols).

B. Characterization of the Model Food Emulsions: Granulometry

A Malvern Mastersizer laser diffrectometer (Malvern Instruments, Orsay, France) was used to determine the particle size distribution in the model food emulsions.

C. Solid-Phase Microextraction Coupled with Gas Chromatography Mass Spectrometry (Multiple Ion Monitoring Mode) Analysis

Samples were placed in 20-mL vials and allowed to equilibrate at 25°C. A solid-phase microextraction (SPME) fiber, polydimethyl siloxane/divinyl-benzene, (PDMS/DVB); 65 μm (Supelco Park, Bellefonte, PA), was used for sampling volatile compounds, using the methodology already described [18].

Volatile compounds were desorbed by inserting the fiber into the gas chromatography (GC) injector, set at 250°C for 10 min; 1 min was allowed for desorption (purge off) and 9 min for cleaning (purge on).

All the SPME operations were automated by using a MPS2 Multi-Purpose Sampler (Gerstel, Applications, Brielle, The Netherlands).

Experiments were done in triplicate at 25°C.

1. Gas Chromatography Mass Spectrometry Analysis

A HP 6890-GC equipped with a split/splitless injector coupled with a mass selective detector 5970 (Hewlett Packard, Palo Alto, CA) was used. A fused-silica capillary column DB-Wax, 50 m, 0.32 mm inner diameter (ID), 1 μm film thickness (J&W Scientific), was employed. The carrier gas was helium (35 cm/sec).

The GC oven heating was started at 50°C, then increased to 220°C at a rate of 5°C/min. The total time of analysis was 39 min. The injector was always maintained at 250°C.

The mass spectrometer was operated in the mass range from 29 to 300 at a scan rate of 1.89 sec/scan. The quantification was realized by multiple ion monitoring (MIM) mode. The selected and specific ions were 43 for diacetyl, 88 for ethyl butanoate, 87 for methyl hexanoate, 101 for ethyl hexanoate, 82 for hexenol, 93 for linalol, 142 for mesifuran, and 85 for gamma-octalactone.

D. Sensory Analysis: Time-Intensity

Time-intensity sensory measurements were conducted on the six types of model food emulsions. Ten external panelists (nine women and one man) performed time-intensity study. Assessors worked 3 days a week in 1- or 1-1/2-hour sessions. Tasters were selected, before this study, for their sensory abilities, their motivation, and their group behavior. The panel was trained as follows.

The first step was a classical profile (individual evaluation, followed by round table discussions to reach consensus). The aim of these sessions was to generate attributes to describe the model food emulsions and to evaluate whether the panelists already perceived differences between samples.

In a second step, we trained the assessors in the method of time-intensity with sounds (varying in intensity, in volume, and in frequency). Using a personal computer (PC) mouse, they reported their sensation in real time along an intensity scale visible on the PC screen.

The third step was a training on individual parameters of time-intensity. Assessors were to score the intensity of the fruity perception on a structured scale at specified times on paper.

The fourth step was practice of full time-intensity in the booths. Assessors were to score the intensity of the fruity perception on an unstructured scale continuously.

At the end of the training program, panel performance was validated.

Time-Intensity measurements were quantified by using Fizz Delphi software developed by Biosystemes (Biosystems, Couternon, Dijon,

France). To record responses, panelists, using a mouse, moved a cursor along a horizontal unstructured scale. For the evaluation of the sensory attribute, the panelists were asked to put a small stainless-steel spoon of model food emulsions into their mouth, followed by an almost instantaneous click on the mouse to start the recording. The panelists were free to choose the beginning point of the scale. Then they moved the cursor according to the evolution of the sensation. The recording duration was 120 sec. Presentation orders of model food emulsions were built with a Latin square design. Individual cups were coded with a three-digit random number. To minimize errors, all evaluations were made in individual booths under white light. Time-intensity evaluation was performed twice for each product.

E. Statistical Analysis

1. Solid-Phase Microextraction

Experiments were done in triplicate. For each flavor compound, a one-way analysis of variance (product effect to evaluate whether there is a difference between products) was performed with Statbox software (Grimmer logiciel, France). Mean comparison was made by a Newman-Keuls test at 5% risk level.

2. Analysis of Time-Intensity Data

The trapezoid method allowed modeling of the experimental time-intensity curves, in order to extract characteristic times and intensities [19]. This is shown in Fig. 1. In particular, this treatment of time-intensity curves allows the calculation of the total duration and area of the perception.

Ten panelists, six products, one attribute, and two replications amounted to 120 time-intensity curves. An analysis of variance and the least significant difference (LSD) were computed for each primary and secondary parameter.

3. Correlation Analysis of Time-Intensity Data and Solid-Phase Microextraction Data

Correlation coefficients (r = correlation coefficient of Pearson) between the data of solid-phase microextraction (SPME) analysis and the data of time-intensity sensory analysis, were calculated with Excel software.

Intensity

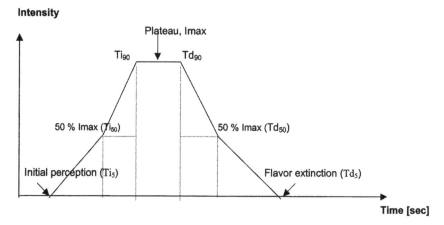

Figure 1 Average skeleton time-intensity curves. Imax, maximal intensity of the sensation; Ti_5, Ti_{50}, Ti_{90}, times corresponding to 5%, 50%, and 90% Imax (increasing phase); Td_{90}, Td_{50}, Td_5, times corresponding to 90%, 50%, and 5% Imax (decreasing phase); Di (duration of increasing part of the curve) = $(Ti_{50} - Ti_5) + (Ti_{90} - Ti_{50})$; Dm (duration of plateau) = $(Td_{90} - Ti_{90})$; Dd (duration of decreasing phase) = $(Td_5 - Td_{50}) + (Td_{50} - Td_{90})$; Dt (total duration) = Di + Dm + Dd; Ai, area of increasing phase; Am, area of plateau; Ad, area of decreasing phase; At (total area) = Ai + Am + Ad.

III. RESULTS AND DISCUSSION

A. Characterization of the Model Food Emulsions: Granulometry

Emulsions are all monomodal; they have closed values of D (v, 0.5) and surface area. Therefore, the structure of these food model emulsions does not seem to be a factor that can explain differences observed in aroma release (see Table 3).

B. Solid-Phase Microextraction Analysis

SPME, coupled with gas chromatography mass spectrometry (GCMS) [multiple ion monitoring (MIM) mode], was used to quantify aroma release in the vapor phase. The polydimethyl siloxane divinylbenzene (PDMS/DVB) coating fiber was chosen because its sensitivity was adapted to the eight aroma compounds (20). The experiments were made at 25°C, in

Table 3 Emulsion Characteristics

Emulsions	D (v, 0.5), µm[a]	Surface area, m^2/mL	Dispersion, µm
X-L monomodal	0.78	8.56	2.21
X-B monomodal	0.58	11.26	0.86
Y-L monomodal	0.67	9.39	1.44
Y-B monomodal	0.78	8.71	1.68
Z-L monomodal	0.78	7.39	2.05
Z-B monomodal	0.58	9.61	2.16

[a] D (v, 0.5): median diameter, (micrometers).

triplicate. The eight aroma compounds were detected and quantified (Fig. 2).

Some aroma compounds were more influenced than the others by the nature of fats. The nature of fat seemed to have more influence on the release of diacetyl, ethyl butanoate, methyl hexanoate, ethyl hexanoate, linalol, and γ-octalactone than of hexenol and mesifuran.

The release of hydrophobic compounds such as esters was more important in emulsions with X or Y fats and less important in Z emulsions. The release of hydrophilic compounds such as diacetyl was greater in Z emulsions and was lower in emulsions with X or Y fats.

To conclude, the vegetable oils, as opposed to the animal fat, seemed to lead to higher flavor release for hydrophobic compounds.

C. Time-Intensity Analysis: Fruity Attribute

By observing Fig. 3, we noticed differences in the maximal intensity and in the total duration of aroma perception (also see Table 4).

For maximal perceived intensity, three groups significantly differed:

X-B and Y-B, which had the highest value,
X-L and Y-L, which were intermediate
Z-B and Z-L, which had the lowest value

A quite similar pattern was observed for total duration of aroma perception.

These results were in agreement with the data of flavor release analysis (the aroma compounds were released in greater quantity from emulsions that contained X or Y oil).

Y-B and X-B emulsions presented higher maximal intensity for the fruity attribute and differed by the total duration of this attribute. Thus,

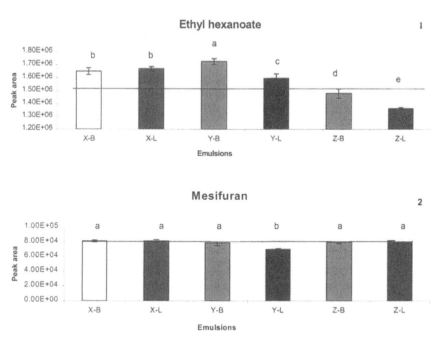

Figure 2 Aroma release (1, ethyl hexanoate; 2, mesifuran) of different emulsions. Means within a row with different letters (a–e) are significantly different ($P = 0.05$).

Y-B emulsion acted on the temporal aspect and on the intensity of flavor perception. Moreover, X-L and Y-L emulsions had a higher maximal intensity than Z-L emulsion.

Thus, emulsions made with X and Y oils had higher maximal intensity and longer duration of perception than emulsions made with Z fat.

In relation to fat, we observed a protein effect for Y oil. The fruity attribute was more intense and more persistent in B emulsion than in L emulsion. No significant difference was observed for Z fat. For emulsions made with X oil, no difference in duration was observed but only in the level of maximal intensity, which was again more important for B emulsion than for L emulsion.

Thus, by looking at time-intensity sensory analysis, we noticed that there was not a general effect (or single effect) of either protein type or fat type, but a significant combined effect of protein and fat on flavor perception.

The time-intensity experiment allowed a description of the differences in flavor perception; it gave more information than a classical profile,

Fruity

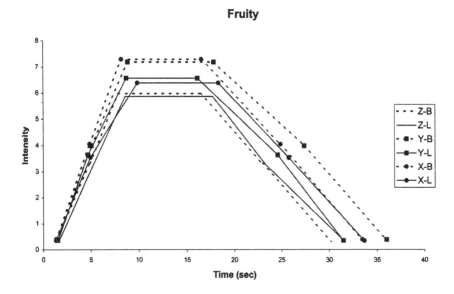

Figure 3 Average skeleton time-intensity curves of "fruity sensory attribute."

notably because it allowed a better understanding of the links between perceptions that occur during emulsion tasting.

Time-intensity allowed the verification of one hypothesis: the presence of vegetable fat instead of animal fat leads to higher intensity and longer duration of the fruity perception. Furthermore, an effect of the nature of protein was observed with vegetable fats.

D. Correlation Between Time-Intensity Analysis and Solid-Phase Microextraction Analysis

Significant positive correlations were found between aroma release of esters and the three parameters of time-intensity for the fruity attribute (see Table 5). Linalol and γ-octalactone were less correlated than esters but more than hexenol and mesifuran. These results were in agreement with SPME analysis: indeed, the nature of fat seemed to have more influence on release of esters than of linalol and γ-octalactone and than of hexenol and mesifuran. Diacetyl behaved differently than the other aroma compounds because it was the only hydrophilic compound. It was also the only one that was released in greater quantity from animal fat than from vegetable fats used in this study.

Table 4 Primary and Secondary Parameters of Time-Intensity Analysis

Parameters	Z-B	Z-L	Y-B	Y-L	X-B	X-L	LSD[h]
Primary parameters							
I_{max}[a]	6.7	6.5	8.0	7.3	8.1	7.1	1.0
TI_5[b]	1.3	1.6	1.4	1.4	1.3	1.3	0.7
TI_{50}[b]	4.9	5.2	5.0	4.7	4.8	4.9	1.2
TI_{90}[b]	7.9	8.6	8.8	8.7	8.1	9.8	1.6
Td_{90}[c]	16.4	17.7	17.8	16.1	16.5	18.3	2.5
Td_{50}[c]	23.0	23.4	27.4	24.6	24.9	25.7	4.0
Td_5[c]	30.2	31.6	35.9	31.5	33.5	33.7	6.2
Secondary parameters							
Di[d]	6.5	7.0	7.4	7.2	6.8	8.5	1.3
Dm[e]	8.6	9.1	9.0	7.4	8.4	8.5	0.9
Dd[f]	13.8	13.9	18.1	15.3	16.9	15.4	0.9
Dt[g]	28.9	30.0	34.5	30.0	32.1	32.4	0.9
Ri[d]	0.8	0.8	1.0	1.0	1.0	0.9	0.1
Rd[f]	0.4	0.5	0.3	0.3	0.4	0.4	1.7
Ai[d]	18.3	19.5	26.2	24.2	23.9	28.4	7.5
Am[e]	48.4	50.7	61.1	46.2	57.7	51.2	16.1
Ad[f]	39.2	36.2	65.1	51.5	59.8	47.0	19.0
At[g]	105.9	106.4	152.4	121.9	141.3	126.6	33.5

[a] I_{max}: maximal intensity of the sensation.
[b] TI_5, TI_{50}, and TI_{90}: times corresponding to 5%, 50%, and 90% Imax (increasing phase).
[c] Td_{90}, Td_{50}, and Td_5: times corresponding to 90%, 50%, and 5% Imax (decreasing phase).
[d] The duration Di, the rate Ri (between 5 and 50 Imax), the area Ai of the increasing part of the curve.
[e] The duration Dm and the area Am of the 90% Imax plateau.
[f] The duration Dd, the rate Rd (between 90% and 50% Imax), and the area Ad of the decreasing part of the curve.
[g] The total area At and the total duration Dt of the shape.
[h] LSD, least significant difference.

IV. CONCLUSIONS

Aroma release of hydrophobic compounds such as esters was higher in emulsions with X or Y (vegetable oils) and was less important in Z emulsions (animal fat). Hydrophilic compound release (diacetyl) was greater in Z emulsions and lower in emulsions with X or Y fats.

Table 5 Correlation Between Time-Intensity Analysis and Solid-Phase Microextraction Analysis

25°C	Diacetyl	Ethyl butanoate	Methyl hexanoate	Ethyl hexanoate	γ-Octalactone	Linalol	Mesifuran	Hexenol
	r[a]	r	r	r	r	r	r	r
Imax[b]	−0.09	0.86	0.84	0.84	0.54	0.55	−0.09	0.14
Dt[c]	0.04	0.83	0.76	0.81	0.66	0.72	0.22	−0.23
At[d]	0.02	0.89	0.84	0.87	0.69	0.70	0.08	−0.01

[a] r, correlation coefficient (Excel software).
[b] Imax, maximal intensity of perception.
[c] Dt, total duration.
[d] At, total area.

Sensory time-intensity allowed us to conclude that the presence of vegetable oils instead of animal fat led to a higher intensity and a longer duration. An effect of the nature of protein was observed in vegetable emulsions.

Thus, we concluded that the evolution in the mouth of the fruity note in emulsions was mainly due to aroma release of esters.

ACKNOWLEDGMENTS

This work was financially supported by the Nestlé Company and the Regional Council of Burgundy. We thank Fis (York, UK) for providing aroma compounds and flavor mixture. We also thank A. Rytz (Nestlé) for assistance in statistical analysis as well as C. Vaccher (Nestlé) and S. Stroh (Nestlé) for technical help.

REFERENCES

1. KL Franzen, JE Kinsella. Parameters affecting the binding of volatile flavor compounds in model food systems. I. Proteins. J Agric Food Chem 22:675–678, 1974.
2. T O'Neill, JE Kinsella. Flavor protein interactions: Characteristics of 2-nonanone binding to isolated soy protein fractions. J Food Sci 52:98–101, 1987.
3. K Sostmann, E Guichard. Immobilised beta-lactoglobulin on a HPLC-column: A rapid way to determine protein/flavor interactions. Food Chem 62:509–513, 1998.
4. AP Hansen, JJ Heinis. Decrease of vanillin flavor perception in the presence of casein and whey proteins. J Dairy Sci 74:2936–2940, 1991.
5. AP Hansen, JJ Heinis. Benzaldehyde, citral, and d-limonene flavor perception in the presence of casein and whey proteins. J Dairy Sci 75:1211–1215, 1992.
6. KB De Roos. How lipids influence food flavour. Food Technol 51:60–62, 1997.
7. RG Buttery, JL Bomben, DG Guadagni, LC Ling. Some considerations of the volatiles of organic flavor compounds in foods. J Agric Food Chem 19:1045–1048, 1971.
8. RG Buttery, DG Guadagni, LC Ling. Flavor compounds: Volatilities in vegetable oil and oil-water mixtures: Estimation of odor thesholds. J Agric Food Chem 21:198–201, 1973.
9. M Harrison, BP Hills, J Bakker, T Clothier. Mathematical models of flavor release from liquid emulsions. J Food Sci 62:653–658, 664, 1997.
10. JP Schirle-Keller, GA Reineccius, LC Hatchwell. Flavor interactions with fat replacers: Effect of oil level. J Food Sci 59:813–815, 1994.

11. SE Ebeler, RM Pangborn, WG Jennings. Influence of dispersion medium on aroma intensity and headspace concentration of menthone and isoamyle acetate. J Agric Food Chem 36:791–796, 1988.

12. C Guyot, C Bonnafont, I Lesschaeve, S Issanchou, A Voilley, HE Spinnler. Effect of fat content and odor intensity of three aroma compounds in model emulsions: d-decalactone, diacetyl, and butyric acid. J Agric Food Chem 44:2348, 1996.

13. JR Piggott. Dynamism in flavour science and sensory methodology. Food Res Int 33:191–197, 2000.

14. T Pyysalo, M Suiko, E Honkanen. Odour thresholds of the major volatiles identified in cloudberry (*Rubus chamaemorus* L.) and arctic bramble (*Rubus arcticus* L.). Lebensm.–Wiss. Technol. 10:36–39, 1977.

15. RG Buttery, JL Bomben, DG Guadagni, LC Ling. Some considerations of the volatiles of organic flavour compounds in foods. J Agric Food Chem 19:1045–1048, 1971.

16. T Schieberle, T Hofmann. Evaluation of the character impact odorants in fresh strawberry juice by quantitative measurements and sensory studies on model mixtures. J Agric Food Chem 45:227–232, 1997.

17. E Guichard, I Lesschaeve, N Fournier, S Issanchou. Séparation et caractérisation sensorielle des énantiomères de molécules chirales très odorantes. Revista italiana EPPOS, numero speciale: 11èmes Journées Internationales des Huiles Essentielles, Digne-les-bains Febbraio 79–89, 1993.

18. M Fabre, V Aubry, E Guichard. Comparison of different methods: Static and dynamic headspace and solid-phase microextraction for the measurement of interactions between milk proteins and flavor compounds with an application to emulsions. J Agric Food Chem 50:1497–1501, 2002.

19. M Lallemand, A Giboreau, A Rytz, B Colas. Extracting parameters from time-intensity curves using a trapezoid model: The example of some sensory attributes of ice cream. J Sens Stud 14:387–399, 1999.

20. DD Roberts, P Pollien, C Milo. Solid-phase microextraction method development for headspace analysis of voltile flavor compounds. J Agric Food Chem 48:2430–2437, 2000

12

The Influence of Random Coil Overlap on Perception of Solutions Thickened Using Guar Gum

Conor M. Delahunty, Siobhan O'Meara, Fiachra Barry, Edwin R. Morris, Persephoni Giannouli, and Katja Buhr
University College Cork, Cork, Ireland

I. INTRODUCTION

Thickened foods require added flavoring to produce the same flavor intensity as that of foods that are more fluid [1–4]. This interaction may take place in the food before its introduction into the mouth, where components are made available to the senses; it may be caused when components of the food reduce access of flavor active compounds to receptors but may also be contributed to by cross-sensory modality interactions, as the senses responsible for perception of taste, aroma (perceived retronasally during consumption), and texture do not respond to stimulus independently of one another [5–7].

Interactions that occur within the food, between food components, can influence sensory perception by determining texture properties and the availability of taste and aroma-active compounds to receptors during consumption [8]. It has also been shown that increasing hydrocolloid concentration reduces the perceived intensity of tastes and aromas [9] and reduces the partition coefficients of volatile compounds [10]. As the concentration of random coil polysaccharides, e.g., guar gum, in solution is increased, c^* marks the transition from a dilute solution of polymer coils free to move independently to an entangled network at higher polymer concentrations. Therefore, a mechanism proposed to explain the interaction that causes taste and aroma change is that increasingly entangled polymer networks that occur at concentrations above c* inhibit movement of taste and aroma-active molecules to the surface, where they become available for perception [4]. However, in the case of taste, increasing viscosity might also reduce access to the receptors [11], and therefore this interaction might occur between food and consumer rather than within the food itself.

On the other hand, cross-modal sensory interactions can be caused by at least two factors: (a) learning of an association between one sensory attribute (or modality) and another, often leading to placing greater reliance on one sensory modality over another, e.g., the dominance bias most easily demonstrated by the way color determines taste expectations [12], or (b) the influence of the function of one sense (threshold measures, concentration-response functions) on a stimulus of another sense, e.g., capsaicin desensitization, which reduces perceived taste intensity [13]. In addition, aroma-taste interactions have been widely documented [5,6,14], but it is as yet unclear what the mechanisms for these interactions are. The physiological characterizations of the senses of olfaction and taste are independent, so a direct interaction is unlikely; however, it is extremely difficult to localize congruent tastes and aromas when they are presented mixed together, and increasing the concentration of any one increases the perceived intensity of both. Most recently it was proposed that changes in texture have a significant influence on flavor intensity when release of aroma-active volatiles was unaffected [7]. Again, a mechanism for such an interaction has yet to be elucidated, although one could speculate that both aroma-taste interactions and texture-flavor interaction described occur at the cognitive level, where stimulus integration takes place.

The objectives of this study were to determine the influence of increasing concentration of the random coil polysaccharide guar gum on odor (perceived orthonasally by smelling), thickness, taste, and aroma (perceived retronasally when sample is placed in the mouth) perception and to study the effect of onset of random coil entanglement above c* on odor, thickness, taste, and aroma perception, with particular emphasis on aroma release and perception. Guar gum was chosen for this study because it is used extensively for thickening products by the food industry and because it is a neutral polymer.

II. MATERIALS AND METHODS

A. Samples

In a hydrocolloid system, the effective hydrodynamic volume of polymer coils may be characterized by intrinsic viscosity [η]. The number of coils present is proportional to concentration. Therefore, c[η] gives a measure of the extent of space occupancy by the polymer. The intrinsic viscosity of the guar gum used for this study was determined at 20°C on a Contraves low-shear 30 rotational viscometer with coaxial cylinder using solutions with relative viscosities in the range 1.2–2.0. It was calculated to be 16.2 dL/g by extrapolation of experimental values of specific viscosity by a combined Huggins, Kraemer, and single-point method [15]. The onset of coil entanglement (c*) for the guar gum was calculated to be close to 0.32% w/w by observing a point of deflection on a plot of log c[η] against log-specific viscosity. Twenty-one solutions were prepared to provide a $7 \times 3 \times 1$ experimental design that included seven concentrations of guar increasing in a 0.25 log series (0.056%, 0.1%, 0.178%, 0.32%, 0.56%, 1.0%, 1.78% w/w), three concentrations of sucrose (5%, 10%, 20% w/w), and one concentration of ethyl butyrate (5 ppm). Therefore, the design studied included three solutions below c*, one close to c*, and three solutions above c*. One drop of a flavorless red food color was added to all samples to mask appearance differences and to make appearance of the samples more acceptable to assessors.

B. Sensory Analyses

Twenty-one sensory assessors used the method of magnitude estimation [16] to measure the intensity of the sensory attributes "fruity odor" (assessed by smelling), "thickness" (assessed during consumption), "sweetness" (assessed during consumption), and "fruity flavor" (assessed during consumption). The middle solution in the design (0.32% guar, 10% sucrose, 5ppm ethyl butyrate) was used as a reference and was assigned an arbitrary score of 100. All 21 solutions were compared to the reference, one attribute at a time, using an incomplete block design that was repeated twice, resulting in 10 individual measures per attribute per solution.

C. Instrumental Analyses

Static headspace analysis of ethyl butyrate release was carried out on the seven guar solutions containing 10% sucrose. Two milliliters of sample was incubated at 37°C in a sealed 10-mL glass vial and agitated at 750 rpm for 10 min using an automated headspace unit (Combipal-CTC Analytics, JVA Analytical Ltd., Dublin, Ireland). One milliliter of equilibrated headspace was injected and analyzed by gas chromatography with flame ionization detection (GC-FID). Analyses were carried out in triplicate.

Proton-transfer reaction mass spectrometry (PTR-MS) (Ionicon Analytik, Innsbruck, Austria) of ethyl butyrate release during guar consumption was also carried out on the seven guar solutions containing 10% sucrose. One participant consumed all solutions. This person swirled 20 g of each guar solution around his mouth for 1 min, at a rate of one chewing oscillation every 2 sec, all the time the quantity of ethyl butyrate released was measured in real time, by displacing the compound through the nose. After 1 min, the sample was swallowed. The instrument requires a constant inlet flow of 100 mL/min, of which 15 mL/min is split into the reaction chamber [17]. A special nosepiece allowed breathing into the instrument, without disturbing the natural breathing pattern of the person, yet maintaining constant flow for the instrument. The masses 89 and 117, which represent the main fragments of ethyl butyrate [18], were monitored with a dwell time of 0.2 sec, respectively. In addition, the masses 21, 32, and 37, which monitor the general performance of the instrument, were monitored. This resulted in one full measuring cycle per 0.9 sec. Resulting data were corrected on the basis of transmission and fragmentation according to the procedure referenced [18]. Each solution was analyzed by this method on five separate occasions.

D. Data Analyses and Representation

Sensory data were normalized to take account of individual scale-use differences. For each sensory attribute log mean perceived intensity was plotted against log c[η]. Static headspace ethyl butyrate release data were also plotted against log c[η]. A noncentered principal components analysis was used to generate a first principal release curve based on the five replications of in-nose measured ethyl butyrate for each concentration of guar gum. Principal components analysis provides a method of averaging replicate analyses of time-intensity curves without distorting the results or losing information on between-replicate variability [19]. The area under each principal curve was calculated, and this value was plotted against log c[η].

III. RESULTS AND DISCUSSION

Perceived thickness increased with increasing coil overlap, increasing abruptly with the onset of coil entanglement at c*. Sucrose concentration had little influence on perceived thickness in the presence of the guar gum (Fig. 1). Perceived sweetness was independent of coil overlap at concentrations of guar below c* and increased somewhat as thickness increased. However, the intensity of perceived sweetness decreased rapidly with the onset of coil entanglement above c* (Fig. 2). The three different concentrations of sucrose were clearly distinguished by perceived sweetness

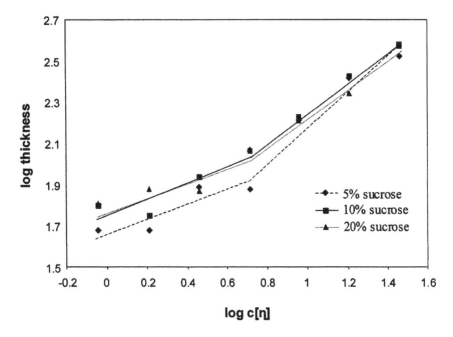

Figure 1 Perceived thickness of model food solutions with increasing coil overlap.

intensity, and the interaction of sweetness with increasing viscosity was similar regardless of sucrose concentration (Fig. 2). In general, perceived fruitiness mirrored perceived sweetness, even though only one concentration of ethyl butyrate was added to all solutions (Fig. 3). This demonstrated an aroma-taste interaction, or halo effect. This might not have occurred for noncongruent taste and aroma mixtures. Perceived fruity odor was found to be the same in all 21 solutions (Fig. 4). Static headspace concentrations of ethyl butyrate were also unaffected by coil overlap and entanglement (Fig. 5). These results showed that the ethyl butyrate was not held in the solution in any way that influenced partitioning. However, concentrations of ethyl butyrate released in the mouth during consumption and measured in-nose were influenced by coil overlap, and concentrations decreased abruptly with the onset of coil entanglement above c* (Fig. 6 and 7). The sensory evaluation results suggest that perception of both sucrose compounds and ethyl butyrate is inhibited at higher degrees of coil entanglement. This conclusion is drawn, even though sucrose release was not measured, for if sucrose release were unaffected by viscosity, we would not expect to see a fall in the perceived intensity of sweetness or of fruity flavor. In addition, if

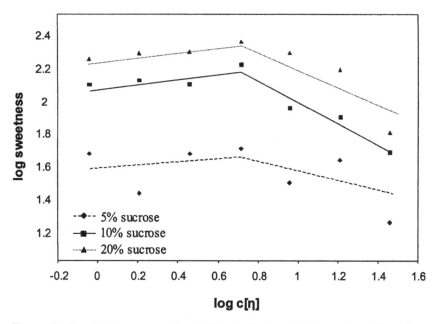

Figure 2   Perceived sweetness of model food solutions with increasing coil overlap.

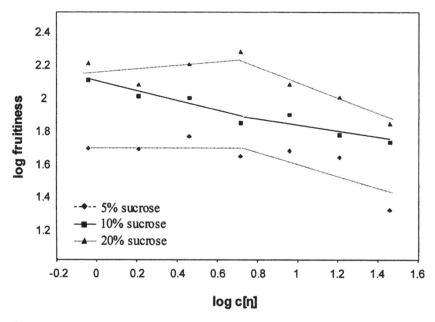

Figure 3   Perceived fruitiness of model food solutions with increasing coil overlap.

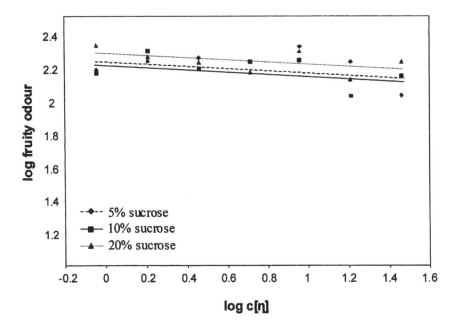

Figure 4 Perceived fruity odor of model food solutions with increasing coil overlap.

ethyl butyrate release were unaffected by viscosity, fruity flavor intensity should not fall to the same extent as sweetness intensity because of the halo effect caused by the aroma. The results of instrumental ethyl butyrate release analysis agree with the sensory evaluation results.

The instrumental analyses were carried out only on solutions containing 10% sugar. However, the concentration of sucrose in solution may have influenced ethyl butyrate release [10]. Sensory analysis results suggest that this interaction, if it occurred, was not large, since there was no difference in perceived odor intensity between solutions with different sugar concentrations (Fig. 4). With regard to the texture-aroma interactions observed, it was thought unlikely that sucrose concentration would have an effect that would change the trends observed.

The results found for the volatile release during consumption in this study contradict those found in 2002 by other researchers [20], who observed no change in either ethyl butyrate or benzaldehyde release during consumption of hydroxypropylmethylcellulose (HPMC) at concentrations in solution above c*. However, this study differs from that referenced in a number of ways that may explain the different findings: (a) In this study the

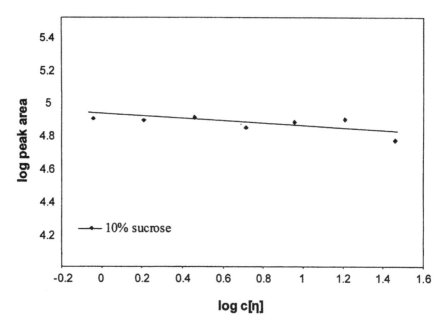

Figure 5 Release of ethyl butyrate into static headspace with increasing coil overlap.

sampling protocol for in-nose measurements allowed the consumer to swirl the sample continuously in the mouth, creating sample movement and mixing that were determined by viscosity, whereas in the study referenced participants held the sample in the mouth without swirling or mixing. (b) In this study the sampling time was 1 min, and ethyl butyrate release data were collected continuously, and therefore (although quantity of compound released fell below instrument sensitivity before 20 sec) the total quantity of compound measured had both time and intensity dimensions. The quantity of compound released was measured as area below the time-intensity release curve. In the comparable study referenced [20], participants held the sample in the mouth only for a short time before swallowing, and the entire sampling procedure was achieved during one breath. (c) In this study, the concentration of volatile compound used (5 ppm) was considerably lower than that used in the study referenced (200 ppm strawberry flavor, 55 ppm benzaldehyde). The reasons for these differences in concentration used are likely to be related to the procedural requirements of each study, which include the different hydrocolloids and flavor systems that were used. However, this difference in volatile concentration may have determined the

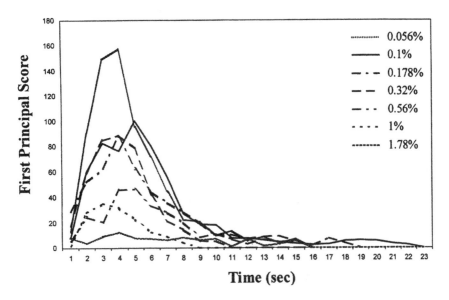

Figure 6 Release of ethyl butyrate during consumption of model food solutions with increasing coil overlap: first principal release curves for each concentration of guar gum in solution.

concentration of compound released into headspace during consumption in each case. (d) In this study, as a result of choice of a log concentration series, the range of concentration of hydrocolloid above c* was wider than that of the study referenced, and at the top end of the range a considerably more viscous solution with a greater degree of coil overlap was studied.

To determine correctly the influence of random coil polysaccharides on aroma and flavor perception, further investigation of the influence of experimental factors such as those considered is required. In addition, a technique to determine release of tastant, e.g., sucrose, must be developed. Finally, further studies with different thickeners, different tastants, and volatile compounds of different aroma character are needed as the interactions found will most likely be compound- and hydrocolloid-specific.

ACKNOWLEDGMENTS

The authors would like to thank Dr. Liz Sheehan, who assisted with the sensory analyses, and Mike Geary, who assisted with the static headspace analyses.

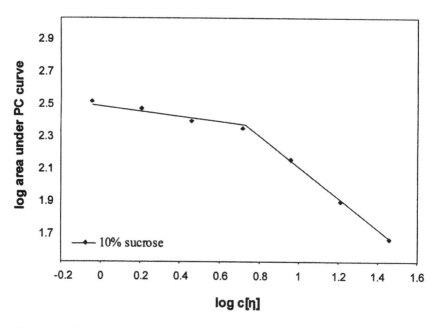

Figure 7 Release of ethyl butyrate during consumption with increasing coil overlap: area below release curve for each concentration of guar gum in solution.

REFERENCES

1. RM Pangborn, AS Szczesniak. Effect of hydrocolloids and viscosity on flavor and odor intensities of aromatic flavor compounds. J Texture Stud 4:467–482, 1974.
2. CM Christensen. Effects of solution viscosity on perceived saltiness and sweetness. Percept Psychophys 28:347–353, 1980.
3. AM Calviño, MR Carcia-Medina, JE Cometo-Muñiz, MB Rodriguez. Perception of sweetness and bitterness in different vehicles. Percept Psychophys 54:751–758, 1993.
4. ZV Baines, ER Morris. Flavour/taste perception in thickened systems: The effect of guar gum above and below c*. Food Hydrocolloids 1(3):197–205, 1997.
5. M Cliff, AC Noble. Time-intensity evaluation of sweetness and fruitiness and their interaction in a model food system. J Food Sci 55:450–454, 1990.
6. RJ Stevenson, J Prescott, R Boakes. Confusing tastes and smells: How odours can influence the perception of sweet and sour tastes. Chem Senses 24:627–635, 1999.

7. KGC Weel, AEM Boelrijk, AC Alting, PJJM van Mil, JJ Burger, H Gruppen, AGJ Voragen, G Smit. Flavor release and perception of flavored whey protein gels: Perception is determined by texture rather than by release. J Agric Food Chem 50(18):5149–5155, 2002.

8. AJ Taylor. Flavour matrix interactions. In: KAD Swift, ed. Current Topics in Flavours and Fragrances. Dordrecht, The Netherlands: Kluwer Academic Publishers, 1999, pp 123–138.

9. RM Pangborn, ZM Gibbs, C Tassan. Effect of hydrocolloids on apparent viscosity and sensory properties of selected beverages. J Texture Stud 9:415–436, 1978.

10. MA Godshall. How carbohydrates influence food flavor. Food Technol 51:63–67, 1997.

11. J Lynch, YH Liu, DJ Mela, HJH MacFie. A time-intensity study of the effect of oil mouthcoating on taste perception. Chem Senses 18(2):121–129, 1993.

12. FM Clydesdale. Color as a factor in food choice. Crit Rev Food Sci Nutr 33(1):83–101, 1993.

13. T Karrer, L Bartoshuk. Effects of capsaicin desensitization on taste in humans. Physiol Behav 57(3):421–429, 1995.

14. RA Frank, J Byram. Taste-smell interactions are tastant and odorant dependent. Chem Senses 13:445–455, 1988.

15. ER Morris. Applications of hydrocolloids. In: GO Phillips, DJ Wedlock, PA Williams, eds. Gums and Stabilizers for the Food Industry 2. Oxford: Pergamon Press, 1984, pp 57–72.

16. HR Moskowitz, BE Jacobs. Magnitude estimation: Scientific background and use in sensory analysis. In: HR Moskowitz, ed. Applied Sensory Analysis of Foods. Vol. 1. Boca Raton, FL: CRC Press, 1988, pp 194–222.

17. W Lindinger, A Hansel, A Jordan. On-line monitoring of volatile organic compounds at pptv levels by means of proton-transfer-reaction mass spectrometry (PTR-MS): Medical applications, food control and environmental research. Int J Mass Spectrom 173(3):191–241, 1998.

18. K Buhr, S van Ruth, C Delahunty. Analysis of volatile flavour compounds by proton transfer reaction mass spectrometry: Fragmentation patterns and discrimination between isomeric and isobaric compounds. Int J Mass Spectrom 221(1):1–7, 2002.

19. GB Dijksterhuis. Principal component analysis of time-intensity bitterness curves. J Sens Stud 8(4):317–328, 1993.

20. TA Hollowood, RST Linforth, AJ Taylor. The effect of viscosity on the perception of flavour. Chem Senses 27(7):583–591, 2002.

13

Differences in the Aroma of Selected Fresh Tomato Cultivars

Florian Mayer, Gary Takeoka, Ron Buttery, and Linda Whitehand
Western Regional Research Center, Agricultural Research Service, U.S. Department of Agriculture, Albany, California, U.S.A.

Yair Bezman, Michael Naim, and Haim Rabinowitch
Faculty of Agricultural, Food and Environmental Quality Sciences, The Hebrew University of Jerusalem, Rehovot, Israel

I. INTRODUCTION

The tomato is the second largest vegetable crop in dollar value in the United States; its fresh market production value was $1.16 billion in 2000 [1]. The United States produced 3.77 billion lb of fresh market tomatoes in 2000. In the 1950s tomato yields were relatively low (8900 lb per acre in 1955 vs. 30,600 lb per acre in 2000) and shelf life was very short. Advances in agrotechniques, genetics, and breeding have led to dramatic increases in both yield and shelf life. The latter resulted from the introduction of ripening inhibitor genes that adversely affect the flavor [2]. Consumers have been complaining about the lack of flavor in commercially available fresh tomatoes for several years now. One well-known reason is that many tomatoes are harvested green and induced to ripen by the use of ethylene before marketing. Although sensory characteristics such as color, texture, and sugar-acid content are important factors in consumer acceptance, the aroma content is considered a major quality trait for which a batch of tomatoes will be accepted or rejected. One objective of the present study was to determine the most important contributors to fresh tomato aroma. Another objective was to investigate the relationship between sensory perception and instrumental analysis. We selected two highly accepted and two less accepted tomato cultivars and analyzed their volatile flavor composition to find out whether the reason for consumer preference or rejection can be related to particular flavor compounds and their concentrations.

II. MATERIALS AND METHODS

A. Tomatoes

Tomatoes (cultivars BR-139, R 144, R 175, and FA-624) grown on a farm near Manteca, California, in 2000, 2001, and 2002 were picked regularly during the months of July through October and stored at room temperature until used for analysis. The properties of the four different tomato cultivars investigated are detailed in Table 1. The Brix value was determined after filtering blended tomatoes through a Kimwipes® EX-L wipe (Kimberly-Clark Corp., Roswell, GA, U.S.A.) and measuring the refractive index of the juice by using a Bellingham + Stanley Inc. (Atlanta, GA, U.S.A.) model RFM 81 automatic digital refractometer.

B. Chemicals

All chemicals and reference compounds used were obtained commercially or synthesized according to published methods. All odorants were purified by preparative gas chromatography (Varian 3700 GC, Walnut Creek, CA) using a glass packed column (250 × 0.5 cm, inner diameter [i.d.] packed with

Table 1 Properties of the Investigated Tomato Cultivars

	Tomato cultivar			
	R 144	R 175	BR-139	FA-624
Fruit weight, g	76–100	51–68	6–8	48–68
pH	4.5–4.7	4.3–4.5	4.3–4.6	4.3–4.5
Brix (early season–late season)	5.4–6.5	4.9–6.6	7.7–10.3	6.4–8.5
Preference	Less		More	

1% Carbowax 20M (Union Carbide, Danbury, CT, U.S.A.) on 120-140 mesh Chromosorb G (Johns Manville, NY, U.S.A.)). *Trans-* and *cis*-4,5-epoxy-(*E*)-2-decenal were synthesized by epoxidation of (*E,E*)- and (*E,Z*)-2,4-decadienal, respectively, using 3-chloroperoxybenzoic acid [3]. (*Z*)-1,5-Octadien-3-one was synthesized, following procedures described by Swoboda and Peers [4]. Diethyl ether was freshly distilled every 4 wk and stored in the dark after addition of antioxidant 330 (1,3,5-trimethyl-2,4,6-tris-[3,5-di-*tert*-butyl-4-hydroxybenzyl]-benzene).

C. Sensory Evaluation

All sensory evaluations were performed under orange light in a separate room providing space for four panelists at a time in four separate booths divided by vertical walls. The samples were presented to the panelists through sliding doors between the sensory preparation room and each of the four booths.

1. Determination of Odor Thresholds

Odor thresholds were determined according to Guadagni and Buttery [5].

2. Training of the Sensory Panel

The sensory panel was trained as previously described [6]. The number of participating panelists was increased from 12 to 16.

3. Flavor Profile Analysis of Fresh Tomato Samples

The flavor profile analysis was performed as previously described [6]. The number of evaluated flavor attributes was reduced from six to five (sweet, green, floral, fruity, and sour), removing the smoky odor quality as it was not important for the tomatoes investigated here. The intensities of the odor attributes were scored on a category scale from 0 (not perceptible) to 3 (strongly perceptible) in increments of 0.5.

4. Flavor Profile Analysis of Aroma Models Compared to Fresh Tomato

Aroma models of all four tomato cultivars were prepared by making concentrated stock solutions of all single odorants in absolute ethanol (200 proof) and adding the corresponding amount of each odorant (according to its concentration determined in tomato) to 500 mL of purified water. The pH of the model was adjusted to 4.3 using citrate buffer. Two milliliters of the aroma model solution and 2 mL of the corresponding tomato juice (prepared immediately before sensory evaluation by squeezing tomato pieces) were filled into two red colored tubes. The panelists were asked to dip an aroma testing paper strip (Measureline™, 6 × 1/4 in, Orlandi Inc., Farmingdale, NY) in each of the solutions, smell the paper strip, and score the intensities of the odor qualities—sweet, green, fruity, floral, and sour—on a category scale from 0 (not perceptible) to 3 (strongly perceptible) in increments of 0.5. The intensity scores of all odor qualities were compared between the model and the real tomato for significant differences by using the F-test and paired t-test.

D. Sample Preparation

Sample preparation was carried out by following three previously described methods [6–9] as summarized in the flowchart shown in Fig. 1.

E. Aroma Extract Dilution Analysis

Aroma extract dilution analysis (AEDA) [10] was performed as previously described [6].

F. Quantification of Fresh Tomato Odorants by Gas Chromatography Mass Spectrometry

Quantification of fresh tomato odorants was performed as previously described [6] using the same analytical instruments, internal standards mass spectrometry (MS) fragment ions, and recovery values. One additional compound, 2-isobutylthiazole, was included in the quantitative analysis. The ion used for quantification was m/z 99, and the recovery value determined for the solvent assisted flavor evaporation (SAFE) sample preparation method was 78%.

III. RESULTS AND DISCUSSION

A. Preference Ranking and Flavor Profile Analysis of Four Different Fresh Tomato Cultivars

After an independent tomato flavor testing with consumers, two highly accepted (BR-139 and FA-624) and two less accepted (R 144 and R 175) cultivars were chosen for detailed investigation to get an insight whether differences in odorant composition can be related to preference ranking. The BR-139 and FA-624 cultivars were preferred to the R 144 and R 175 cultivars. Furthermore, our sensory panel was asked to score the intensities

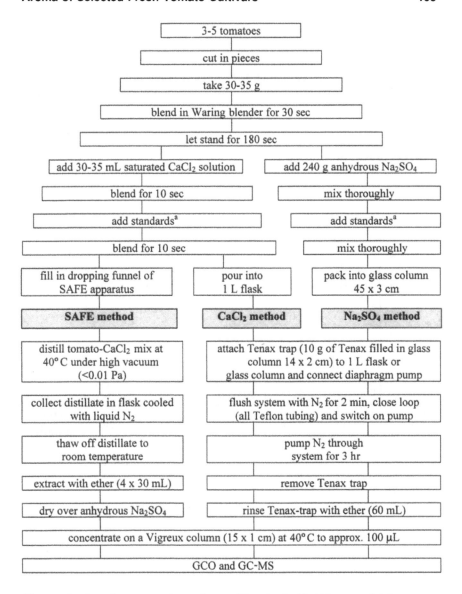

Figure 1 Sample preparation scheme. [a]Standard: 10–100 ppm of 3-hexanone, 2-octanone, anethole, maltol in 1 mL H_2O.

of certain tomato flavor attributes (see Sec. II.C.3) to produce a flavor profile of each of the four different cultivars. The flavor profiles are shown in Fig. 2. Statistical evaluations of the data revealed significant differences in the intensities of some odor qualities between some tomato cultivars. The R 144 and R 175 tomato cultivars were rated significantly less sweet than the BR-139 and FA-624 cultivars. The intensity scores for green odor were significantly higher for the R 144 than the BR-139 and FA-624 cultivars. No significant differences were found for the fruity note. The R 144 and BR-139 cultivars were rated more floral than the other cultivars, but the only significant differences were between the R 144 and the R 175 and FA-624 cultivars. The intensities of the sour odor quality of the R 144 and R 175

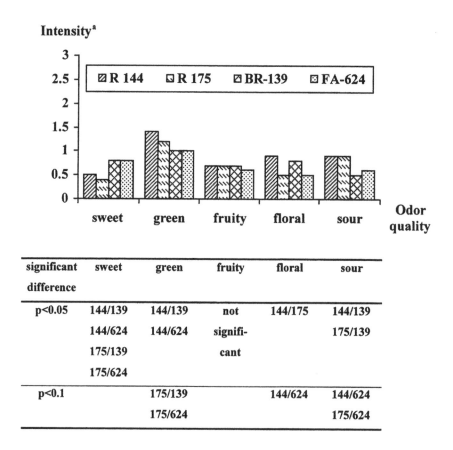

significant difference	sweet	green	fruity	floral	sour
p<0.05	144/139	144/139	not	144/175	144/139
	144/624	144/624	signifi-		175/139
	175/139		cant		
	175/624				
p<0.1		175/139		144/624	144/624
		175/624			175/624

Figure 2 Flavor profiles of the four investigated tomato cultivars. [a]Intensity scale: 0, not perceptible; 3, strongly perceptible.

cultivars were significantly higher than those of the BR-139 and FA-624 cultivars. In summary, the less preferred tomato cultivars R 144 and R 175 were rated less sweet, more green, and more sour than the other cultivars, the R 144 tomato also had a higher floral note. In contrast, the more preferred cultivars, BR-139 and FA-624, showed higher intensities of the sweet and lower intensities of the green and sour odor attributes. Next we investigated whether the results of the sensory evaluation could be correlated to the volatile aroma composition of the four cultivars.

B. Aroma Extract Dilution Analysis

Aroma extract dilution analysis [10] was performed on all four fresh tomato cultivars and the results compared to those previously published (7,11–13). The AEDA results of one of the more preferred tomato cultivars (FA-624) is presented in Table 2. We found the same odorants in all four investigated cultivars that other researchers had previously found in fresh tomato, with three exceptions. Of the 19 flavor compounds perceived in the present samples, 10 were already reported by Buttery and associates [7,11]. Among those are (Z)-3-hexenal, β-ionone, hexanal, β-damascenone; 1-penten-3-one, 3-methylbutanal, (E)-2-hexenal, phenylacetaldehyde, 2-phenylethanol, and 2-isobutylthiazole. Four other compounds, 1-octen-3-one, methional, 4-hydroxy-2,5-dimethyl-3(2H)-furanone (Furaneol®, Aldrich Chemical Company, Inc., Milwaukee, WI, U.S.A.) and (E,E)-2,4-decadienal, were reported by Krummbein and Auerswald [12]. Additionally, Guth and Grosch [13] found 3-methylbutyric acid and trans-4,5-epoxy-(E)-2-decenal among the most important odorants in fresh tomato. The three compounds not previously mentioned as important contributors to fresh tomato aroma, but that we could perceive, were (E,Z)-2,4-decadienal, cis-4,5-epoxy-(E)-2-decenal, and (Z)-1,5-octadien-3-one. The latter compound was identified in 2002 as a fresh tomato odorant for the first time [6]. In contrast, some other compounds previously reported as fresh tomato odorants could not be detected (by AEDA) in our samples, e.g., 1-nitro-2-phenylethane, 6-methyl-5-hepten-2-one, and methyl salicylate [7,11]. AEDA revealed no major differences in flavor dilution (FD) factors among the four cultivars with the exceptions of 4-hydroxy-2,5-dimethyl-3(2H)-furanone (Furaneol®), which had higher dilution factors in the more preferred cultivars BR-139, and FA-624; 2-isobutylthiazole, which had a higher FD factor in the R 175 cultivar, and phenylacetaldehyde and 2-phenylethanol, which had higher FD factors in the R 144 tomato than in the other tomato cultivars.

Table 2 Aroma Extract Dilution Analysis of Fresh Tomato (FA-624)

Odorant	Odor quality	Kovats index (DB-Wax)	Flavor dilution factor
3-Methylbutanal	Sweet, malty	914	2
1-Penten-3-one	Pungent	1016	32
Hexanal	Green, grassy	1077	64
(Z)-3-Hexenal	Green, grassy	1135	2048
(E)-2-Hexenal	Fruity, green apple	1214	8
1-Octen-3-one	Mushroom	1297	32
(Z)-1,5-Octadien-3-one	Metallic, geranium	1380	32
2-Isobutylthiazole	Green, tomato leaves	1396	1
Methional	Cooked potato	1446	128
Phenylacetaldehyde	Honey, beeswax	1636	16
3-Methylbutyric acid	Sweaty	1680	8
(E,Z)-2,4-Decadienal	Fatty	1765	16
(E,E)-2,4-Decadienal	Deep fried	1808	16
β-Damascenone	Fruity, cooked apple	1819	1024
2-Phenylethanol	Floral, rose	1910	16
β-Ionone	Floral, violet	1939	16
cis-4,5-Epoxy-(E)-2-decenal	Fatty, metallic	2000	16
trans-4,5-Epoxy-(E)-2-decenal	Fatty, metallic	2020	2048
4-Hydroxy-2,5-dimethyl-3(2H)-furanone	Sweet, caramel, cotton candy	2037	512

C. Determination of Odorant Concentrations

The concentrations of the 19 odorants perceived by AEDA in the four cultivars were determined. The quantitative results for the odorant concentrations are presented in Table 3. The quantitative results confirmed the AEDA studies but also showed some other differences. The amount of 4-hydroxy-2,5-dimethyl-3(2H)-furanone (Furaneol) in the BR-139 and FA-624 cultivars was 220 μg/kg, 4 to 12 times higher than in the R 144 and R 175 cultivars. The concentration of 2-isobutylthiazole was highest in the R 175 cultivar. The R 144 cultivar contained up to five times more phenylacetaldehyde and up to four times more 2-phenylethanol than the other cultivars. Other large differences in odorant concentrations between the four cultivars were found for 3-methylbutanal, 3-methylbutyric acid, and the (E,E)- and (E,Z)-2,4-decadienal isomers. The R 175 cultivar contained more 3-methylbutanal and 3-methylbutyric acid than the other cultivars. In the FA-624 tomato the concentrations of the two decadienal isomers were up to 20 times higher than in the R 144 tomato. The β-ionone content was 10 times higher in the FA-624 tomato than in the R 144 tomato. The FA-624 and BR-139 tomatoes had two to three times higher concentrations of (Z)-3-hexenal than the R 144 and R 175 cultivars. The first two cultivars also contained more (E)-2-hexenal and 1-penten-3-one than the latter two. For the other odorants the differences in concentration among the four cultivars were rather small. The hexanal content in all four cultivars ranged from 0.7 to 1.7 mg/kg, the amount of 1-octen-3-one varied between 3 and 7 μg/kg, and the concentration of β-damascenone was between 1 and 3 μg/kg. All cultivars contained similar amounts of methional. The concentration of the cis- and trans-4,5-epoxy-(E)-2-decenals in all four cultivars was between 90 and 140 μg/kg for the cis isomer and between 330 and 510 μg/kg for the trans isomer. (Z)-1,5-Octadien-3-one, which was clearly perceived as a fresh tomato odorant by gas chromatography olfactometry (GCO) in AEDA in all cultivars, could not be quantified with the usual amount of sample used for analysis (35 g) because of its low concentration. For its identification, we needed about 900 g of tomatoes and preparative GC to get a clear mass spectrum of (Z)-1,5-octadien-3-one. The estimated amount of (Z)-1,5-octadien-3-one in fresh tomato is less than 0.1 μg/kg [6].

On the basis of the preference ranking of the four fresh tomato cultivars, we may conclude that preference was positively influenced by higher concentrations of 1-penten-3-one, (Z)-3-hexenal, (E,Z)- and (E,E)-2,4-decadienal, and 4-hydroxy-2,5-dimethyl-3(2H)-furanone (Furaneol) (BR-139 and FA-624 cultivars) and negatively influenced by high amounts of phenylacetaldehyde and 2-phenylethanol (R 144 cultivar) and

Table 3 Concentrations of Potent Odorants in Four Fresh Tomato Cultivars

Odorant	Concentration, ppb			
	R 144	R 175	BR-139	FA-624
3-Methylbutanal	55	97	21	14
1-Penten-3-one	200	210	500	530
Hexanal	1200	670	860	1700
(Z)-3-Hexenal	3200	2100	6000	6800
(E)-2-Hexenal	79	46	210	130
1-Octen-3-one	3	6	7	4
2-Isobutylthiazole	140	410	200	110
Methional	8	9	5	9
Phenylacetaldehyde	740	410	150	450
3-Methylbutyric acid	640	1800	200	200
(E,Z)-2,4-Decadienal	8	20	29	160
(E,E)-2,4-Decadienal	2	5	7	33
β-Damascenone	1	3	1	3
2-Phenylethanol	1700	1100	460	880
β-Ionone	2	3	6	21
cis-4,5-Epoxy-(E)-2-decenal	93	110	88	140
trans-4,5-Epoxy-(E)-2-decenal	330	420	420	510
4-Hydroxy-2,5-dimethyl-3(2H)-furanone	18	56	220	220

3-methylbutanal, 3-methylbutyric acid, and 2-isobutylthiazole (R 175 cultivar).

The quantitative results were in good accord with the results of the flavor profile analyses (Fig. 2). The higher intensity of the sweet odor in the more preferred tomato cultivars, BR-139 and FA-624, could be correlated to the higher amounts of 4-hydroxy-2,5-dimethyl-3(2H)-furanone (Furaneol). The more intense floral note in the R 144 tomato was likely caused by distinctly higher concentrations of phenylacetaldehyde and 2-phenylethanol. The R 144 and R 175 cultivars were perceived to have higher intensities of green odor than the BR-139 and FA-624 cultivars, although the latter two cultivars contained more (Z)-3-hexenal. This may be explained by lower concentrations of some other odorants in the first two cultivars mentioned, which caused the green note to be perceived more intensely, whereas in the latter two cultivars higher amounts of other odorants masked the green note, making it harder to perceive. In the R 175 tomato the higher concentration of 2-isobutylthiazole may have also contributed to the higher intensity of the green odor.

D. Calculation of Odor Units

To get additional insight into which compounds are the most important contributors to fresh tomato aroma and which odorants are responsible for the aroma differences among the four cultivars, odor units (odor activity values) were calculated by dividing each odorant's concentration by its odor threshold in water (water content of tomato is about 95%). The thresholds are listed in Table 4. The calculated odor units are shown in Table 5. In all fresh tomato cultivars investigated (Z)-3-hexenal and trans-4,5-epoxy-(E)-2-decenal were, by far, the most potent odorants, with odor units ranging from 8000 to 20,000. β-Ionone, β-damascenone, and 1-octen-3-one followed, with odor units between 300 and 3000. In the more preferred FA-624 cultivar (E,Z)-2,4-decadienal had an odor unit value of 1450, whereas in the less preferred R 144 cultivar it had an odor unit value of only 70. (E,E)-2,4-Decadienal had higher odor units in the FA-624 cultivar than in the other cultivars. The odor units of 1-penten-3-one were higher in the FA-624 and BR-139 tomatoes than in the R 144 or R 175 tomatoes. 3-Methylbutanal had quite high odor units in the R 175 cultivar with values seven times higher than in the FA-624 cultivar. Hexanal had odor units of 190 to 380 in all four cultivars. In the R 144 cultivar phenylacetaldehyde had an odor unit value of 185, which was five times higher than in the BR-139 cultivar. 2-Phenylethanol had two to four times higher odor units in the R 144 cultivar than in the other cultivars. The odor units of methional were similar in all cultivars. In the R 175 cultivar 2-isobutylthiazole had higher

Table 4 Odor Thresholds of Important Fresh Tomato Odorants in Water

Odorant	Odor threshold in water, µg/L[a]
3-Methylbutyric acid	250
2-Phenylethanol	68[b]
4-Hydroxy-2,5-dimethyl-3(2H)-furanone	21
(E)-2-Hexenal	17
2-Isobutylthiazole	4.6[b]
Hexanal	4.5
Phenylacetaldehyde	4
1-Penten-3-one	1
(Z)-3-Hexenal	0.25
3-Methylbutanal	0.2
Methional	0.2
(E,Z)-2,4-Decadienal	0.11[b]
(E,E)-2,4-Decadienal	0.11[b]
$trans$-4,5-Epoxy-(E)-2-decenal	0.02[b]
β-Ionone	0.007
1-Octen-3-one	0.005
β-Damascenone	0.002
cis-4,5-Epoxy-(E)-2-decenal	Not yet determined

[a] According to Ref. 14.
[b] Own result.

odor units when compared to the other cultivars. 4-Hydroxy-2,5-dimethyl-3(2H)-furanone (Furaneol) had odor units of only 1 to 10, but they were higher in the more preferred BR-139 and FA-624 cultivars than in the less preferred R 144 and R 175 cultivars. The odor units of (E)-2-hexenal were also low, but there were no large differences between the cultivars. 3-Methylbutyric acid had odor units less than 1 in the BR-139 and FA-624 tomatoes (indicating that its concentration did not exceed its odor threshold), whereas in the R 175 cultivar it was 7.

In summary, we may conclude that the basic fresh tomato aroma is composed of the same compounds in all four cultivars, especially (Z)-3-hexenal and $trans$-4,5-epoxy-(E)-2-decenal, and that the aroma differences between the cultivars that are responsible for differences in preference are caused by concentration differences of odorants with less impact, i.e., lower odor units. The more preferred tomatoes contained more 1-penten-3-one, (E,E)- and (E,Z)-2,4-decadienal, and 4-hydroxy-2,5-dimethyl-3(2H)-furanone (Furaneol). High concentrations of phenylacetaldehyde and 2-phenylethanol (as in the R 144 cultivar) or 3-methylbutanal and 2-isobutylthiazole (as in the R 175 tomato) had a negative influence on preference.

Table 5 Comparison of Odor Units of Potent Fresh Tomato Odorants in Four Cultivars

Odorant	Odor units[a]			
	R 144	R 175	BR-139	FA-624
3-Methylbutanal	275	485	105	70
1-Penten-3-one	200	210	500	530
Hexanal	267	149	191	378
(Z)-3-Hexenal	12800	8400	24000	27200
(E)-2-Hexenal	5	3	12	8
1-Octen-3-one	600	1200	1400	800
2-Isobutylthiazole	30	89	43	24
Methional	40	45	25	45
Phenylacetaldehyde	185	103	38	113
3-Methylbutyric acid	2	7	<1	<1
(E,Z)-2,4-Decadienal	73	182	264	1455
(E,E)-2,4-Decadienal	18	45	64	300
β-Damascenone	500	1500	500	1500
2-Phenylethanol	25	16	7	13
β-Ionone	286	429	857	3000
cis-4,5-Epoxy-(E)-2-decenal	n.d.[b]	n.d.[b]	n.d.[b]	n.d.[b]
trans-4,5-Epoxy-(E)-2-decenal	16500	21000	21000	25500
4-Hydroxy-2,5-dimethyl-3(2H)-furanone	<1	3	10	10

[a] Odor units = concentration / odor threshold in water.
[b] n.d., Not determined, no odor threshold available.

E. Preparation of Aroma Models

To confirm the identification of all important fresh tomato odorants—that the concentrations of the flavor compounds were determined correctly and that the differences between the cultivars were caused by concentration differences—we prepared aroma model solutions of all cultivars (by mixing together the odorants in the concentrations determined) and compared each aroma model to its corresponding real tomato. The aroma models were prepared in citrate buffer at pH 4.3. The odorant composition of an aroma model of one of the more preferred tomatoes compared to that of a previously published aroma model is given in Table 6 [15]. When it is compared to the earlier model, there are many more compounds in the

Table 6 Comparison of an Aroma Model of Fresh Tomato Prepared According to the Results of the Present Investigation and a Published Recipe[a]

	Concentration in H_2O, ppb	
Odorant	Present model[b]	Previous model[a]
3-Methylbutanal	14	200
1-Penten-3-one	530	200
Hexanal	1700	600
(Z)-3-Hexenal	6800	3500
(E)-2-Hexenal	130	160
1-Octen-3-one	4	—[c]
2-Isobutylthiazole	110	10
Methional	9	—
Phenylacetaldehyde	450	—
(E,Z)-2,4-Decadienal	160	—
(E,E)-2,4-Decadienal	33	—
β-Damascenone	3	—
2-Phenylethanol	880	—
β-Ionone	21	10
cis-4,5-Epoxy-(E)-2-decenal	140	—
trans-4,5-Epoxy-(E)-2-decenal	510	—
4-Hydroxy-2,5-dimethyl-3(2H)-furanone	220	—
(Z)-3-Hexenol	—	1500
6-Methyl-5-hepten-2-one	—	100
Methyl salicylate	—	50

[a] From Ref. 15.
[b] pH Adjusted to 4.3 with citrate buffer.
[c] —, Not added.

present model. However, there are also three compounds missing that could never be perceived by us by AEDA. The flavor profile of the aroma model of one of the more preferred tomatoes (FA-624) compared to that of the corresponding real tomato is shown in Fig. 3. It can be noted that the flavor profile of the model is similar to the flavor profile of the real sample. Statistical evaluation of the results revealed no significant differences between the model and the real tomato in any of the odor qualities. The problems of comparing a liquid aroma model to a real tomato for similarity leave some room for further improvement, especially for finding ways to simulate tomato matrix to create realistic conditions for the distribution of odorants between the matrix and the vapor phase [16].

IV. CONCLUSIONS

Sensory evaluation of the aroma model confirmed the results of our instrumental analyses. All important odorants of fresh tomato aroma could be identified and their concentrations determined. The differences in preference ranking among the four cultivars under the experimental conditions were due to variations in the concentrations of certain flavor compounds. Higher amounts of the (E,E)- and (E,Z)-2,4-decadienal

Intensity[a]

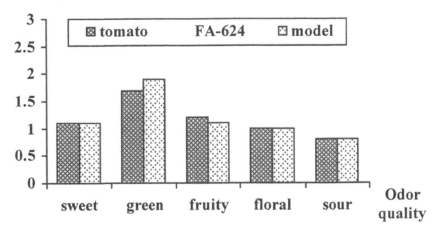

Figure 3 Comparison of the flavor profile of a fresh tomato and its corresponding aroma model. [a]Intensity scale: 0, not perceptible; 3, strongly perceptible.

isomers, 1-penten-3-one, and 4-hydroxy-2,5-dimethyl-3(2*H*)-furanone (Furaneol) had a positive influence on preference, whereas high concentrations of phenylacetaldehyde, 2-phenylethanol, and 2-isobutylthiazole had a negative influence.

ACKNOWLEDGMENT

This research was supported by Research Grant Award IS-2980-98 from BARD, the United States–Israel Binational Agricultural Research and Development Fund.

REFERENCES

1. Agricultural Statistics 2002, U.S. Department of Agriculture, National Agricultural Statistics Service, Washington, DC: United States Government Printing Office, 2002.
2. WB McGlasson, JH Last, KJ Shaw, SK Meldrum. Influence of the non-ripening mutants *rin* and *nor* on the aroma of tomato fruit. Hortic Sci 22:632–634, 1987.
3. PAT Swoboda, KE Peers. *trans*-4,5-Epoxyhept-*trans*-2-enal: The major volatile compound formed by the copper and α-tocopherol induced oxidation of butterfat. J Sci Food Agric 29:803–807, 1978.
4. PAT Swoboda, KE Peers. Metallic odour caused by vinyl ketones formed in the oxidation of butterfat: The identification of octa-1,*cis*-5-dien-3-one. J Sci Food Agric 28:1019–1024, 1977.
5. DG Guadagni, RG Buttery. Odor threshold of 2,3,6-trichloroanisole in water. J Food Sci 43:1346–1347, 1978.
6. F Mayer, GR Takeoka, RG Buttery, Y Nam, Y Bezman, M Naim, H Rabinowitch. Aroma of fresh field tomatoes. In: K Cadwallader, H Weenen, eds. Freshness and Shelflife of Foods. ACS Symposium Series No. 836. Washington, DC: American Chemical Society, 2002, pp 144–161.
7. RG Buttery, R Teranishi, LC Ling. Fresh tomato aroma volatiles: A quantitative study. J Agric Food Chem 35:540–544, 1987.
8. RG Buttery, GR Takeoka, M Naim, H Rabinowitch, Y Nam. Analysis of furaneol in tomato using dynamic headspace sampling with sodium sulfate. J Agric Food Chem 49:4349–4351, 2001.
9. W Engel, W Bahr, P Schieberle. Solvent assisted flavour evaporation—a new and versatile technique for the careful and direct isolation of aroma compounds from complex food matrices. Eur Food Res Technol 209:237–241, 1999.
10. F Ullrich, W Grosch. Identification of the most intense volatile flavour compounds formed during autoxidation of linoleic acid. Z Lebensm Unters Forsch 184:277–282, 1987.

11. RG Buttery, R Teranishi, RA Flath, LC Ling. Fresh tomato volatiles. In: R Teranishi, RG Buttery, F Shahidi, eds. Flavor Chemistry, Trends and Developments. ACS Symposium Series No. 388. Washington, DC: American Chemical Society, 1989, pp 213–222.

12. A Krummbein, H Auerswald. Characterization of aroma volatiles in tomatoes by sensory analyses. Nahrung 42/6:395–399, 1998.

13. H Guth, W Grosch. Evaluation of important odorants in foods by dilution techniques. In: R Teranishi, EL Wick, I Hornstein, eds. Flavor Chemistry: 30 Years of Progress. New York, NY: Kluver Academic/Plenum Publishers, 1999, pp 377–386.

14. M Rychlik, P Schieberle P, W Grosch. Compilation of Odor Thresholds, Odor Qualities and Retention Indices of Key Food Odorants. Garching, Germany: Deutsche Forschungsanstalt für Lebensmittelchemie und Institut für Lebensmittelchemie der Technischen Universität München, 1998.

15. RG Buttery. Quantitative and sensory aspects of flavor of tomato and other vegetables and fruits. In: TE Acree, R Teranishi, eds. Flavor Science, Sensible Principles and Techniques. ACS Professional Reference Book. Washington, DC: American Chemical Society, 1993, pp 259–286.

16. Y Bezman, F Mayer, GR Takeoka, RG Buttery, G Ben-Oliel, H Rabinowitch, M Naim. Differential effects of tomato (*Lycopersicon esculentum*, Mill) matrix on the volatility of important aroma compounds. J Agric Food Chem 51:722–726, 2003.

14

New Flavor Compounds from Orange Essence Oil

Sabine Widder, Marcus Eggers, Jan Looft,
Tobias Vössing, and Wilhelm Pickenhagen
DRAGOCO Gerberding & Co., Holzminden, Germany

I. INTRODUCTION

Today citrus fruit is the most important fruit crop in the world. Its production exceeds by far that of all other fruits: apples, pears, etc. The main reasons for this outstanding position of citrus fruits are their desirable and refreshing taste and flavor, as well as their positive effect on human health and vitality. Therefore, it is not surprising that citrus fruits are a significant item in modern diet. Additionally citrus is one of the most appreciated flavors for soft drinks and confectionery and is accepted all over the world. In 2001 about 71 million tons of citrus fruits was produced globally. The most important species by far are oranges, accounting for more than two-thirds of the whole production. Approximately 60% of the orange crop is consumed as fresh fruits, the remaining 40% are processed to juice, mainly concentrated juice, and several by-products, such as peel oil, essence oil, and aqueous essence. These by-products exhibit the characteristic flavor of orange. Therefore, they are interesting raw materials for the flavor and fragrance industries. Orange essence oil is obtained during production of concentrated orange juice. During the concentration step, flavor compounds are stripped from the juice together with the water. These vapors are condensed and collected as aroma essence, which is subsequently separated into aqueous essence and essence oil. Essence oil contains the oil-soluble volatiles and is characterized by fresh, floral, sweet, and fruity flavor attributes. The yield of orange essence oil is about 100 mg/kg orange fruit.

Because of its excellent flavor properties orange essence oil is a valuable raw material for the creation of flavors and perfumes. Numerous studies on orange essence oil have already been performed to determine the composition of the volatile fraction [1,2]. These studies indicated hydrocarbons, especially limonene, as the most abundant compounds, accounting for more than 97% of the volatile fraction. The other 3% are aldehydes, esters, ketones, and some other oxygenated compounds (Table 1). However, none of these studies tried to evaluate the flavor impact of single-flavor compounds identified in the orange essence oil.

One systematic approach to estimation of the contribution of single volatiles in the overall odor is gas chromatography olfactometry (GCO) of stepwise diluted aroma extracts [aroma extract dilution analysis (AEDA)] [3]. This method has already been used successfully in numerous studies to characterize key odorants in food [4].

This study describes the evaluation of the most potent flavor compounds of orange essence oil by AEDA and the enrichment, purification, and identification of new flavor compounds that have not been described for orange flavor.

II. EXPERIMENTAL PROCEDURES

A. Materials

Different orange essence oils originating from Florida and Brazil were sensorially evaluated. The oil with the highest quality in freshness, sweet fruitiness, and juiciness originating from Brazil was selected for the investigation. For analysis the essence oil was diluted with diethylether by a ratio of 1:25.

Table 1 Composition of the Volatile Fraction of Orange Essence Oil

Components	Amount, g/100 g
Limonen	91.7
Valencen	1.44
Other terpenes	3.70
Aldehydes	1.02
Alcohols	1.13
Ester	0.04
Others	0.97

B. High-Resolution Gas Chromatography Olfactometry and Mass Spectrometry

High-resolution gas chromatography olfactometry (HRGCO) and high-resolution gas chromatography mass spectrometry (HRGCMS) were performed in parallel by means of a GCD type mass spectrometer (Hewlett-Packard) equipped with a cold injection system (CIS, Gerstel GmbH, Mülheim, Germany) and a sniffing port (Gerstel GmbH, Mülheim, Germany). The following columns were used: DB-1 (30 m × 0.25 mm fused silica, 0.25 µm d_f, J&W Scientific, Fisons, Mainz, Germany) and DB-Wax (30 m × 0.25 mm fused silica, 0.25 µm d_f, J&W Scientific, Fisons, Mainz, Germany). Gas chromatographic conditions were the same as described previously [5]. Mass spectra in the electron ion impact mode were generated at 70 eV. For identification of the chemical structures, MS data, retention indices, and odor quality of reference materials were used.

C. Aroma Extract Dilution Analysis

Aroma extract dilution analysis was performed as described by Schieberle [4]. The essence oil was diluted stepwise with diethylether (1:1), and an aliquot of each dilution step was subsequently analyzed by GC. This procedure was repeated until no odorant was detectable.

D. Column Chromatography

Orange essence oil was fractionated on a glass column (61 × 3.7 cm) packed with silica gel (40 µm, J. T. Baker BV, Deventer, The Netherlands) in pentane to remove hydrocarbons (500 g batches). Chromatography was performed under nitrogen pressure, by maintaining a constant flow of 5 mL/min and by using the following solvents: n-pentane (fraction I), diethylether (fraction II). Each fraction was concentrated and analyzed by HRGCO to localize the odor-active compounds for mass spectral measurements.

E. Preparative Capillary Gas Chromatography

For enrichment and isolation of unknown trace compounds located in fraction II, a mass flow-controlled automated multidimensional switching system (MCS, Gerstel GmbH, Mühlheim, Germany) was employed in combination with an HP 6890 gas chromatograph (Hewlett Packard, Waldenbronn, Germany) equipped with a cold injection system (CIS) (Gerstel GmbH, Mühlheim, Germany). A combination of a DB 1 column (15 m × 0.53 mm, 3 µm d_f, J&W Scientific) and a DB-Wax column

(30 m × 0.53, 2 μm d_f, J&W Scientific) was used. Gas chromatographic conditions were the same as described previously [5].

F. Odor Thresholds in Water

A defined amount of each compound dissolved in 0.1 mL of ethanol was added to water (1 L). After stirring for 10 min, this stock solution was diluted stepwise with water (1:1, v/v) and stirred for 5 min. Odor thresholds were determined by triangle tests with a trained panel of at least 15 panelists.

III. RESULTS AND DISCUSSION

A. Identification of Odor-Active Volatiles

In preliminary experiments different orange essence oils originating from Florida and Brazil were compared by hedonic sensory evaluation. One essence oil of Brazilian origin with the highest quality in freshness, sweet fruitiness, and juiciness gained the highest acceptance by a panel of five flavorists. On the basis of this result, this essence oil was selected for further investigations.

First HRGCO was applied to detect odor-active regions and AEDA was performed to select the most potent volatiles contributing to the overall flavor of orange essence oil. In the flavor dilution (FD) range of 2 to 512, 27 odor-active compounds could be perceived. The results in Table 2 reveal four odorants, linalool (flowery, sweet), decanal (citruslike, soapy), octanal (citruslike, green), and ethyl butyrate (fruity), as key contributors. Most of the volatiles listed in Table 2 had already been described for orange essence oil [1,2,6] or freshly squeezed orange juice [7].

Additionally, five odor-active regions with interesting flavor qualities could be detected. These aroma compounds did not yield an MS signal, indicating low odor thresholds. The odor qualities of these compounds perceived at the sniffing port are listed in Table 3. The unknown compounds I, II, and III, all giving a fresh and fruity impression, seemed especially to contribute to the overall flavor.

B. Identification of Unknown Volatiles

Numerous investigations on orange flavor volatiles showed that their main components are hydrocarbons [1,2,6]. Because of their high odor thresholds, however, their contribution to the overall flavor is rather low. Extensive studies on the flavor of freshly squeezed orange juice have confirmed that the overall flavor impression is predominantly caused by oxygenated

Table 2 Most Potent Compounds of Orange Essence Oil

RI DB5	Odorant	FD factor
1101	Linalol	512
1207	Decanal	512
1004	Octanal	256
802	Ethyl butyrate	256
933	*a*-Pinene	128
1033	Limonene	64
1077	Unknown I	64
1196	4-Decenal	64
1382	*trans*-4,5-Epoxy-(E)-2-decenal	64
799	(Z)-3-Hexenal	32
986	Myrcene	32
1109	Nonanal	32
799	Hexanal	16
849	Ethyl-2-methyl-butanoate	16
1319	(E,E)-2,4-Decadienal	16
1597	Unknown II	16
1136	Ethyl-3-hydroxy-hexanoate	8
967	Unknown III	8
1248	Neral	8
1199	Ethyl hexanoate	4
1266	(E)-2-Decenal	4
1275	Geranial	4
1491	*β*-Ionone	4
1172	Unknown IV	2
1280	Unknown V	2
1498	Valencene	2
1815	Nootkatone	2

RI, retention index; FD, flavor dilution.

odorants [7]. Therefore, for the identification of the unknown compounds, hydrocarbons were removed from about 1 kg of orange essence oil by column chromatography on silica gel. All of the target compounds could be localized in fraction II–containing oxygenated compounds. On the basis of mass spectral data, four of the five odorants could be identified as 6-methyl octanal (no. I in Table 3), 6-methyl heptanal (no. III in Table 3), 8-methyl nonanal (no. IV in Table 3), and 8-methyl decanal (no. V in Table 3). For

Table 3 Unknown Odorants of Orange Essence Oil

RI DB5	Odorant	Odor quality	FD factor
1077	Unknown I	Fruity, orange peel-like, fresh, green, orange	64
1597	Unknown II	Fresh, sweet, fruity, orange peel-like, albedo-like	16
967	Unknown III	Fatty, fruity, fresh, green, juicy, sweet	8
1172	Unknown IV	Fatty, soapy, orange peel-like	2
1280	Unknown V	Fatty, soapy, orange peel-like	2

RI, retention index; FD, flavor dilution.

unknown compound no. II, no unequivocal mass spectral data were obtained.

The proposed structures were confirmed by synthesis. Figure 1 shows the synthetic route, which yields 8-methyl decanal. According to a literature procedure, in the first step a Wittig reaction of 2-methyl butanal with the triphenyl phosphonium salt of W-bromo ethyl hexanoate was performed. Lithium alanate reduction of the resulting ester led to the unsaturated alcohol, and subsequent hydrogenation using a palladium catalyst finally resulted in the formation of the 8-methyl decanol, which in the last step was oxidized with pyridinium chlorochromate to the final product. 6-Methyl octanal was prepared accordingly, starting with the triphenyl phosphonium

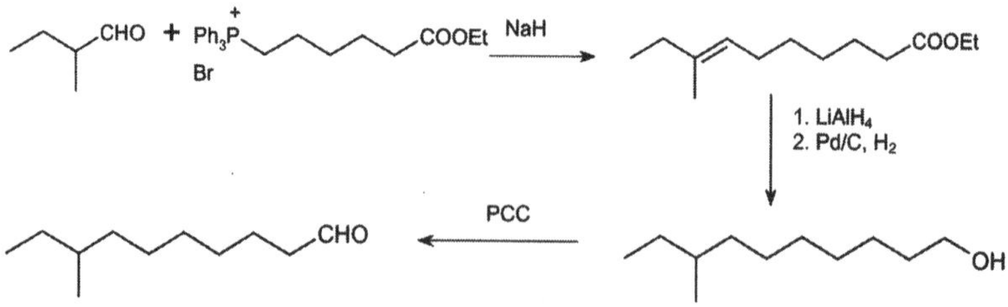

Figure 1 Synthesis of 8-methyl decanal.

salt of W-bromo ethyl butanoate. For the synthesis of 8-methyl nonanal the same procedure, using 3-methyl butanal and the triphenyl phosphonium salt of W-bromo ethyl pentanoate, was performed.

6-Methyl octanal, 8-methyl nonanal, and 8-methyl decanal have not been reported for oranges, but have been reported for green and yellow yuzu peel oil [8]. Yuzu (*Citrus junos Tanaka*) is a kind of sour orange mainly grown in Japan, where it is used as flavorful acidulant in products such as seasonings, pickled vegetables, seafood, and vinegar [9]. 6-Methyl heptanal has been reported in heated beef fat, *Zanthoxylum rhetsoides* Drake, and *Helichrysum* species [10–12]. The sensory properties of the methyl branched aldehydes, flavor quality, and odor thresholds in water are summarized in Table 4. The odor description was performed by a panel of five flavorists in mixtures containing sugar, citric acid, and 3 ppm of the aldehyde. The odor thresholds in water were determined by a trained panel of at least 15 judges. 6-Methyl heptanal and 6-methyl octanal were especially judged as very interesting because of their fresh, green, juicy, and orangelike odor qualities and their low odor thresholds of 1.2 µg/L and 5.6 µg/L, respectively.

For the enrichment and purification of unknown compound II, the hydrocarbons of about 20 kg of orange essence oil were removed by column chromatography on silica gel. Subsequently the compound occurring in fraction II was further purified by two-dimensional preparative capillary gas chromatography. Interpretation of the resulting mass spectra finally led to the identification of 8-tetradecenal. The mass spectrum is shown in Fig. 2.

8-Tetradecenal had not been described as a flavoring compound. Its structure was confirmed by synthesis. The two possible diasteroisomers, 8-(Z)-tetradecenal and 8-(E)-tetradecenal, were stereoselectively synthesized. Both isomers could be detected in orange essence oil. Figure 3 shows the synthetic route, which yields 8-(Z)-tetradecenal. In the first step an oxidative cleavage of cyclooctene by ozonolysis was performed, leading to methyl-8-

Table 4 Odor Properties of Methyl Branched Aldehydes

Compound	Odor quality	Odor threshold water, ppb
6-Methyl heptanal	Fresh, green, juicy, sweet, orangelike	1.2
6-Methyl octanal	Fatty, green, juicy, orangelike	5.6
8-Methyl nonanal	Fatty, melon-like, soapy	42
8-Methyl decanal	Fatty, soapy, peel-like	28

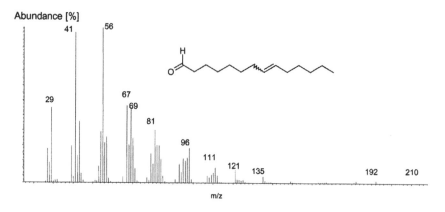

Figure 2   Mass spectra of 8-tetradecenal.

oxo-octanoate. Subsequent Z-selective Wittig reaction with *n*-hexyl-triphe-nyl phosphonium bromide resulted mainly in the formation of methyl 8-(Z)-tetradecenoate, which in the last step was reduced to 8-(Z)-tetradecenal. The synthesis of 8-(E)-tetradecenal are published elsewhere. The sensory properties of 8-(Z)- and 8-(E)-tetradecenal are summarized in Table 5. The results demonstrate a considerable difference in the odor quality as well as in the odor threshold between both isomers. The cis isomer exhibits a

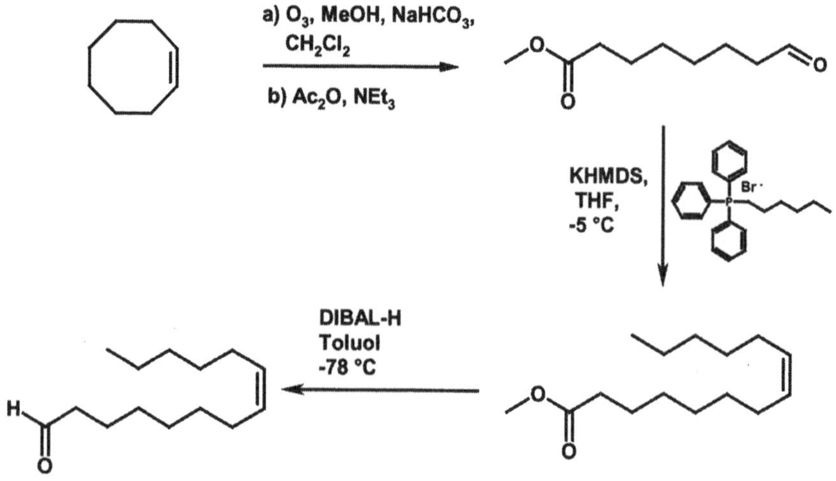

Figure 3   Synthesis of 8-(Z)-tetradecanal.

Table 5 Sensory Properties of 8-Tetradecenal Isomers

Isomer	Odor quality	Odor threshold water, ppb
(E)-8-Tetradecenal	Fatty, waxy, green, chemical, sweet	1.24
(Z)-8-Tetradecenal	Citruslike, fruity, sweet, fresh, orange peel-like, albedolike, fatty	0.009

pleasant fruity, fresh, citruslike flavor quality, whereas the trans isomer has a fatty, waxy, green, and chemical flavor. Additionally, the odor threshold of (Z)-8-tetradecenal in water (0.009 µg/L) is lower by a factor of about 140. These data show that 8-(Z)-tetradecenal has high-impact flavor qualities as well as high odor strength.

IV. CONCLUSION

The combination of HRGCO and AEDA provides suitable tools to detect and select compounds responsible for the overall flavor impression of edible materials. In addition, the odor quality and the sensory value of each odorant eluting from the GC column can be assessed. Even odorants having low FD values but interesting odor qualities should thus be selected as targets for identification experiments.

The results of these studies have shown that by using this approach, high-impact flavor chemicals can be identified.

REFERENCES

1. PE Shaw. Review of quantitative analysis of citrus essential oils. J Agric Food Chem 27:246–257, 1979.
2. MG Moshonas, PE Shaw. Flavor and compositional comparison of orange essences and essence oils produced in the United States and in Brazil. J Agric Food Chem 38:799–801, 1990.
3. TE Acree. Bioassays for flavor. In: TE Acree, R Teranishi, eds. Flavor Science: Sensible Principles and Techniques. Washington, DC: ACS Professional Reference Book, 1993, pp 1–10.

4. P Schieberle. New developments in methods for analysis of volatile flavor compounds and their precursors. In: AG Gaonkar, ed. Characterization of Food: Emerging Methods. Amsterdam: Elsevier, 1995, pp 403–423.

5. S Widder, C Sabater Lüntzel, T Dittner, W Pickenhagen. 3-Mercapto-2-methylpentan-1-ol, a new powerful aroma compound. J Agric Food Chem 48:418–423, 2000.

6. RL Swaine. Citrus oils: Proscessing, technology, and applications. Perf Flav 13:2–20, 1988.

7. A Hinterholzer, P Schieberle. Identification of the most odour-active volatiles in fresh, hand-extracted juice of Valencia late oranges by odour dilution techniques. Flavour Fragrance J 13:49–55, 1998.

8. K Tajima, S Tanaka, T Yamaguci, M Fujita. Analysis of green and yellow yuzu peel oils (*Citrus junos Tanaka*) novel aldehyde components with remarkably low odor thresholds. J Agric Food Chem 38:1544–1548, 1990.

9. SM Njoroge, H Ukeda, H Kusunose, M Sawamura. Volatile compounds of Japanese yuzu and lemon oils. Flavour Fragrance J 9:159–166, 1994

10. S Ohnishi, T Shibamoto. Volatile compounds from heated beef fat and beef fat with glycine. J Agric Food Chem 32:987–992, 1984.

11. P Weyerstahl, H Marschall, U Splittgerber, PT Son, PM Giang, VK Kaul. Constituents of the essential oil from the fruits of *Zanthoxylum rhetsoides Drake* from Vietnam and from the aeral parts of *Zanthoxylum alatum Roxb.* from India. Flavour Fragrance J 14:225–229, 1999.

12. J Kuhnke, F Bohlmann. Synthesis of naturally occurring phloroglucinol derivatives. Tetrahedron Lett 26:3955–3958, 1985.

15
Heat-Induced Changes in Aroma Components of Holy Basil (*Ocimum sanctum* L.)

Sompoche Pojjanapimol and Siree Chaiseri
Kasetsart University, Bangkok, Thailand

Keith R. Cadwallader
University of Illinois, Urbana, Illinois, U.S.A.

I. INTRODUCTION

The genus *Ocimum*, Lamiaceae, collectively called basil and comprising over 30 species, is one of the most popular culinary herbs. The fresh leaves of holy basil (*Ocimum sanctum* Linn.) provide a rich, spicy, and pungent aroma that is much appreciated in Thai cuisine. The herb is most often used in stir-fried meat dishes including beef, pork, chicken, and prawn that are called Pad Gaprao. The herb also is added to curries, e.g., country-style curry. Additionally, deep fried holy basil is used to garnish some Thai dishes. In general, holy basil leaves are added near the end of the cooking process, to help in minimizing heat-induced changes and loss in the characteristic flavor of this herb.

The distilled essential oil of holy basil is sometimes used as a food flavoring agent, e.g., in instant noodles. However, the essential oil is relatively expensive compared to essential oils of other herbs because of its limited production, which is partially due to low yield of only about 0.29% of the fresh herb weight [1]. Additionally the flavor of the steam-distilled or hydrodistilled essential oil is considered inferior to that of the fresh herb and has led to limited success of the essential oil in the marketplace.

There have been a number of reports on the essential oil composition of *O. sanctum* [1–9]; however, there are no reports on the characteristic aroma-active components of the fresh herb and distilled essential oil. Furthermore, no study has specifically addressed the volatile changes that occur during cooking of the herb or during manufacture of the distilled essential oil. The following investigation was conducted to identify compounds responsible for heat-induced aroma changes in holy basil herb.

II. ESSENTIAL OIL COMPOSITION OF *OCIMUM SANCTUM*

In Thailand there are three forms (green, purple, and hybrid) of *O. sanctum* available for commercial and culinary use. All three forms have been reported to belong to the same chemotype independently of color [5]. The published literature reports that *O. sanctum* contains a high proportion of eugenol (24%–83%), methyl eugenol (3%–46%), and sesquiterpenes [1–9]. In sesquiterpene-rich oils, β-caryophyllene is usually reported as the main constituent. In 2002 Modello and associates [6] reported on *O. sanctum* L. (green and purple) essential oils isolated by hydrodistillation (Clevenger-type apparatus, 2-hr extraction). Both the green and purple types contained eugenol as the major constituent, 41.7% and 77.5%, respectively, and only trace levels of methyl eugenol were found. β-Caryophyllene was the most abundant sesquiterpene in both essential oils.

In the present study, we evaluated the composition of a commercial sample (obtained from a local manufacturer in Bangkok, Thailand) of holy basil distilled essential oil. Volatile composition (based on percentage peak area) was determined by gas chromatography mass spectrometry (GCMS). After drying over anhydrous magnesium sulfate, oil was injected neat (1 µL) by cold split mode using a programmed temperature vaporizer (PTV) inlet [− 50°C (0.1 min initial time) to 240°C (10 min final time) at 12°C/sec, 1:200 split ratio] into an HP6890 GC/5973 GCMS system (Agilent Technologies, Inc., Palo Alto, CA) equipped with a Stabilwax DA [30 m × 0.25 mm inner diameter (i.d.) × 0.5-µm film, Restek Corp., Bellefonte, PA] capillary column. Helium was carrier gas at a constant linear velocity of 35 cm/sec. Oven temperature was programmed from 100°C to 200°C at 2°C/min and then ramped to 230°C at 4°C/min (30-min final hold time).

The volatile profile of the commercial holy basil essential oil sample is shown in Fig. 1. The composition is similar to that previously reported by Modello and colleagues [6] for *O. sanctum* (green and purple varieties) grown in Bangladesh. Major components (based on peak area percentage) were eugenol (36.7%), (*E*)-caryophyllene (30.4%), and β-elemene (14%). Minor components consisted mainly of monoterpene hydrocarbons (e.g., α-

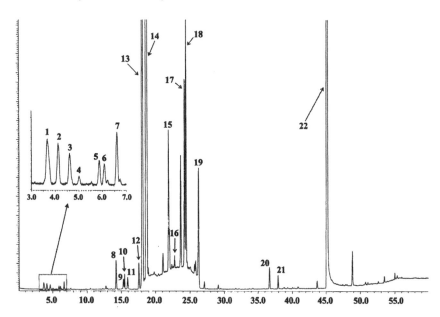

Figure 1   GCMS total ion chromatogram of a commercial hydrodistilled essential oil of *Ocimum sanctum*. L. Peak identifications: 1, α-pinene; 2, camphene; 3, β-pinene; 4, β-myrcene; 5, limonene; 6, 1,8-cineol; 7, (Z)-ocimene; 8, coapaene*; 9, β-bourbonene*; 10, linalool; 11, β-cubebene*; 12, (Z)-caryophyllene; 13, β-elemene; 14, (E)-caryophyllene; 15, α-humulene*; 16, borneol; 17, β-selinene*; 18, α-selinene*; 19, germacrene A*; 20, caryophyllene oxide; 21, methyl eugenol; 22, eugenol., * Compound tentatively identified on the basis of mass spectral database match. GCMS, gas chromatography mass spectrometry.

and β-pinene), sesquiterpene hydrocarbons, and various oxygenated compounds (e.g., methyl eugenol, 1,8-cineol, linalool, borneol, and caryophyllene oxide).

III. CHARACTERISTIC AROMA COMPONENTS OF *OCIMUM SANCTUM*

The predominant aroma-active components of the essential oil sample described and of fresh leaves of *O. sanctum* (purple type) obtained from a local market (Bangkok, Thailand) were compared by gas chromatography olfactometry and aroma extract dilution analysis. During the isolation of volatile constituents from the fresh herb, the influence of extraction technique on the composition must be carefully considered. It is often difficult to obtain representative aroma extracts from fresh plant material because of possible alteration of the volatile profile due to enzyme action (e.g., lipoxygenase). Figure 2 compares the volatile profiles obtained by dynamic headspace sampling (DHS) GC analysis of fresh (disrupted tissue) and water blanched leaves of *O. sanctum*. Lipoxygenase (LOX) action led to the generation of numerous volatile C6 and C9 aldehydes in the fresh herb, whereas these compounds were found in low abundance in the blanched herb. However, use of any heat treatment, such as blanching, would undoubtedly affect the aroma composition. Therefore, in order to reduce the formation of LOX-derived artifacts and eliminate the effect of heating of product, we employed a cold direct solvent extraction technique [10] followed by high-vacuum distillation cleanup of the extract [11]. A typical GCMS total ion chromatogram of a volatile extract prepared in this manner demonstrates the effectiveness of this procedure in minimizing LOX activity during volatile extraction (Fig. 3, note that the figure caption contains sample preparation and analysis details). Except for the higher proportion of methyl eugenol, the volatile composition of fresh herb was comparable to that of the distilled oil sample.

A. Comparison of Essential Oil and Fresh Herb

For AEDA, the essential oil sample mentioned previously was first diluted in dichloromethane (1:50 w/v), and then serial dilutions (1:3 v/v) were prepared in dichloromethane. Likewise, the aroma extract of the fresh herb underwent ternary (1:3 v/v) dilutions in dichloromethane. Aroma extract dilution analysis (AEDA) was conducted by two experienced panelists using a published procedure [11]. On-column injection was used during AEDA to minimize possible generation of thermally derived artifacts.

Despite their overall high abundance, the sesquiterpene hydrocarbons (*E*)-caryophyllene and *β*-elemene were not detected by GCO and AEDA. This indicates that these compounds may play either no role or only a minor role in the overall aroma of *O. sanctum*. Results of AEDA (Table 1) indicate that eugenol (*cloves*) is the predominant odorant in the essential oil, followed by *β*-damascenone (*floral, applesauce*), 2-isopropyl-3-methoxypyrazine (*earthy, potato*), and 1,8-cineol (*minty, eucalyptus*). The fact that eugenol and 1,8-cineol are key aroma components is not surprising and is supported by previous studies [1–9]. However, *β*-damascenone and 2-isopropyl-3-methoxypyrazine had not been previously reported as components of *O. sanctum*. The AEDA procedure was conducted on the fresh herb to determine whether these latter components are indeed natural constituents of the herb or whether they are generated during the production of the

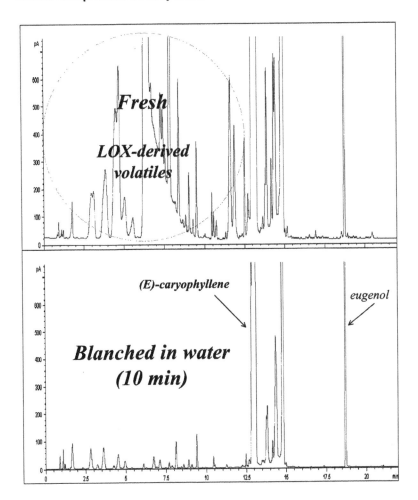

Figure 2 Dynamic headspace sampling GCMS comparison of fresh versus blanched (in water for 10 min) leaves of *O. sanctum*. Sample preparation: 1, fresh leaves (10 g) were blended in 50 mL of deodorized-distilled water using a hand-held mixer or, 2, fresh leaves (10 g) were blanched in 50 mL of boiling deodorized-distilled water and then the mixture was blended. For analysis, 10 g of mixture 1 or 2 was transferred to a 250-mL purge and trap apparatus and the volatiles purged at 23°C for 20 min with nitrogen (50 mL/min) onto a preconditioned Tenax TA trap (Tekmar-Dohrmann, Cincinnatti, OH, U.S.A.) (200 mg). Volatiles were thermally desorbed at 220°C (TDS, Gerstel, Germany) and analyzed by GC using a DB-FFAP (15 m × 0.53 mm i.d. × 1-μm film; J&W Scientific, Folsom, CA) capillary column and a flame ionization detector. GCMS, gas chromatrography mass spectrometry; LOX, lipoxygenase.

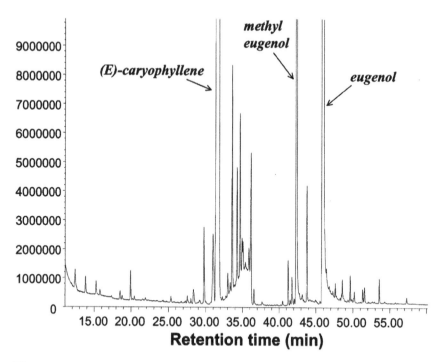

Figure 3 GCMS total ion chromatogram of volatile constituents isolated by cold direct solvent extraction of fresh leaves of *O. sanctum* L. (purple type). Leaves (10 g) plus 10 g of NaCl and 10 μL of internal standard solution (see text) were frozen and ground under liquid nitrogen in a 250-mL glass centrifuge bottle. Fifty milliliters of dichloromethane was added and the contents warmed to 0°C in an ice-water bath. Contents were blended using a hand-held mixer, flitered (No. 40 Whatman), and the filtrate subjected to high-vacuum distillation. Aroma extract was dried over anhydrous sodium sulfate and concentrated to 200 μL before analysis. Conditions for GCMS analysis were the same as previously described for the essential oil except that on-column injection and a DB-FFAP (30 m × 0.25 mm i.d. 0.25-μm film, J&W Scientific, Folsom, CA, U.S.A.) capillary column were used. GCMS, gas chromatography mass spectrometry.

oil. Results of AEDA (Table 1) showed that except β-damascenone, which was not detected in the aroma extract of the fresh herb, and linalool (*floral, honeysuckle*), which was markedly more intense in the fresh herb, the predominant odorants including 2-isopropyl-3-methoxypyrazine were similar for the two aroma extracts. The fresh herb contained additional minor aroma constituents that were not detected in the essential oil, such as lipoxygenase-derived compounds 1-octen-3-one and 2,6-nonadienal. On the basis of these findings we hypothesized that β-damascenone was likely a

Table 1 Aroma Extract Dilution Analysis Comparison of a Commercial Hydrodistilled Essential Oil and Fresh Leaves (Purple) of *Ocimum sanctum*

Compound	RI[a]		Odor description	$\log_3$ FD factor[b]	
	FFAP	DB-5		Oil[c]	Fresh[d]
2-/3-Methylbutanal[e]	929	652	*Malty, Chocolate*	1	2
α-Pinene[f]	1025	937	*Piney*	—[g]	1.5
Camphene[f]	1062	950	*Sweet, solvent*	—	1
1-Hexen-3-one[e]	1090	777	*Plastic*	—	1
Unknown	1189	808	*Malty, chocolate*	—	1
1,8-Cineol[f]	1208	1031	*Minty, eucalyptus*	2	3
Octanal[f]	1284	1006	*Orange oil*	—	2
1-Octen-3-one[f]	1305	980	*Mushroom*	—	1
2-Isopropyl-3-methoxypyrazine[e]	1433	1096	*Earthy, potato*	3	3.5
3-(Methylthio)propanal[e]	1451	905	*Potato*	2	2.5
Linalool[f]	1534	1103	*Floral, honeysuckle*	1	4
(E,Z)-2,6-Nonadienal[e]	1592	1154	*Cucumber*	—	1
Borneol[f]	1712	1190	*Soil, camphorous*	—	2
Unknown	1717	1250	*Catty, green, pungent*	—	2
β-Damascenone[e]	1824	1487	*Floral, applesauce*	3.5	—
Guaiacol[e]	1867	1085	*Smoky*	—	1.5
Eugenol[f]	2188	1354	*Spicy, cloves*	6	7
Unknown	2297	1704	*Woody, chicory*	1	—
Vanillin[f]	2566	1411	*Vanilla*	1	2

[a] Retention index based on GCO data; FFAP = DB-FFAP column (15 m × 0.32 mm i.d. × 0.25 μm film; J&W Scientific), DB-5 = DB-5 ms column (15 m × 0.32 mm i.d. × 0.5 μm film; J&W Scientific (Folsom, CA, U.S.A.)). RI, retention index; FD, flavor dilution; GCO gas chromatography olfactometry; DP-FFAP.

[b] Average $\log_3$ FD factor ($n = 2$).

[c] Commercial *O. sanctum* essential oil.

[d] Aroma extract prepared from fresh leaves of *O. sanctum* (purple) as described in Fig. 2 caption.

[e] Compound tentatively identified by comparison of RI values and odor properties with those of reference compound.

[f] Compound positively identified by comparison of RI values, odor values, and mass spectum with reference compound.

[g] No odor detected.

thermally generated compound originating from a glycoside precursor(s), during manufacture of the hydrodistilled essential oil. This is supported by results of a 2001 study in which it was demonstrated that glycosides precursors of β-damascenone are released/transformed during thermal processing of black tea [12]. In subsequent experiments we focused on verifying the thermal generation of β-damascenone by monitoring the aroma changes that occurred during the "controlled" cooking of the fresh leaves of *O. sanctum*.

IV. HEAT-INDUCED AROMA CHANGES IN *OCIMUM SANCTUM*

A. Experimental Approach

Ocimum sanctum plants (green type) were grown at the University of Illinois, Urbana, from seeds obtained from the Department of Horticulture, Kasetsart University, Bangkok, Thailand. Plants were harvested during July–August 2002. Fresh leaves were either immediately extracted or cooked (in water) just before extraction.

Cold direct solvent extraction of fresh leaves was conducted as previously described except that $10 \mu L$ of an internal standard solution (containing $11.2 \, mg/mL$ of *tert*-butylbenzene, $10.1 \, mg/mL$ of 2-undecanol, $10.5 \, mg/mL$ of tridecane, $11.3 \, mg/mL$ of 2-methoxy-4-propylphenol in methanol) was added to the sample before extraction.

The apparatus shown in Fig. 4 was used for cooking of fresh leaves. Ten grams of fresh leaves plus $50 \, mL$ of deodorized-distilled water were raised to a boil ($100°C$) and then refluxed while stirring ($\sim 30 \, rpm$) for 5 or 10 min. The mixture was immediately cooled in an ice-water bath, and internal standard solution ($10 \mu L$), $10 \, g$ of NaCl, and $50 \, mL$ of dichoromethane were added to the flask. The mixture was thoroughly blended using a hand-held mixer, shaken for 30 min, and the dichoromethane layer recovered and subjected to a high-vacuum distillation cleanup step as described earlier. Aroma extract was dried over anhydrous sodium sulfate and concentrated to $200 \mu L$ before analysis. Methods for GCMS and AEDA were the same as described earlier.

B. Aroma of Fresh Versus Cooked Herb

Under the cooking conditions used in the present study no noticeable losses due to evaporation were observed for the major volatile constituents (Fig. 5, Table 2). In fact, a slightly elevated level of eugenol was detected in the cooked herb. Occurrence of glycosidically bound eugenol in *O. sanctum* has been previously reported [7]. Therefore, the observed increase in eugenol could have been due to the hydrolysis of glycosidically bound eugenol during cooking.

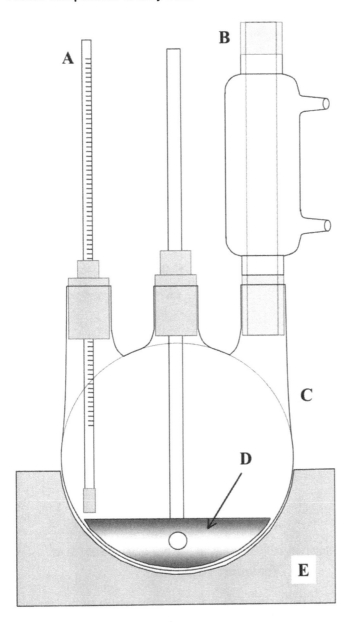

Figure 4 Apparatus used for water cooking (boiling) of fresh leaves. A, thermometer; B, condenser (0°C); C, three-neck round bottom flask; D, polytetra-fluoroethylene stirrer blade assembly (Kontes, Vineland, NJ, U.S.A.); E, heating mantle.

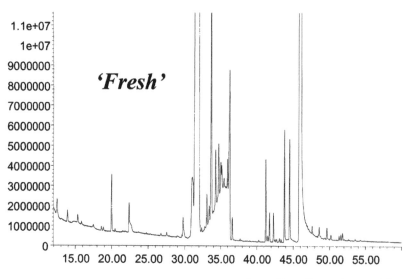

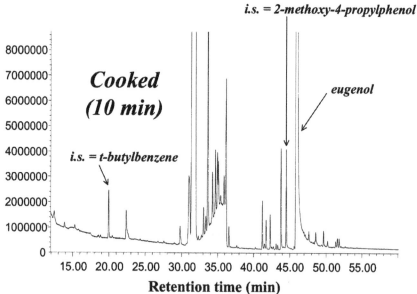

Figure 5 GCMS total ion chromatogram comparison of volatile extracts prepared from fresh versus cooked (10 min) leaves of *O. sanctum* (green type). GC analysis conditions were the same as described in Fig. 3 legend. GCMS, gas chromatography mass spectrometry.

Table 2 Relative Concentrations of Selected Volatile Constituents of Fresh and Cooked Leaves of *Ocimum sanctum* (Green Type)

Compound	Relative concentration, µg/g[a]		
	Fresh leaves	C-5	C-10
α-Pinene[b]	3.2	3.8	3.7
Camphene[b]	2.1	2.2	1.4
β-Pinene[b]	1.4	1.4	1.0
1,8-Cineol[c]	0.60	0.79	0.68
Linalool[c]	4.1	4.4	4.4
Borneol[c]	8.1	9.2	7.3
Methyl eugenol[d]	2.1	2.1	4.6
Eugenol[d]	297	392	436

[a] Average concentration ($n = 2$), fresh weight basis, relative to the internal standard (i.s.; C-5 and C-10 represent leaves cooked for 5 and 10 min, respectively.
[b] i.s., *tert*-butylbenzene.
[c] i.s., 2-undecanol.
[d] i.s., 2-methoxy-4-propylphenol.

Results of AEDA of fresh and cooked leaves of *O. sanctum* are given in Table 3. The predominant odorants eugenol, borneol, 1,8-cineol, linalool, 2-isopropyl-3-methoxypyrazine, and 3-(methylthio)propanal did not differ for the fresh and the cooked herb. Both also contained comparable levels of lipoxygenase-derived compounds (Z)-3-hexenal, (Z)-4-heptenal, 1-octen-3-one, (Z)-1,5-octadien-3-one, and (E,Z)-2,6-nonadienal. In agreement with our earlier findings, β-damascenone intensity increased as a result of cooking of the fresh herb. In addition, $\log_3$ flavor dilution factors for two unknown compounds having *skunky, pungent* ($RI_{FFAP} = 1107$) and *catty, green, pungent* ($RI_{FFAP} = 1718$) odors increased after cooking. The Maillard reaction products 2,3-butanedione and 2-acetyl-1-pyrroline were two other compounds found only in the cooked herb.

V. CONCLUSIONS

Results of this study indicated that eugenol, 1,8-cineol, linalool, and 2-isopropyl-3-methoxypyrazine are characteristic aroma compounds of the essential oil and fresh leaves of *O. sanctum*. Meanwhile, heating or cooking of fresh herb caused formation of additional thermally derived aroma compounds, such as β-damascenone and other thermally derived com-

Table 3 Aroma Extract Dilution Analysis Comparison of Fresh and Cooked Leaves of *Ocimum sanctum* (Green Type)

Compound	RI[a]		Odor description	Log₃ FD factor[b]		
	FFAP	DB-5		Fresh	C-5	C-10
Methylpropanal[c]	821	n.a.	*Malty, chocolate*	2	<1	<1
2-/3-Methylbutanal[c]	927	655	*Malty, chocolate*	3.5	2	3
2,3-Butanedione[d]	985	582	*Buttery*	—[e]	1.5	1.5
α-Pinene[d]	1025	937	*Piney*	<1	1.5	1.5
Unknown	1042	950	*Sweet, solvent*	2.5	2.5	2.5
1-Hexen-3-one[c]	1105	777	*Plastic*	1	1.5	1
Unknown	1107	766	*Skunky, pungent*	1	2	4
(Z)-3-Hexenal[d]	1147	801	*Green, cut-leaf*	2	1	1
Unknown	1188	808	*Malty, chocolate*	3.5	4	2.5
1,8-Cineol[d]	1200	1031	*Minty, eucalyptus*	5	5	5
(Z)-4-Heptenal[c]	1247	902	*Rancid*	<1	<1	<1
Octanal[d]	1286	1006	*Orange oil*	2.5	2.5	2.5
1-Octen-3-one	1299	980	*Mushroom*	2.5	4	3
2-Acetyl-1-pyrroline[c]	1330	920	*Popcorn, baked*	—	2.5	2
(Z)-1,5-Octadien-3-one[c]	1364	928	*Metallic*	3.5	4	3.5
2-Isopropyl-3-methoxypyrazine[c]	1427	1091	*Earthy, potato*	4.5	5	4.5
3-(Methylthio)propanal[c]	1451	907	*Potato*	5.5	5.5	5.5

Compound	RI	RI	Odor description			
Linalool[d]	1534	1103	*Floral, honeysuckle*	5	5	5
Phenylacetaldehyde[d]	1643	1051	*Rosy, plastic*	4.5	1.5	2
(E,Z)-2,6-nonadienal[c]	1592	1154	*Cucumber*	1	1.5	1
Borneol[d]	1697	1190	*Soil, comphorous*	3.5	5.5	5
Unknown	1718	1250	*Catty, green, pungent*	4.5	6	6
Unknown	1727	n.a.	*Hay, dry grass*	1	1.5	1.5
β-Damascenone[c]	1819	1487	*Floral, applesauce*	3	6.5	7
Guaiacol[c]	1859	1090	*Smoky*	<1	2	1
Eugenol[c]	2166	1354	*Spicy, cloves*	7.5	8	7.5
o-Aminoacetophenone[c]	2297	1704	*Grape, corn tortilla*	2.5	3	2.5
Indole[c]	2450	1291	*Fecal, mothballs*	2.5	1.5	1.5
Vanillin[d]	2566	1411	*Vanilla*	1.5	3	1.5

[a] Retention index based on GCO data; FFAP = Stabilwax DA column (15 m × 0.32 mm i.d. × 0.5 μm film, Restek Corp., Bellefonte, PA, U.S.A.), DB-5 = DB-5ms column (15 m × 0.32 mm i.d. × 0.5 μm film, J&W Scientific, Folsom, CA, U.S.A.), GCO, gas chromatography olfactometry; DB, FFAP, FD, flavor dilution; RI, retention index.

[b] Average $\log_3$ FD factor (n = 2), C-5 and C-10 represent leaves cooked for 5 and 10 min, respectively.

[c] Compound tentatively identified by comparison of RI values and odor properties with those of reference compound.

[d] Compound positively identified by comparison of RI values, odor values, and mass spectum with reference compound.

[e] No odor detected.

pounds, which imparted atypical cooked notes to the herb. No dramatic change due to cooking, especially any decrease, was observed for the major volatile constituents of the herb. Results of our study may be useful to anyone considering the use of the hydrodistilled essential oil in place of the fresh herb or possibly the use of the fresh herb in a thermally processed food product.

REFERENCES

1. D Choochoat, Chemical Composition of Essential Oils from Thai Lamiaceous Plants. MS Thesis, Chulalongkorn University, Thailand, 1998.
2. I Laakso, T Seppänen-Laakso, B Herrmann-Wolf, N Kühnel, K Knobloch. Constituents of the essential oil from the holy basil or tulsi plant, *Ocimum sanctum*. Planta Med 56:527, 1990.
3. MI Lacerda Machado, MC de Vasconcelos Silva, FJ Abreu Matos, AA Craverio, JW Alencar. Volatile constituents from leaves and inflorescence oil of *Ocimum tenuiflorum* L.f. (syn *O. sanctum* L.) grown in northeastern Brazil. J Essent Oil Res 11:324–326, 1999.
4. S Laskar, S Gosh Majumdar. Variation of major constituents of essential oil of the leaves of *Ocimum sanctum* Linn. J Indian Chem Soc 65: 301–302, 1988.
5. ML Maheshwari, BM Singh, R Gupta, M Chien. Essential oil of sacred basil (*Ocimum sanctum*). Indian Perfumer 31:137–145, 1987.
6. L Modello, G Zappia, A Cotroneo, I Bonaccorsi, JU Chowdhuri, M Yusuf, G Dugo. Studies on the essential oil-bearing plants of Bangladesh. Part VIII. Composition of some *Ocimum* oil *O. basilicum* L. var. *purpurascens*; *O. sanctum* L. green; *O. sanctum* L. purple; *O. americanum* L., citral type; *O. americanum* L., camphor type. Flavour Fragr J 17:335–340, 2002.
7. H Nörr, H Wagner. New constituents from *Ocimum sanctum*. Planta Med 58:574, 1992.
8. A Pino, A Rosado, M Rodriquez, D Garcia. Composition of the essential oil of *Ocimum tenuiflorum* L. grown in Cuba. J Essent Oil Res 10:437–438, 1998.
9. PM Raju, M Ali, A Velasco-Negueruela, MJ Perez-Alonso. Volatile constituents of the leaves of *Ocimum sanctum* L. J Essent Oil Res 11:159, 1998.
10. KR Cadwallader, R Surakarnkul, SP Yang, TE Webb. Character-impact aroma components of coriander (*Coriandrum sativum* L.) herb. In F. Shahidi, C.-T. Ho, ed. Flavor Chemistry of Ethnic Foods. New York: Kluwer Academic/Plenum Publishers, 1999, pp. 77–84.
11. Q-X Zhou, CL Wintersteen, KR Cadwallader. Identification and quantification of aroma-active components that contribute to the distinct malty flavor of buckwheat honey. J Agric Food Chem 50:2016–2021, 2002.
12. K Kumazawa, H Masuda. Changes in the flavor of black tea drink during heat processing. J Agric Food Chem 49:3304–3309, 2001.

16

Characterization of Flavor Compounds During Grinding of Roasted Coffee Beans

Masayuki Akiyama, Kazuya Murakami, Noboru Ohtani, Keiji Iwatsuki, and Kazuyoshi Sotoyama
Morinaga Milk Industry Co., Ltd., Zama, Kanagawa, Japan

Akira Wada, Katsuya Tokuno, Hisakatsu Iwabuchi, and Kiyofumi Tanaka
San-Ei Gen F.F.I., Inc., Toyonaka, Osaka, Japan

I. INTRODUCTION

Numerous volatile compounds of coffee have been found in brewed and ground coffee by many studies [1]. In the 1990s, gas chromatography olfactometry (GCO) was applied to determine which of these volatile compounds are the most potent compounds and likely to contribute to the characteristic aroma of brewed and ground coffee [2,3]. Some studies of the headspace volatile compounds arising from roasted ground coffee used a gastight syringe to sample and GCO to analyze [4]; the others used a static headspace sampler to investigate the effects of time and temperature on the volatile compounds released from ground roasted Arabica coffee [5]. In 1999 and 2001, the influence of origin and roast degree on 28 potent odorants in Arabica coffee beans using stable isotope dilution assays [6] and the changes in volatile composition above that of ground coffee sampled by collecting the headspace on Tenax® traps from a novel sampling apparatus [7] were reported. Solid-phase microextraction (SPME) [8], involving both exposure to the gas phase above a sample and submersion in the liquid phase of a liquid sample, has been applied to the flavor analyses of both brewed and ground coffee under static, no-gas-flow conditions [9,10].

The pleasant aroma arising from roasted coffee beans during grinding is as attractive as the aroma of fresh brewed coffee for coffee flavored prepared foods. However, because these pleasant aromas are especially highly volatile and unstable compounds, these are easily lost during the industrialized processing and storage of coffee products such as beverages. Therefore, the study of potent odorants released during the grinding of roasted coffee beans could help food chemists and flavorists make new and more desirable coffee flavors for prepared foods. This chapter reports the development of a new SPME-based dynamic headspace sampling method useful for the investigation of volatile compounds arising from grinding various types of roasted Arabica coffee beans (Ethiopia, Tanzania, and Indonesia) and their analytical results and aroma profiles obtained by gas chromatography mass spectrometry (GCMS) and gas chromatography olfactometry (GCO) using this sampling method. Also, aroma characterizations of each roasted coffee beans achieved by applying principal component analysis (PCA) to the results obtained by dynamic SPME GCO are reported.

II. MATERIALS AND METHODS

A. Samples and Chemicals

Arabica coffee beans (Mocha, Kilimanjaro, and Mandheling) originating from Ethiopia, Tanzania, and Indonesia were roasted using a Probat G-12 roaster (Emmerich, Germany), packed in 1-kg portions, and stored at $-20\,°C$ until used. Degree of roasting was represented as an L value. The L value was determined by measuring ground roasted coffee (particle sizes $<500\,\mu m$) using a color difference meter (Nippon Denshoku, Tokyo, Japan). All roasted coffee beans (L26, L23, and L18) were held for 2 hr to reach room temperature before grinding. Reference standards were purchased from Tokyo Kasei Kogyo Co., Ltd. (Tokyo, Japan), CTC Organics (Atlanta, GA), and Oxford Chemicals Ltd. (Hartlepool, England).

B. Solid-Phase Microextraction Device

The solid-phase microextraction device (SPME) was purchased from Supelco Co. (Bellefonte, PA). The following types of SPME fiber were used: polydimethylsiloxane (PDMS) with 100-μm thickness, PDMS/divinylbenzene (DVB) with 65-μm thickness, and Carboxen™ (Supelco Co., Bellefonte, PA, U.S.A.)/PDMS (CAR/PDMS) with 75-μm thickness.

Unless otherwise noted, PDMS/DVB fiber was used for sampling headspace volatile compounds.

C. Gas Chromatography Mass Spectrometry and Gas Chromatography Olfactometry Parameters

Gas chromatography mass spectrometry analyses were performed on an Hewlett-Packard (HP) 5973 (Palo Alto, CA, U.S.A.) with a fused silica capillary column DB-WAX (60 m × 0.25 mm, 0.25-μm film thickness). The temperature program was set at an initial 50°C for 2 min, followed by an increase of 3°C/min to 220°C and held at 220°C for 20 min. The injection port equipped with an 0.75-mm i.d. liner (Supelco Co.) was maintained at 250°C and the injection purge on the GC was off for the initial 1 min. GCO (CharmAnalysis^{TM}) was conducted in triplicate on an HP6890 GC modified by DATU, Inc. (Geneva, NY) [11]. A fused silica capillary column DB-WAX (15 m × 0.32 mm, 0.25-μm film thickness) was used. The flow of the carrier gas helium was 3.2 mL/min. The temperature program was set at an initial temperature of 40°C, followed by an increase of 6°C/min to 230°C and held at 230°C for 20 min. The injection and detector ports were maintained at 250°C, and the injection purge on the GC was off for the initial 1 min. The retention time of each compound was converted to Kovats indices using C_6–C_{28} n-alkanes.

D. Identification of the Volatile Compounds

Volatile compounds were identified by comparing their mass spectra and Kovats indices using C_6–C_{28} n-alkanes to those of standard compounds and to those from the literature [12]. Also, identifications of some potent odorants found only by the GCO analyses were made by comparing their Kovats indices and aroma properties to those of standard compounds and to those from the literature [2].

E. Apparatus Used for Dynamic Headspace Sampling

The sampling apparatus (5L, Fig. 1) was designed and built for analyses. The electronic grinder (Model CG-4B, Melitta Japan, Tokyo) was purchased from the commercial market and modified for continuous grinding.

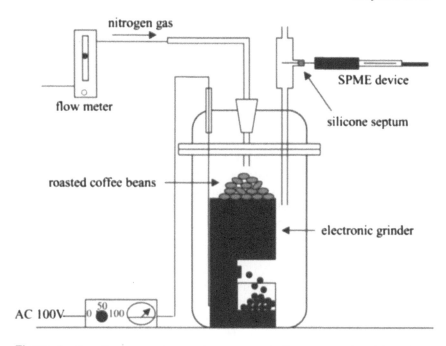

Figure 1 Sampling apparatus used to trap volatile compounds arising from roasted coffee beans during grinding.

F. Isolation of Headspace Volatile Compounds from Samples

1. General Solid-Phase Microextraction Procedure

Before headspace sampling, the SPME fiber was reconditioned according to the recommendations (Data sheet T7941231, Supelco Co.) in the GC injection port. After sampling, the fiber was placed into the injection port of the GCMS or GCO and thermally desorbed for 10 min at 250°C. Each SPME sampling was carried out at room temperature (25°C) and conducted in triplicate.

2. General Static Headspace Solid-Phase Microextraction Sampling

Ground Ethiopia coffee (L23, particle sizes 400–800 μm) suitable for the so-called paper drip mode was hermetically stored in polyester/aluminum/polyethylene package at − 20°C after sealing under vacuum and used within 24 hr for headspace sampling. Before headspace sampling, the package containing ground coffee was allowed to stand for 2 hr at room temperature. The headspace volatile compounds were collected by using three fibers, PDMS, PDMS/DVB, and Carboxen/PDMS, under the same static condition used for the choice of fiber suitable for coffee volatile sampling. Ground coffee (200 g) was transferred into a 1-L round glass bottle sealed with a Teflon™-coated silicone cap, where it equilibrated for 1.5 hr at room temperature (25°C) before 8 min of SPME headspace sampling. After sampling, the SPME device was treated as described.

3. General Dynamic Headspace Solid-Phase Microextraction Sampling

Roasted coffee beans were placed on the hopper of the electronic grinder in the sampling apparatus (Fig. 1). Grinding speed of the electronic grinder in the sampling apparatus was controlled by varying the voltage to obtain ground coffee. After inserting the stainless steel housing of the SPME device into the sampling port, the nitrogen gas was passed into the glass vessel. The SPME fiber was pushed out of its stainless steel housing immediately after starting the grinding and exposed to the effluent gas.

4. Sampling Parameters for Dynamic Headspace Solid-Phase Microextraction Sampling

Ethiopia coffee beans (L23) were used to determine suitable sampling parameters for headspace volatile compounds arising from roasted coffee beans during grinding. To determine the nitrogen gas flow rate, the actual air volume rate during sniffing was measured by using the Digital Flow Meter (Agilent Technologies, Inc., Wilmington, DE, U.S.A.) ($n = 5$). The GCMS analyses were conducted on the headspace volatile compounds of roasted coffee beans (150 g) adsorbed on the SPME fiber under different nitrogen gas flow rates (200, 400, 600, 800, and 1000 mL/min) for 8 min to investigate the effect of effluent gas flow rate, and to those of roasted coffee beans (220 g) adsorbed on the SPME fiber with different exposure times (1, 2, 4, 8, and 12 min) at the same gas flow rate (600 mL/min) to determine appropriate sampling time.

G. Dilution Analysis

The SPME fiber was fully exposed (1.0 cm), approximately half exposed (0.5 cm), approximately one-fourth exposed (0.25 cm), and approximately one-eighth exposed (0.12 cm) to the headspace gas (SH-SPME sampling) and the nitrogen gas stream (DH-SPME), respectively. Exposure length was controlled by creating three additional notches in the SPME holder [13].

H. Gas Chromatography Olfactometry Evaluation of Samples

Aroma activities of volatile compounds obtained by GCO dilution analyses were represented as Charm values, and the relative intensities of component odorants were represented in terms of the odor spectrum value (OSV) [14]. Each Charm value was rounded off to two significant figures in order to reflect the actual resolution of the dilution analysis. Acidic, buttery-oily, green-black currant, green-earthy, nutty-roast, phenolic, smoke-roast, soy sauce, sweet-caramel, and sweet-fruity were the aroma descriptions used in all GCO experiments to describe potent odorants. These descriptions were chosen from the results of a single preliminary free choice GCO analysis using a lexicon of words commonly used for coffee evaluation.

I. Multivariate Analysis Using Dynamic Headspace Solid-Phase Microextraction Gas Chromatography Olfactometry

Principal component analysis (PCA) was carried out by using an SPSS program. Thirty-five odorants with an OSV of 50 or above, calculated by using mean Charm values of each odorant, were selected from the volatile compounds of the aroma released during grinding and categorized into 10 aroma descriptions; the total of Charm values of each description was applied to the PCA.

III. RESULTS AND DISCUSSION

A. Parameters for Dynamic Headspace Solid-Phase Microextraction Sampling

The SPME technique has been developed by Arthur and Pawliszyn [8] and now is well known as a simple, rapid, and sensitive sampling method for liquid or gaseous volatile samples. However, the analytes adsorbed on the SPME fiber generally depend on their polarities and the SPME fiber affinity. Because coffee aroma consists of compounds having many kinds of functional groups, proper selection of the type of fiber to use is important to obtaining accurate and reproducible results [15]. In 2000, in a comparison of three fibers (PDMS, PDMS/DVB, and Carboxen/PDMS) for brewed coffee headspace volatile compounds, it was reported that PDMS/DVB gave the overall best sensitivity, especially for phenols such as 2-methoxyphenol, 4-ethyl-2-methoxyphenol, 4-ethenyl-2-methoxy-phenol, and polar compounds such as 4-hydroxy-2,5-dimethyl-3(2H)-

furanone, and 3-hydroxy-4,5-dimethyl-2(5*H*)-furanone [10]. Therefore, the GCMS analyses were conducted to investigate volatile compounds adsorbed on three fibers (PDMS, PDMS/DVB, Carboxen/PDMS) under the static, no-gas-flow, "equilibrium" condition using ground Ethiopia coffee (L23). Typical total ion chromatograms (TICs) of volatile compounds adsorbed by the three SPME fibers are shown in Fig. 2 and the compounds identified listed in Table 1. The Carboxen/PDMS fiber was the most sensitive to small molecules such as 2-methylfuran, 2- and 3-methylbutanals, 2,3-butanedione, pyrrole, and pyridine, but its sensitivity to molecules such as 2-methoxyphenol, 4-ethenyl-2-mehoxyphenol, and 4-hydroxy-2,5-dimethyl-3(2*H*)-furanone was lower than that of the other two fiber coatings. Also, considerable peak tailings of the small molecules mentioned previously were observed. The traditional nonpolar fiber, PDMS, is known to have good stability and has many applications for flavor analysis. However, the PDMS fiber showed the lowest sensitivity to coffee volatiles. The more polar volatile compounds mentioned earlier

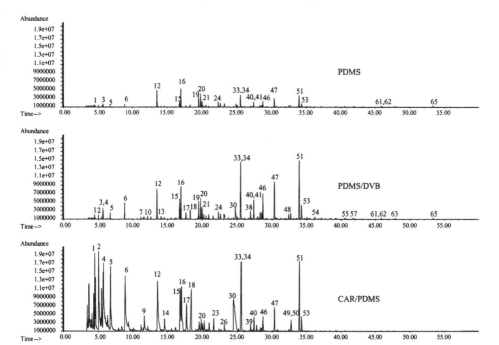

Figure 2 Typical total ion chromatograms of the static headspace volatiles of ground roasted Ethiopia coffee beans (roast degree; L23) obtained by three different solid-phase microextraction fibers. PDMS, polydimethyl siloxane; DVB, divinylbenzene; CAR, carboxen.

Table 1 Volatile Compounds Found in the Headspace of Roasted Ethiopian Coffee Beans (L23) Under Static and Dynamic Conditions

No.	Compound	Peak area × 10⁵ Static	Peak area × 10⁵ Dynamic
1	2-Methylpropanal	56.49	91.62
2	2-Methylfuran	35.89	73.68
3	2-Methylbutanal	113.8	170.76
4	3-Methylbutanal	95.41	134.33
5	2,3-butanedione	71.41	119.39
6	2,3-Pentanedione	407.57	448.84
7	2,3-Hexanedione	62.12	59.92
8	3,4-Hexanedione	50.86	44.43
9	1-Methyl-1H-pyrrole	65.64	60.55
10	2-Vinyl-5-methylfuran	95.19	80.95
11	Myrcene	66.61	136.17
12	Pyridine	992.14	1280.28
13	Limonene	88.38	144.96
14	Pyrazine	63.36	63.36
15	Dihydro-2-methyl-3(2H)-furanone	986.83	787.95
35	2-Ethyl-3,5-dimethylpyrazine	4.92	36.43
36	Cis-linalool oxide (furanoid)	13.74	80.05
37	2-((Methylthio)methyl)furan	22.5	65.88
38,39[a]	2-Methyl-5-vinylpyrazine, furfuryl formate	201.33	355.29
40,41[a]	2-Acetylfuran, 2,5-dimethyl-3(2H)-furanone	521.55	1023.69
42	Pyrrole	38.67	50.62
43	1-(2-Furyl)-2-propanone	62.24	177.85
44,45[a]	2,4,5-Trimethyl-3(2H)-furanone, 1-(Acetyloxy)-2-butanone	184.42	569.52
46	Furfuryl acetate	588.23	1638.28
47	5-Methyl-2-furancarboxaldehyde	1065.89	2599.84
48	1-Methyl-1H-pyrrole-2-carboxaldehyde	133.53	360.16
49,50[a]	Acetylpyrazine, dihydro-2(3H)-furanone	226.96	586.77
51	2-Furanmethanol	2019.36	4213.11
52	3-Mercapto-3-methyl-1-butanol	11.29	88.5

No.			
16	2-Methylpyrazine	1184.87	1118.24
17	3-Hydroxy-2-butanone	284.97	255.93
18	1-Hydroxy-2-propanone	526.85	406.03
19	2,5-Dimethylpyrazine	470.14	844.05
20	2,6-Dimethylpyrazine	583.66	973.1
21	2-Ethylpyrazine	288.1	442.43
22	2,3-Dimethylpyrazine	81.77	152.99
23	1-Hydroxy-2-butanone	146.38	117.3
24	2-Ethyl-6-methylpyrazine	164.96	424.71
25	2-Ethyl-5-methylpyrazine	146.47	392.74
26,27[a]	2-Methyl-3(2H)-furanone, trimethylpyrazine	219.73	617.36
28,29[a]	5-Methyl-2(3H)-furanone, 2-furanmethanethiol	23.36	76.06
30,31,32[a]	Acetic acid, trans-linalool oxide (furanoid), 3-ethyl-2,5-dimethylpyrazine	815.2	889.23
33,34[a]	1-(Acetyloxy)-2-propanone, 2-furancarboxaldehyde	1994.4	3223.88

No.			
53	3-Methylbutyric acid	441.6	1316.8
54	1-(1H-pyrrol-1-yl)-1-propanone[b]	56.26	199.6
55	1-(2-Furanylmethyl)-1H-pyrrole	20.22	112.65
56	2-Hydroxy-3-methyl-2-cyclopenten-1-one	20.68	148.75
57	2-Methoxyphenol	10.16	71.8
58	2,3-Dihydro-5-hydroxy-6-methyl-4H-pyran-4-one	6.53	84.07
59	1-(5-Methyl-2-furanyl)-1,2-propanedione	10.4	49.63
60	3-Ethyl-2-hydroxy-2-cyclopenten-1-one	4.27	37.07
61,62[a]	1-(1H-pyrrol-2-yl)-ethanone, 3-Hydroxy-2-methyl-4H-pyran-4-one	20.41	183.45
63	1H-pyrrole-2-carboxaldehyde	26.96	97.97
64	4-Hydroxy-2,5-dimethyl-3(2H)-furanone	4.1	73.6
65	4-Ethenyl-2-methoxyphenol	22.43[b]	92.85
66	1-(2-Furanylmethyl)-1H-pyrrole-2-carboxaldehyde[b]	trace	13.14

[a] Volatile compounds detected as overlapping peaks on TIC.
[b] Identifications were achieved by comparing mass spectra and retention indices with data from Ref. 12.

were adsorbed by the PDMS/DVB fiber. From these analytical results, PDMS/DVB fiber was selected for the DH-SPME sampling during the grinding of roasted coffee beans.

The amount of volatile compounds adsorbed on the SPME fiber was assumed to be dependent on the nitrogen gas flow rate [16]. The airflow rate during inhalation through the external nares has been reported to be 100 mL/sec [17], but the average of the actual measurement of the airflow rate of an adult quiet sniff using the digital flow meter was about 665 mL/min. Then, headspace volatile compounds released during the grinding of Ethiopia coffee bean (L23) were adsorbed on the SPME fiber (PDMS/DVB) using our new apparatus at 200, 400, 600, 800, and 1000 mL/min, respectively. The GCMS analyses showed similar gas chromatograms, and the coefficients of variation of the total peak area of the 47 volatile compounds (Table 1) separated on TIC were less than 5%. In the case of the small molecules, such as 2-methylfuran, 2- and 3-methylbutanals, 2,3-butanedione, and 2,3-pentanedione, each peak area decreased gradually with increasing nitrogen gas flow rate, and each peak area of the relative polar molecules, such as 1-(2-furanylmethyl)-1H-pyrrole, 2-hydroxy-3-methyl-2-cyclopenten-1-one, 1H-pyrrole-2-carboxaldehyde, 4-hydroxy-2,5-dimethyl-3(2H)-furanone, and 4-ethenyl-2-methoxyphenol, increased gradually with increasing nitrogen gas flow rate, but their coefficients of variation were less than 12%.

These results showed that the adsorption of volatile compounds by the fiber depended little on the nitrogen gas flow rate within the 200- to 1000-mL/min range. This was attributed to the existence of abundant volatile compounds released from the surface of a large number of coffee grains. On the basis of these results, the operating nitrogen gas flow rate was determined to be 600 mL/min.

To determine appropriate exposure time, the SPME fiber (PDMS/DVB) was exposed to nitrogen gas discharged from the apparatus at 1, 2, 4, 8, and 12 min, respectively, during the grinding of Ethiopia coffee beans (220 g, L23). Average total peak area ($n = 3$) of the 47 volatile compounds reached an almost saturated level within 12 min, and total peak area reached about 97% of the almost saturated level after 8 min, a period that corresponds to the time for finishing grinding 150 g of roasted coffee beans.

According to these results, the DH-SPME sampling parameters were determined as follows: SPME fiber, PDMS/DVB; roasted coffee beans used, 150 g; nitrogen gas flow rate, 600 mL/min; SPME sampling time; 8 min. The GCMS analyses conducted under these conditions showed that the coefficients of variation of peak area of the 47 volatile compounds, which corresponded to about 68% of total peak area detected on GCMS analysis, were less than 5% [18].

B. Gas Chromatography Mass Spectrometry Analyses of Volatile Compounds of Roasted Coffee Beans

1. Comparison of Static Headspace and Dynamic Headspace Solid-Phase Microextraction Samplings

The TICs of the headspace volatiles of Ethiopia coffee bean (L23) adsorbed on the SPME fiber under static and dynamic conditions under the same GCMS conditions are shown in Fig. 3. The DH-SPME sampling resulted in acceptable peak intensity and a chromatographic profile, and the ratios of components adsorbed were different from those obtained by the SH-SPME sampling. The amount of peak area of 47 volatile compounds trapped under the dynamic condition was about 1.8 times that of volatiles trapped under the static condition (Table 1). Also, minor compounds reported as potent odorants of coffee, such as 4-hydroxy-2,5-dimethyl-3(2H)-furanone, 2-methoxyphenol, 4-ethenyl-2-methoxyphenol, and 2-ethyl-3,5-dimethylpyrazine, were found to be present in higher concentrations in the dynamic headspace than in the static headspace. This chromatographic difference was considered to have resulted from additional volatilization of compounds induced by flowing inert gas above the sample [10].

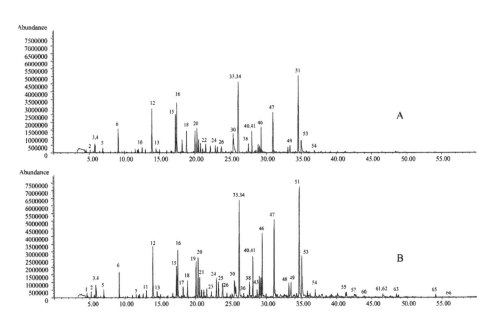

Figure 3 Typical total ion chromatograms of the headspace volatiles of roasted Ethiopia coffee beans obtained by using static (A) and dynamic (B) solid-phase microextraction sampling.

2. Gas Chromatography Mass Spectrometry Analyses of Volatile Compounds in Arabica Coffee Beans with Different Roast Degrees

The DH-SPME sampling method was applied to different roast degrees of different Arabica coffee beans, including Ethiopia coffee beans (L23), to investigate changes in the contents of volatile compounds. Compounds identified and their peak areas under the same GCMS conditions are shown in Table 2.

In 1999, Mayer and associates reported the influence of origin and roast degree on the composition of 28 potent odorants in the headspace or extracts of Arabica coffees using stable isotope dilution assays [6]. Comparing our results with those from reported data, some similarities and differences in aroma changes with roasting and coffee bean origins were found as follows: Aliphatic saturated aldehydes are 2- and 3-methylbutanals, whose aroma properties are reported to be pungent, malty, fermented, and sweaty [2]. These saturated aldehydes decreased gradually beyond the medium roasting range, especially in the case of Ethiopia and Indonesia coffee beans, and the changes of their peak area with roasting were similar to results reported for Tipica from Colombia. On the other hand, the peak area changes of these aldehydes with roasting in Tanzania coffee beans were similar to results reported for Colombia and Kenya coffee beans. The concentrations in coffee beans of 2,3-butanedione and 2,3-pentanedione have been reported to be affected by the differences in coffee bean origins. In our experiments, the peak area changes of 2,3-pentanedione exhibited significant differences between the different regional coffee beans we used, especially in the medium roast. Alkyl pyrazines are predominant volatile compounds in roasted coffee beans. Peak area changes of alkyl pyrazines contained in Tanzania and Indonesia coffee beans were similar: that is, major alkyl pyrazines such as 2-methylpyrazine and 2,5- and 2,6-dimethylpyrazines tended to decrease gradually with roasting, but those of other alkyl pyrazines were largely unchanged. On the other hand, peak area changes of major alkyl pyrazines contained in Ethiopia coffee beans were significant: that is, these pyrazines showed their maximal peak area at the vicinity of roast degree L23 and decreased significantly with roasting. 4-Hydroxy-2,5-dimethyl-3(2*H*)-furanone is known as an aroma-active compound possessing sweet and caramel-like aromas. The peak area change of this furanone with roasting in Indonesia coffee beans was similar to results reported for Colombia and Kenya coffee beans. Mayer and coworkers reported that the concentrations of 2-methoxyphenol, 4-ethyl-2-methoxyphenol, and 4-ethenyl-2-methoxyphenol increased with roasting. As shown in Table 2, the peak area of 2-methoxyphenol and 4-ethenyl-2-methoxyphenol increased gradually with roasting in all the roasted coffee bean types used in our study but decreased beyond the medium roasting range in all the roasted coffee bean types used in our experiments. Also, the concentration of 4-ethenyl-2-methoxyphenol was reported to be affected by the differences in coffee bean origins. This phenol was contained in larger amounts in Indonesia coffee beans than in Ethiopia and Tanzania coffee beans at each roast degree we tested, and it seemed to be one of the key compounds indicating regional differences of Arabica coffee beans. Major furan derivatives are 2-furancarboxaldehyde, furfuryl acetate, 5-methyl-2-furan-carboxaldehyde, and 2-furanmethanol. 2-Furancarboxaldehyde tended to decrease gradually with roasting, and 5-methyl-2-furancarboxaldehyde showed a maximal peak area at the vicinity of roast degree L23 in all coffee beans. 2-Furanmethanol and furfuryl acetate containing Tanzania and Indonesia coffee beans tended to increase gradually with roasting, but the latter two furan derivatives in Ethiopia coffee beans showed different behavior with roasting: that is, reached their maximal peak area at the vicinity of roast degree L23. The content changes with roasting found in coffee beans of different origins were very complicated, but, in general, that of Ethiopia coffee beans was significant, and those of Tanzania and Indonesia coffee beans showed similar tendencies.

C. Gas Chromatography Olfactometry Evaluation of Volatile Compounds of Roasted Coffee Beans

1. Comparison of Static Headspace and Dynamic Headspace Solid-Phase Microextraction Samplings

Aroma chromatograms of volatile compounds of Ethiopia coffee beans (roast degree, L23) using SH- and DH-SPME samplings are shown in Fig. 4. Charm values and odor spectrum values (OSVs) [14] of 20 potent odorants with an OSV of 50 or greater detected in the headspace under dynamic and static conditions are listed in Table 3. The totals of the Charm values including these 20 potent odorants detected in the headspace under dynamic and static conditions are shown in Fig. 5. The total of the Charm values for these 20 potent odorants comprised 72% (dynamic condition) and 88% (static condition) of the total of Charm values, respectively. OSVs are independent of concentration and approximate the relative intensity of odorants by accounting for the exponential nature of olfactory psychophysics; Charm values indicate the true odor activity measurement (potency) and are a linear function of concentration [14]. Compared with those of the static SPME headspace sampling, the total of Charm values of the dynamic SPME sampling indicated higher values in all aroma

Table 2 Volatile Compounds Found in the Headspace of Roasted Coffee Beans (L26, 23, 18) of Three Different Origins

No.	Compound	Ethiopia			Tanzania			Indonesia		
		L26	L23	L18	L26	L23	L18	L26	L23	L18
1	2-Methylpropanal	73.36	85.78	29.21	86.87	72.90	95.12	108.37	163.07	75.65
2	2-Methylfuran	72.28	81.28	134.46	73.49	105.67	238.60	94.46	173.05	284.65
3	2-Methylbutanal	261.53	261.14	198.25	274.31	271.56	297.48	448.01	466.80	357.24
4	3-Methylbutanal	186.10	204.31	180.14	207.33	208.50	235.25	283.54	346.22	283.86
5	2,3-Butanedione	177.12	166.32	108.89	183.45	171.51	152.49	130.98	187.50	140.63
6	2,3-Pentanedione	191.21	480.02	177.24	370.22	398.89	220.21	510.21	577.55	260.99
7	2,3-Hexanedione	66.12	101.47	78.27	89.73	104.14	128.88	106.44	137.74	135.73
8	3,4-Hexanedione	44.13	63.55	26.31	48.24	55.41	57.23	68.97	78.62	53.62
9	1-Methyl-1*H*-pyrrole	66.30	86.30	93.22	81.00	105.89	184.44	69.18	134.70	197.23
10	2-Vinyl-5-methylfuran	89.78	115.94	62.04	93.90	102.93	114.12	119.87	146.53	109.87
11	Myrcene	201.27	255.78	63.81	39.81	37.77	27.30	98.46	69.63	47.52
12	Pyridine	1179.62	1479.81	1732.85	1159.12	1788.33	3134.59	1252.44	2031.23	3292.64
13	Limonene	210.77	388.95	101.71	73.78	113.24	109.81	126.69	188.04	138.71
14	Pyrazine	94.64	115.48	47.41	99.78	104.31	108.54	107.89	115.69	113.91
15	Dihydro-2-methyl-3(2*H*)-furanone	960.30	1162.19	491.27	976.35	1016.32	1023.91	1350.04	1244.44	995.58
16	2-Methylpyrazine	1927.16	1910.64	918.47	1895.49	1893.97	1795.56	2283.22	2186.17	1826.19
17	3-Hydroxy-2-butanone	254.75	368.15	132.98	296.48	337.95	335.28	309.89	365.23	276.48
18	1-Hydroxy-2-propanone	910.94	817.02	171.70	958.54	731.90	411.24	733.53	585.36	354.04
19	2,5-Dimethylpyrazine	1687.51	1673.66	617.93	1528.52	1432.21	1220.88	2113.14	1732.95	1335.14
20	2,6-Dimethylpyrazine	1748.37	1828.21	757.93	1642.58	1603.94	1464.43	2339.34	2059.91	1742.51
21	2-Ethylpyrazine	769.92	815.72	312.83	775.14	736.84	654.58	1014.59	904.38	739.69

Peak area, × 10^5

22	2,5-Dimethylpyrazine	304.65	326.89	170.39	326.81	325.52	360.82	452.54	442.97	457.61
23	1-Hydroxy-2-butanone	187.67	185.05	37.96	221.44	201.60	104.21	180.93	158.68	57.38
24	2-Ethyl-6-methylpyrazine	847.63	883.25	304.52	892.10	846.40	748.29	1238.12	1159.29	752.80
25	2-Ethyl-5-methylpyrazine	617.08	757.66	308.59	657.41	628.58	655.62	894.42	805.57	762.74
34	2-Furancarboxaldehyde[a]	904.34	724.59	122.07	806.75	588.67	210.77	743.42	534.53	155.58
35	2-Ethyl-3,5-dimethylpyrazine	74.27	87.51	52.86	75.36	95.87	65.98	81.25	135.85	110.62
36	Cis-linalool oxide (furanoid)	98.77	155.54	114.91	50.10	69.33	87.78	86.64	125.71	137.16
37	2-((Methylthio)methyl)furan	70.63	81.98	145.21	80.65	110.54	105.32	102.27	158.69	98.58
42	1H-Pyrrole	69.53	82.46	85.22	86.48	105.32	177.15	68.38	111.66	192.65
43	1-(2-Furyl)-2-propanone	178.33	287.05	190.45	208.29	281.86	391.84	294.52	415.08	460.63
46	Furfuryl acetate	1546.87	2650.52	2689.68	2124.39	2863.09	4873.20	2586.22	3788.55	5328.70
47	5-Methyl-2-furancarboxamide	3689.36	4442.63	1407.45	3548.50	3652.65	2329.60	3611.44	4076.17	1865.42
48	1-Methyl-1H-pyrrole-2-carboxaldehyde	371.59	610.30	363.49	379.05	484.25	620.55	493.67	712.65	696.77
51	2-Furanmethanol	5670.16	6278.92	4440.45	6272.13	6613.01	7631.46	5714.67	6732.94	7462.72
52	3-Mercapto-3-methyl-1-butanol	121.23	110.09	61.81	110.25	115.89	70.21	115.23	198.58	92.32
53	3-Methylbutyric acid	1784.97	1887.13	772.21	1814.32	1944.22	1326.00	1943.77	2030.00	1255.31
54	1-(1H-pyrrol-1-yl)-2-propanone[b]	293.36	403.64	234.78	384.42	430.02	453.79	315.49	471.47	520.37
55	1-(2-Furanylmethyl)-1H-pyrrole	117.05	235.01	219.00	155.53	208.12	304.72	174.86	322.65	415.51
56	2-Hydroxy-3-methyl-2-cyclopenten-1-one	147.39	125.23	145.56	110.54	351.21	180.24	148.85	171.28	201.89
57	2-Methoxyphenol	74.98	210.28	190.56	84.41	281.58	210.66	100.03	298.65	205.32
58	2,3-Dihydro-5-hydroxy-6-methyl-4H-pyran-4-one	255.37	118.84	9.96	188.91	40.46	8.54	198.32	28.35	8.77
59	1-(5-Methyl-2-furanyl)-1,2-propanedione	56.75	96.75	49.06	78.42	97.89	78.91	77.97	134.38	64.33
60	3-Ethyl-2-hydroxy-2-cyclopenten-1-one	55.09	77.47	29.56	60.41	55.75	82.05	70.52	107.05	112.91
63	1H-pyrrole-2-carboxaldehyde	143.41	217.13	97.80	159.68	180.39	158.26	159.64	232.96	148.29
64	4-Hydroxy-2,5-dimethyl-3(2H)-furanone	103.37	155.49	70.65	101.15	159.07	89.88	131.42	139.42	95.61
65	4-Ethenyl-2-methoxyphenol	157.59	219.63	130.71	209.11	247.25	218.75	251.61	331.80	311.07
66	1-(2-Furanylmethyl)-1H-pyrrole-2-carboxaldehyde[b]	16.98	27.58	24.23	13.23	25.66	19.49	27.65	29.86	19.65

[a] Peak areas of 2-furancarboxaldehyde are those of selected ion (m/z 96).
[b] Identifications were achieved by comparing mass spectra and retention indices with data from Ref. 12. L, roast degree.

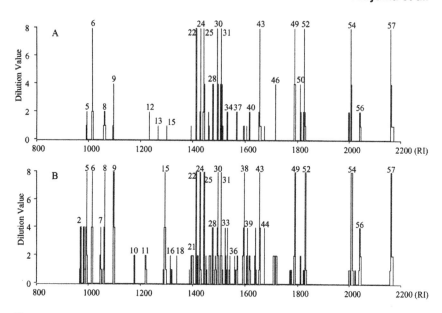

Figure 4 Aroma chromatograms of the headspace volatiles of roasted Ethiopia coffee beans (roast degree, L23) using static (A) and dynamic (B) solid-phase micro-extraction sampling methods.

descriptions, and significant differences were found in nutty-roast and smoke-roast aromas.

Highly volatile compounds 5, 6, and 8 shown in Table 3, which have a buttery-oily aroma, were more abundant in the dynamic headspace than the static headspace, as indicated by their Charm values. Compounds 24, 30, and 31, which were difficult to detect in this GCMS analysis, indicated high OSVs in both static and dynamic conditions, but these compounds were more abundant in the dynamic headspace than in the static headspace. The nutty-roast aroma, compounds 15, 25, and 38 was more abundant in the dynamic headspace than in the static condition, as indicated by their Charm values. Also, these odorants had high OSVs in the dynamic condition and were significant candidates for the nutty-roast aroma released during grinding. Compound 9 was abundant in the dynamic condition, as indicated by its Charm values, and this odorant had a high OSV and seemed to contribute a strong smoke aroma of the volatile during grinding of roasted coffee beans. Compound 49 had high Charm values in both static and dynamic headspace; the polar compounds 43, 52, 54, 56, and 57 were also abundant in the dynamic condition, as predicted by the GCMS analytical

Table 3 GC/O Results of 20 Potent Odorants Detected in the Headspace of Roasted Ethiopia Coffee Beans (L 23) Under Dynamic and Static Conditions

No.	Retention indices	Description	Charm values[a]		OSVs[b]		Compound
			Dynamic condition	Static condition	Dynamic condition	Static condition	
5	994	Buttery-oily	240	47	70	37	2- and 3-Methylbutanals
6	1016	Buttery-oily	340	210	83	79	2,3-Butanedione
8	1062	Buttery-oily	180	96	61	53	2,3-Pentanedione
9	1098	Smoke-roast	360	95	86	53	3-Methyl-2-butene-1-thiol[c]
15	1296	Nutty-roast	230	16	69	22	2-Methyl-3-furanthiol[c]
22	1415	Smoke-roast	320	340	81	100	2-Furanmethanethiol
24	1431	Soy sauce	310	160	80	69	3-(Methylthio)propanal[c]
25	1445	Nutty-roast	230	150	69	66	2-Ethyl-3,5-dimethylpyrazine
28	1478	Nutty-roast	120	90	49	51	2,3-Diethyl-5-methylpyrazine
30	1497	Green-black currant	190	130	62	62	3-Mercapto-3-methylbutyl formate[c]
31	1510	Green-earthy	200	99	64	54	2-Methoxy-3-(2-methylpropyl)pyrazine[c]
32	1514	Buttery-oily	39	93	28	52	(E)-2-nonenal
38	1596	Nutty-roast	170	18	59	23	6,7-Dihydro-5-methyl-5H-cyclopentapyrazine
43	1656	Acidic	280	180	76	73	3-Methylbutyric acid
49	1789	Sweet-fruity	340	340	83	100	(E)-Beta-damascenone
51	1824	Sweet-caramel	38	90	28	51	Unknown
52	1831	Phenolic	210	110	65	57	2-Methoxyphenol
54	2008	Sweet-caramel	410	210	91	79	4-Hydroxy-2,5-dimethyl-3(2H)-furanone
56	2039	Sweet-caramel	170	57	59	41	2-Ethyl-4-hydroxy-5-methyl-3(2H)-furanone
57	2163	Phenolic	490	290	100	92	4-Ethenyl-2-methoxyphenol

[a] Each Charm value is rounded off to two significant digits in order to reflect the actual resolution of the dilution analysis and represented as the average value of three measurements.

[b] Odor spectrum value (OSV) is the normalized Charm value modified with an approximate Stevens's law exponent ($n = 0.5$).

[c] Identifications were achieved by comparing Kovats indices to those of standard compounds and with data from Ref. 2.

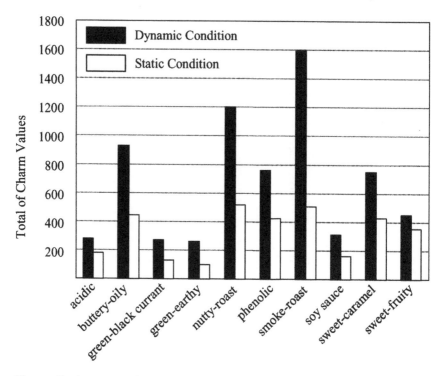

Figure 5 Aroma profiles based on gas chromatography olfactometry (GCO) results of the headspace volatiles of Ethiopia coffee beans (roast degree, L23) obtained by using static and dynamic solid-phase microextraction (SPME) sampling methods.

results mentioned previously. However, 3-hydroxy-4,5-dimethyl-2(5*H*)-furanone and 5-ethyl-3-hydroxy-4-methyl-2(5*H*)-furanone, found in the solvent extracts of roast powder and brewed Arabica coffee as potent odorants possessing seasoning-like aromas, were not detected in this study.

2. Gas Chromatography Olfactometry Evaluations of Volatile Compounds in Arabica Coffee Beans with Different Roast Degrees

This new sampling method was also applied to GCO evaluations to investigate characteristic aroma profiles of coffee beans of three different origins (Ethiopia, Tanzania, and Indonesia) roasted to three different degrees. Some differences were observed in the changes in aroma profiles caused by roasting in each of the three types of coffee beans, as shown in Fig. 6, and potent odorants (above 50% OSV) are listed in Table 4.

Significant changes in the smoke-roast and nutty-roast aromas with roasting were common to all the roasted coffee bean types. As shown in Fig. 6, significant change with roasting in the smoke-roast aroma was common to all the roasted coffee bean types. Especially in the case of Ethiopia coffee beans, the smoke-roast aroma, mainly contributed by 3-methyl-2-buten-1-thiol and 2-furanmethanethiol, intensified with roasting, as has been reported in Colombia and Kenya coffee beans [6]. The nutty-roast aroma increased significantly with roasting but decreased beyond the medium roasting range, especially in the case of Tanzania coffee beans. The Charm values of 2-methyl-3-furanthiol and 2-ethyl-3,5-dimethylpyrazine, which contributed to the nutty-roast aroma, showed similar changes with roasting in all the roasted coffee bean types used in our experiments, but their changes with roasting were not consistent with reported data [6]. The sweet-caramel aroma was mainly contributed by 4-hydroxy-2,5-dimethyl-

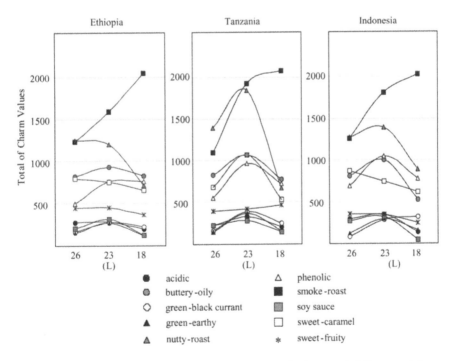

Figure 6 Changes of aroma profiles of roasted coffee beans of different origins with roasting. The numbers (26, 23, and 18) refer to roast degrees (L).

Table 4 Potent Odorants (Above 50% Odor Spectrum Value) Found in the Headspace Volatiles of Roasted Coffee Beans Using the Dynamic Solid-Phase Microextraction Sampling Method

No.	Description	Compound	Retention indices	Charm values[a] (OSVs[b])[e]								
				E26	T26	I26	E23	T23	I23	E18	T18	I18
1	Acidic	3-Methylbutyric acid	1657	270 (83)	220 (77)	300 (80)	280 (76)	320 (81)	350 (85)	190 (65)	200 (70)	140 (60)
2	Buttery-oily											
−1		2- and 3-Methylbutanals	994	230 (77)	200 (74)	190 (64)	240 (70)	310 (80)	100 (46)	290 (80)	290 (84)	120 (55)
−2		2,3-Butanedione	1016	240 (78)	260 (84)	160 (58)	340 (83)	400 (90)	320 (82)	310 (83)	290 (84)	230 (77)
−3		2,3-Pentanedione	1062	61 (40)	99 (52)	140 (55)	180 (61)	150 (55)	170 (60)	46 (32)	35 (29)	18 (21)
−4		(E)-2-Nonenal	1515	71 (43)	88 (49)	95 (45)	39 (28)	110 (47)	130 (52)	110 (49)	65 (40)	86 (47)
−5		Unknown	1975	49 (35)	39 (32)	160 (58)	0 (0)	44 (30)	74 (39)	0 (0)	0 (0)	26 (26)
3	Green-black currant	3-Mercapto-3-methylbutyl formate[d]	1497	83 (46)	80 (46)	74 (40)	190 (62)	250 (71)	260 (74)	120 (52)	130 (56)	160 (64)
4	Green-earthy											
−1		2-Methoxy-3-(1-methylethyl)pyrazine[d]	1420	64 (41)	75 (45)	54 (34)	62 (36)	170 (59)	97 (45)	46 (32)	53 (36)	55 (38)
−2		2-Methoxy 3-(2-methylpropyl)pyrazine[d]	1509	78 (45)	61 (41)	69 (38)	200 (64)	190 (62)	200 (65)	68 (39)	97 (49)	110 (53)
5	Nutty-roast											
−1		2-Methyl-3-furanthiol[d]	1296	120 (55)	220 (77)	150 (56)	230 (69)	390 (89)	220 (68)	56 (35)	13 (18)	39 (32)
−2		Unknown	1318	0 (0)	140 (62)	66 (37)	36 (27)	35 (27)	22 (21)	29 (25)	35 (29)	54 (37)
−3		2-Ethyl-3,5-dimethylpyrazine[d]	1445	160 (64)	190 (72)	170 (60)	230 (69)	340 (83)	250 (72)	120 (52)	87 (46)	180 (68)
−4		2,3-Diethyl 5-methylpyrazine	1479	130 (58)	97 (51)	120 (51)	120 (49)	180 (61)	180 (61)	98 (47)	150 (60)	110 (53)
−5		Unknown	1534	220 (75)	130 (59)	78 (41)	100 (45)	220 (67)	110 (48)	73 (40)	80 (44)	100 (51)
−6		3-Methoxypyridine[c]	1571	93 (49)	61 (41)	78 (41)	65 (36)	110 (47)	110 (48)	46 (32)	70 (41)	97 (50)

	No.	Compound	RI	1	2	3	4	5	6	7	8	9	10
	−7	6,7-Dihydro-5-methyl-5H-cyclopentapyrazine	1594	130 (58)	130 (59)	160 (58)	170 (57)	130 (52)	54 (35)			87 (46)	72 (43)
	−8	3-Mercapto-3-methyl-1-butanol	1641	100 (51)	120 (57)	100 (46)	76 (39)	140 (49)	140 (54)	56 (35)	0 (0)	0 (0)	59 (39)
6 Phenolic	−1	2-Methoxyphenol	1830	120 (55)	110 (55)	110 (48)	210 (48)	350 (65)	350 (85)	350 (85)	210 (68)	240 (77)	220 (75)
	−2	4-Ethyl-2-methoxyphenol	1999	97 (50)	68 (50)	110 (43)	110 (48)	120 (35)	220 (49)	120 (49)	110 (49)	130 (56)	200 (72)
	−3	4-Ethenyl-2-methoxyphenol	2165	280 (85)	370 (85)	470 (100)	490 (100)	480 (100)	490 (100)	440 (100)	350 (92)	440 (99)	360 (96)
7 Smoke-roast	−1	Unknown	985	57 (38)	76 (38)	100 (45)	95 (46)	140 (44)	0 (53)	0 (53)	140 (56)	210 (72)	140 (60)
	−2	Unknown	987	0 (0)	0 (0)	0 (0)	33 (26)	79 (40)	220 (67)	0 (0)	0 (0)	0 (0)	0 (0)
	−3	Unknown	1049	70 (42)	85 (48)	120 (51)	79 (51)	220 (40)	44 (30)			160 (62)	140 (60)
	−4	4-Vinylfuran[c]	1074	77 (44)	0 (0)	0 (0)	0 (0)	0 (0)	0 (0)	0 (0)	0 (0)	300 (86)	240 (78)
	−5	3-Methyl-2-butene-1-thiol[d]	1099	120 (55)	110 (55)	170 (55)	360 (60)	400 (86)	390 (90)			410 (100)	390 (100)
	−6	2-Pentylfuran[c]	1221	32 (29)	69 (43)	32 (43)	64 (60)	120 (36)	130 (49)			110 (52)	78 (45)
	−7	Unknown	1304	0 (0)	0 (0)	0 (0)	0 (0)	0 (0)	0 (0)	0 (0)	0 (0)	99 (49)	51 (36)
	−8	2-furanmethanethiol	1414	320 (91)	300 (90)	340 (90)	320 (85)	400 (81)	320 (90)			320 (88)	350 (95)
	−9	((Methylthio)methyl)furan	1466	100 (51)	93 (51)	75 (50)	110 (40)	140 (47)	130 (53)			99 (49)	34 (30)
	−10	Unknown	1525	51 (36)	66 (42)	67 (42)	120 (38)	160 (49)	120 (57)			65 (40)	92 (49)
8 Soy sauce	−1	3-(Methylthio)propanal[d]	1430	200 (72)	220 (77)	280 (77)	310 (77)	280 (80)	320 (76)			140 (58)	47 (35)
9 Sweet-caramel	−1	2-Hydroxy-3-methyl-2-cyclopenten-1-one	1813	71 (43)	45 (35)	59 (35)	50 (35)	180 (32)	59 (61)			68 (41)	90 (48)
	−2	4-Hydroxy-2,5-dimethyl-3(2H)-furanone	2009	390 (100)	350 (97)	460 (97)	410 (99)	400 (91)	450 (90)			220 (73)	320 (91)
	−3	2-Ethyl-4-hydroxy-5-methyl-3(2H)-furanone	2040	120 (55)	120 (57)	140 (57)	170 (55)	290 (59)	130 (77)			140 (58)	120 (55)
10 Sweet-fruity		(E)-beta-damascenone	1790	290 (86)	310 (92)	260 (92)	340 (74)	350 (83)	350 (85)			370 (95)	160 (64)

[a] Each Charm value was rounded off to two significant digits in order to reflect the actual resolution of the dilution analysis and represented as the average value of three measurements.

[b] Odor spectrum value (OSV) was the normalized Charm value modified with an approximate Stevens' law exponent ($n = 0.5$).

[c] Tentatively identified by gas chromatography mass spectrometry.

[d] Identifications were achieved by comparing Kovats indices to those of standard compounds and with data from Ref 2.

[e] E, Ethiopia; T, Tanzania; I, Indonesia; Numbers (26, 23, and 18) represent roast degrees.

3(2*H*)-furanone and 2-ethyl-4-hydroxy-5-methyl-3(2*H*)-furanone, in which the latter furanone affected the intensity of the sweet-caramel aroma in the medium roasted Tanzania coffee beans, as shown in Fig. 6 and Table 4. Furthermore, the phenolic aroma, mainly contributed by 2-methoxyphenol and 4-ethenyl-2-methoxyphenol, exhibited maximal Charm values in the vicinity of roast degree L23 in Tanzania and Indonesia coffee beans. The Charm values of 3-(methylthio)propanal, which contributed to the soy sauce aroma, in all the roasted coffee bean types increased with roasting and decreased significantly beyond the medium roasting range. These results were consistent with the finding of Mayer and associates in Tipica coffee beans from Colombia [6]. Also, the Charm values of the sweet-fruity aroma in Ethiopia and Indonesia coffee beans decreased gradually with roasting, and the Charm values of (*E*)-beta-damascenone, which mainly contributed to the sweet-fruity aroma, in these coffee beans were similar to that reported by Mayer and colleagues in Tipica coffee beans from Colombia. Aroma profiles of coffee beans of different origins roasted to the same degree are shown in Fig. 7. The lightly roasted coffee beans (L26) exhibited very similar aroma profiles, and there was no significant difference between those of different origins. The most deeply roasted coffee beans (L18), as was the case with the lightly roasted coffee beans (L26), exhibited almost the same aroma profiles in all three kinds of coffee bean, except that Indonesia coffee beans showed lower intensity in the buttery-oily aroma and higher intensity in the nutty-roast aroma than those of other origins. However, the GCO results indicated that aroma characteristics, such as nutty-roast, phenolic, smoke-roast, and sweet-caramel aromas, of roasted coffee beans with different origins were most discernible in the medium roasted coffee beans (L23), as shown in Fig. 7. In the medium roasted Tanzania coffee beans, 2-methyl-3-furanthiol and 2-ethyl-3,5-dimethylpyrazine contributed to the nutty-roast aroma, and 3-methyl-2-butene-1-thiol and 2-furanmethanethiol affected the smoke-roast aroma. As mentioned, the significant intensity found in the medium roasted Tanzania coffee beans was caused by the aroma potency of 2-ethyl-4-hydroxy-5-methyl-3(2*H*)-furanone. The differences found in the phenolic aroma intensities of the medium roasted Tanzania coffee beans were attributed to the Charm values of 2-methoxyphenol and 4-ethyl-2-methoxyphenol. Significant aroma differences caused by 2,3-butanedione, 2,3-pentanedione, 2-methoxy-3-(2-methylpropyl)pyrazine, and 3-mercapto-3-methylbutyl formate were not found in the medium roasted coffee beans.

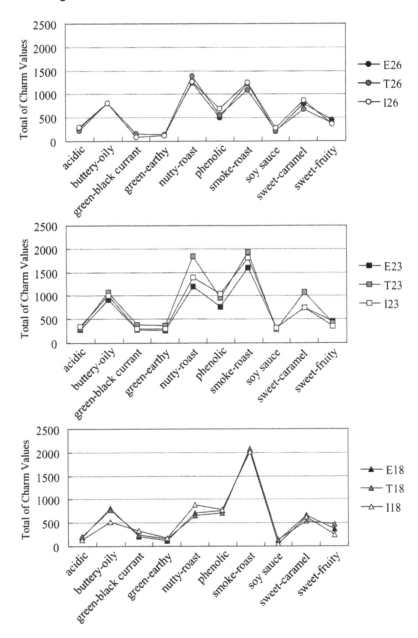

Figure 7 Changes of aroma profiles of coffee beans of different origins roasted to the same degrees. E, Ethiopia; T, Tanzania; I, Indonesia. Numbers (26, 23, and 18) refer to roast degrees.

D. Aroma Investigation Using Principal Component Analysis

The GCO results of each description of roasted coffee beans were applied to the principal component analysis to evaluate objectively how the aroma profiles varied with roast degrees and between origins. The PCAs were conducted as follows: 35 potent odorants with an OSV of 50 or greater shown in Table 4, calculated by using mean Charm values of each odorant, were selected from the volatile compounds of the aroma released during grinding and categorized into 10 aroma descriptions. The two-dimensional scatter plots of the factor loadings and principal component score are shown in Fig. 8. The first principal component (PC1) and the second principal component (PC2) explained 58.4% and 22.0% of the total GCO information, respectively. The descriptions other than smoke-roast aroma were included in PC1. On the other hand, the PC2 indicated the highest contribution to the smoke-roast aroma of roasted coffee beans. This PCA

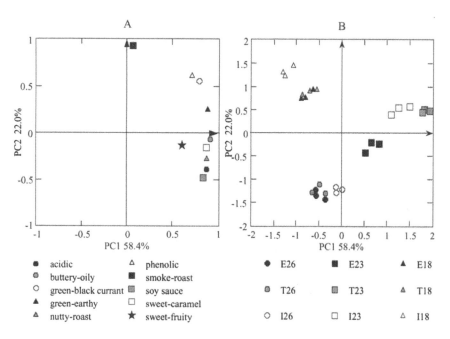

Figure 8 Two-dimensional scatter plots of the factor loading (A) and principal component score (B) using the total Charm values of 10 aroma descriptions (above 50% OSV). E, Ethiopia; T, Tanzania; I, Indonesia. Numbers (26, 23, and 18) refer to roast degrees. OSV, odor spectrum value.

result suggested that the medium roasted coffee beans (L23) of different origins released total flavor most abundantly, except for the smoke-roast aroma. And the most deeply roasted coffee beans (L18) released the least total flavor. On the other hand, the aroma activity of the smoke-roast aroma included in the PC2 increased with roasting. The plot of the principal component score indicated clearly that coffee beans of different origins that were roasted to the same degree released similar aromas. However, the difference in aroma characteristics among the three types of coffee beans was most marked with medium roasting (L23).

In place of Charm values of 10 aroma descriptions mentioned, the PCA was conducted by using the mean Charm values of 35 potent odorants, which showed an OSV of 50 or above; listed in Table 4. The two-dimensional scatter plots of the factor loading and principal component score are shown in Fig. 9. The PC1 and PC2 explained 38.4% and 21.7% of the total GCO information, respectively. The factor loading of odorants (Fig. 9A) was plotted similarly to that of the 10 aroma descriptions into

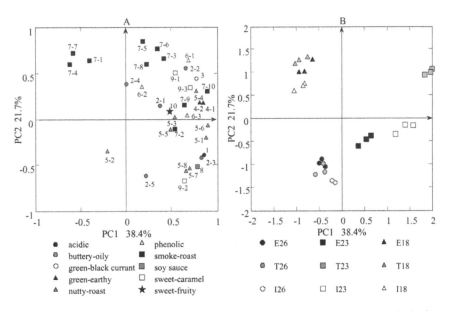

Figure 9 Two-dimensional scatter plots of the factor loading (A) and principal component score (B) using the Charm values of 35 potent odorants (above 50% OSV). A, Numbers correspond to the numbers in Table 4; B, numbers (26, 23, and 18) refer to roast degrees; E, Ethiopia; T, Tanzania; I, Indonesia; OSV, odor spectrum valve.

which the odorants are classified. The results of this factor loading suggest that aroma characters of volatile compounds are similar to those of their aroma descriptors (Fig 9A). Furthermore, the two-dimensional scattered plot of the principal component score (Fig. 9B) is similar to that of the aroma descriptor data (Fig. 8B).

IV. CONCLUSIONS

A DH-SPME sampling method was developed for flavor analyses and evaluation of the characteristic coffee aroma generated during grinding. This method was found to be reproducible and convenient for flavor analyses. When compared with those of SH-SPME sampling using medium roasted coffee beans (L23) that originated in Ethiopia, volatile compounds adsorbed on SPME fiber under the dynamic condition were more abundant, their relative composition differed especially among the more polar and highly volatile compounds, and GCO evaluation showed that the nutty- and smoke-roast aromas were more potent under the dynamic condition. Also, this method was applied to three kinds of Arabica coffee (Ethiopia, Tanzania, and Indonesia) with three different roast degrees (L26, L23, and L18), and their aroma characterizations were investigated by using GCMS and GCO. These studies showed that only smoke-roast aroma increased significantly throughout the roasting process and suggested that the medium roasted coffee beans released the most abundant aromas during grinding and exhibited the most discernible differences in aroma profiles between beans of different origins. In comparing our results with those reported by Mayer and associates [6], some similarities and differences in aroma changes with roasting and coffee bean origins were found. These differences can be attributed to differences in sampling methods and coffee beans used. Furthermore, the PCA showed that the aroma differences were greater in coffee beans roasted to different degrees than in coffee beans of different origins. In other words, it was shown that coffee beans of the same origin and roasted to different degrees showed significantly different aroma characters, whereas coffee beans from different origins and roasted to the same degree exhibited similar aroma characters.

ACKNOWLEDGMENT

We gratefully acknowledge Professor Terry Edward Acree of Cornell University for his useful suggestion about the GCO data and their significance, and also thank Assistant Professor Mitsuya Shimoda of Kyushu University for his useful suggestion about PCA.

REFERENCES

1. LM Nijssen, CA Visscher, H Maarse, LC Willemsens, MH Boelens. Volatile Compounds in Food: Qualitative and Quantitative Data, 7th ed. Zeist, The Netherlands: TNO Nutrition and Food Research Institute, 1996, pp 72.1– 72.23.
2. W Holscher, OG Vitzthum, H Steinhart. Identification and sensorial evaluation of aroma-impact-compounds in roasted Colombian coffee. Café Cacao The 34:205–212, 1990.
3. I Blank, A Sen, W Grosch. Potent odorants of the roasted powder and brew of Arabica coffee. Z Lebensm Unters Forsch 195:239–245, 1992; P Semmelroch, W Grosch. Analysis of roasted coffee powders and brews by gas chromato-graphy-olfactometry of headspace samples. Lebensm Wiss Technol 28:310–313, 1995; P Semmelroch, W Grosch. Studies on character impact odorants of coffee brews. J Agric Food Chem 44:537–543, 1996; M Czerny, F Mayer, W Grosch. Sensory study on the character impact odorants of roasted Arabica coffee. J Agric Food Chem 47:695–699, 1999.
4. W Holscher, H Steinhart. Investigation of roasted coffee freshness with an improved headspace technique. Z Lebensm Unters 195:33–38, 1992.
5. C Sanz, D Ansorena, J Bello, C Cid. Optimizing headspace temperature and time sampling for identification of volatile compounds in ground roasted Arabica coffee. J Agric Food Chem 49:1364–1369, 2001.
6. F Mayer, M Czerny, W Grosch. Influence of provenance and roast degree on the composition of potent odorants in Arabica coffees. Eur Foods Res Technol 209:242–250, 1999.
7. F Mayer, W Grosch. Aroma simulation on the basis of the odourant composition of roasted coffee headspace. Flavour Fragrance J 16:180–190, 2001.
8. CL Arthur, J Pawliszyn. Solid phase microextraction with thermal desorption using fused silica optical fibers. Anal Chem 62:2145–2148, 1990.
9. X Yang, T Peppard. Solid-phase microextraction for flavor analysis. J Agric Food Chem 42:1925–1930, 1994; CP Bicchi, OM Panero, GM Pellegrino, AC Vanni. Characterization of roasted coffee and coffee beverages by solid phase microextraction-gas chromatography and principal component analysis. J Agric Food Chem 45:4680–4686, 1997.
10. DD Roberts, P Pollien, C Milo. Solid-phase microextraction method development for headspace analysis of volatile flavor compounds. J Agric Food Chem 48:2430–2437, 2000.
11. TE Acree, J Barnard, DG Cunningham. A procedure for the sensory analysis of chromatographic effluents. Food Chem 14:273–286, 1984.

12. W Baltes, G Bochmann. Model reaction on roast aroma formation. III. Mass spectrometric identification of pyrroles from the reaction of serine and threonine with sucrose under the condition of coffee roasting. Z Lebensm Unters Forsch 184:478–484, 1987.

13. KD Deibler, TE Acree, EH Lavin. Solid phase microextraction application in gas chromatography/olfactometry dilution analysis. J Agric Food Chem 47:1616–1618, 1999.

14. TE Acree. GC/olfactometry. Anal Chem News Features 69:170A–175A, 1997.

15. R Marsill. Flavor, Fragrance, and Odor Analysis. New York: Marcel Dekker, 2002, pp 205–227.

16. AJ Matich, DD Rowan, NH Banks. Solid phase microextraction for quantitative headspace sampling of apple volatiles. Anal Chem 68:4114–4118, 1996.

17. E Voirol, N Daget. Comparative study of nasal and retronasal olfactory perception. Lebensm Wiss Technol 19:316–319, 1986.

18. M Akiyama, K Murakami, N Ohtani, K Iwatsuki, K Sotoyama, A Wada, K Tokuno, H Iwabuchi, K Tanaka. Analysis of Volatile Compounds released during the grinding of roasted coffee beans using solid-phase microextraction. J Agric Food Chem 51:1961–1969, 2003.

17

Interactions of Selected Flavor Compounds with Selected Dairy Products

Klaus Gassenmeier
Givaudan Schweiz AG, Dübendorf, Switzerland

I. INTRODUCTION

The specific flavor of food is one of its most important properties. Customers often base their preferences on aroma. Ideally the flavor of a product is stable throughout its shelf life. Because of their nature, dairy products often show dramatic changes in terms of flavor profile over time. This variation might be due to interactions with living microorganisms, active enzymes, or specific binding of aroma compounds to milk proteins. Knowledge of the mechanisms involved is important for selection of flavors and the manufacturing of products. The results published previously [1,2] are reviewed here.

II. RESULTS AND DISCUSSION REVIEW

A. Interactions of Aroma Compounds with Yogurt

Flavor changes (e.g., loss of intensity, loss of freshness, change of profile, off-flavor formation) occur frequently during fermentation and storage of yogurts with fruit. Radical changes in the flavor profile of strawberry flavor have been reported [3]; however, the cause of the sensorial phenomenon was not studied in detail. The basic process of yogurt production is outlined in Fig. 1. The milk, milk powder, and a starter culture were mixed; the mass

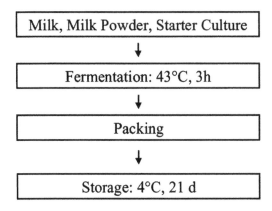

Figure 1 Yogurt production scheme.

was fermented at 43°C for 3 hr, packed, and stored at 4°C. The shelf life of yogurt is approximately 21 days. Added aroma compounds may undergo modifications and degradation during production and/or storage. To get an overview of the relevant processes, flavor model mixtures were added to yogurt before fermentation, as is typical for the production of set-type yogurts. Samples were stored at 4°C for 21 days before analysis. The aroma compounds were quantified by gas chromatography (GC) after isolation by simultaneous distillation extraction (SDE), as described in Ref. 1. No significant degradation was found for esters, ketones, and alcohols. However, huge losses were detected for aldehydes. More than 95% of octanal was degraded during fermentation. The unsaturated aldehydes were significantly more stable when compared to the saturated ones. Only 49% of the initial concentration of (E)-2-hexenal remained after fermentation and 21-day storage. Identification experiments indicated that octanal was converted into the corresponding alcohol octanol. The results suggest that nonspecific hydrogenases produced by the microorganisms are active and can reduce aldehydes but not, e.g., ketones.

Since the microbial activity is highest in the fermentation phase, one might assume that the stability of aldehydes can be improved when they are added after fermentation. Such a process can be realized in the production of a stir-type yogurt. In subsequent experiments, the concentration of hexanal, (E)-2-hexenal, and (E,Z)-2,6-nonadienal as well as the corresponding alcohols in yogurt was monitored over time. The aroma compounds were added after the fermentation process, and the yogurt was stored for 21 days at 4°C. The results are displayed in Fig. 2.

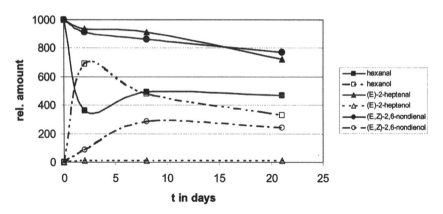

Figure 2 Degradation of hexanal, (E)-2-heptenal, and (E,Z)-2,4-nonadienal during storage in yogurt and formation of the respective alcohols.

Hexanal was degraded quickly, and the corresponding alcohol hexanol was formed. The conversion occurred at the beginning of storage. At the end of the shelf life ca. 50% of hexanal remained in the product. In contrast, the unsaturated aldehyde (E)-2-heptenal was more stable and degradation was linear over time. After 21 days, 70% remained intact. Surprisingly, the corresponding unsaturated alcohol 2-heptenol was not detected. Possibly 2-heptenol was isomerized into heptanal, which is further reduced to heptanol. The (E,Z)-2,6-nonadienal was even more stable than (E)-2-heptenal. Only 20% was degraded during 21 days of storage. Obviously it was converted into (E,Z)-2,6-nonadienol. In previous studies this alcohol was associated with a metallic off-note in buttermilk and exhibited a very low odor threshold of 0.07 ng/L air [4].

On the basis of the results, we hypothesize that instability of aldehydes may contribute to a change of the aroma profile in yogurt.

B. Interactions of Vanillin with Milk Enzymes

When vanilla ice creams were described by quantitative aroma profiling techniques the term *cardboard* is regularly used as a flavor descriptor [5,6]. This note is present, but not desirable. Generally, "cardboard" or "oxidized" notes in dairy products are described to originate from fat autooxidation or packing material [7]. The aroma profile of selected ice creams was described by an expert panel (five panelists), using the consensus technique [8]. The "cardboard" note was found in vanilla flavored ice cream only, but surprisingly not in other types of ice cream. This gave rise to the hypothesis that a specific mechanism is active.

When the concentration of vanillin was measured in market samples of vanilla ice cream exhibiting a cardboard off-note, it was found that vanillin was partially degraded into vanillic acid. Oxidation of vanillin and p-hydroxybenzaldehyde into the corresponding acids was reported in ice cream mass stored at 4°C [9]. After 14 days, significant amounts of the acids are formed. Ehlers [10] observed the same oxidation phenomena in commercial milk products. Similar results were published by Kempe and Kohnen [11] and Anklam and associates [12]. In lab experiments using ultra high-temperature-(UHT) milk or milk heated to 90°C, they found no oxidation; however, in a kitchen-style preparation with pasteurized milk, oxidation occurred. The data suggest that enzymes, which are deactivated during heating, might be involved. It seemed crucial to understand the properties and reaction mechanisms of the enzymes to explain the oxidation of vanillin.

Indeed, milk contains several redox enzymes. The most important are lactoperoxidase and xanthine oxidase. In pasteurized milk these are still active. Experiments by Baumgartner and Neukom [13] indicate that enzymatic oxidation of vanillin by lactoperoxidase yields di-vanillin as the main product. Vanillic acid is not formed. The ability of xanthine oxidase (XO) to oxidize vanillin into vanillic acid is described in the literature. This reaction was even used by Demott and Praepanitchai [14] to measure XO activity in milk. Anklam and colleagues [12] demonstrated that XO added to UHT milk is able to oxidize vanillin to vanillic acid. In the course of the reaction of XO with xanthine, hydrogen peroxide (H_2O_2) or superoxide radical anion ($\cdot O_2^-$) is formed. We assume that the same oxygen species are formed when vanillin is oxidized by XO to vanillic acid. There has been some discussion in the literature whether XO may contribute to fat autooxidation in milk; however, no clear conclusion has been drawn [15]. Under physiological conditions, the superoxide radical anion ($\cdot O_2^-$) is preferably formed [16]. Whether hydrogen peroxide (H_2O_2) or superoxide radical anion ($\cdot O_2^-$) is formed depends also on the reducing status of the enzyme [17]. It was shown that this molecule hardly reacts with linoleic acid [18]. However, in acidic medium it adds a proton to yield a perhydroxy radical ($\cdot O_2H$). In ice cream at pH 6.5 to 7 approximately 1% of the superoxide radical anion ($\cdot O_2$) is transformed into the reactive perhydroxy radical ($\cdot O_2H$, pKs $= 4.7$), which readily reacts with unsaturated fatty acids and forms peroxides.

In medical studies the xanthine/xanthineoxidase system (X/XO) is used to generate in vivo superoxide radicals, and therefore it may be involved in formation of various diseases (e.g., see Ref. 17). Furthermore, it was shown that X/XO resulted in an increase of hydroperoxides in human spermatozoa [19].

In order to clarify whether the oxidation of vanillin is related to the cardboard off-note in ice cream, a model system was developed as outlined in Fig. 3. A cardboard off-note developed in the model system consisting of pasteurized milk, pasteurized cream, sugar, and vanillin within 24 hr. When the system was heated before addition of vanillin, no cardboard note developed. When vanillin was omitted in the not-heated sample, no cardboard note developed. These experiments clearly show that vanillin is involved in the formation of a cardboard off-note in vanilla ice cream. Because no off-notes are formed when the mixture is heated, it can be assumed that enzymes are also involved. When milk is sufficiently heated the enzymes are deactivated. Furthermore, instrumental analyses of the model system revealed that (E)-2-nonenal and heptanal were strongly increased in the not-heated sample with vanillin compared to a not-heated control experiment without vanillin. During gas chromatography olfactometry a cardboard aroma impression was assigned to (E)-2-nonenal. These aldehydes can be generated during lipid autooxidation. To clarify whether they contribute the cardboard off-note quantitative data and odor thresholds will be needed.

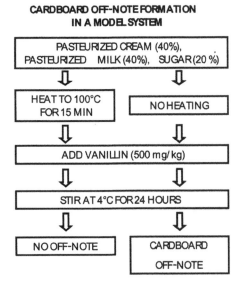

Figure 3 Generation of a cardboard off-note in a model system.

A hypothesis for the generation of a cardboard off-note in ice cream based on the literature data is outlined in Fig. 4.

Vanillin is oxidized into vanillin acid by XO. Thereby reactive oxygen species are released. These react with unsaturated fatty acids to yield hydroperoxides. Upon decomposition of the hydroperoxides aroma-active lipid oxidation products are formed; they cause the cardboard off-note.

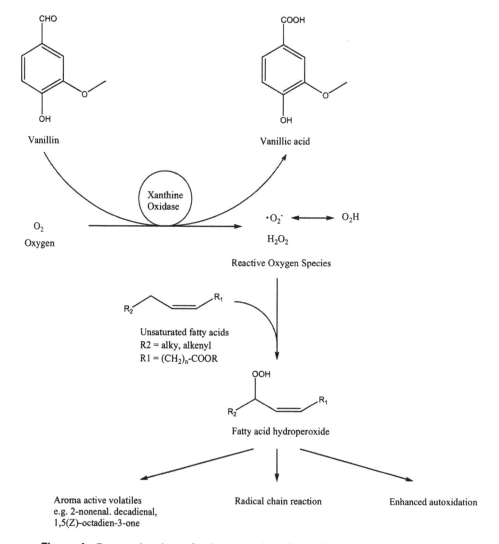

Figure 4 Proposed pathway for the generation of a cardboard off-note in vanilla ice cream.

Furthermore, these hydroperoxides can start a radical chain reaction, leading to additional aroma-active fat degradation products and enhanced autooxidation. The hydroperoxide degradation can happen spontaneously and may also be catalyzed by enzymes. Finally it should be emphasized that XO contributes to lipid oxidation in milk fat only if an appropriate substrate is present.

III. CONCLUSIONS

A fermented dairy product (yogurt) and a nonfermented one (ice cream) have been investigated. In dairy products with living microorganisms aldehydes can be degraded partially or completely during fermentation and/ or storage. The major degradation mechanism is the reduction to the corresponding alcohols. This might be the cause of a loss of aroma impact, change in the aroma profile, and/or formation of unwanted notes. The reported problems can be overcome when the aroma is added to a pasteurized product, when a flavor that avoids unstable compounds is selected, or when microorganisms that do not produce unspecific hydrogenases are used. Surprisingly, in nonfermented dairy products specific interactions of aroma compounds were also found. One reaction is the oxidation of vanillin into vanillic acid, which is catalyzed by XO. In the course of such oxidation reactions, XO can release reactive oxygen species, which may generate aroma-active compounds and trigger autooxidation. The reaction can be prevented by heating the product sufficiently to deactivate the enzyme, before addition of an aroma.

The unwanted interactions of aroma compounds with dairy products can be minimized when the mechanisms involved are known and the flavor and production process are adapted accordingly.

REFERENCES

1. K Gassenmeier. Stability of flavor compounds in yogurt. In: HP Kruse, M Rothe, eds. Flavour Perception—Aroma Evaluation. Potsdam: Eigenverlag Universität Potsdam, 1997, pp 495–498.
2. K Gassenmeier. Vanillin and xanthineoxidase—key factors for the generation of a cardboard off-note in vanilla ice-cream. Lebensm-Wiss u-Technol 36:99–103, 2003.
3. P Marion, S Chardon-Fayard. Incorporation of flavors in foodstuffs. In: G. Charalambous, ed. Flavors and Off-Flavors. Amsterdam: Elsevier Science, 1990, pp 687–699.

4. C Heiler, P Schieberle. Studies on the metallic off flavor in buttermilk: Identification of potent aroma compounds. Lebensm-Wiss u-Technol 29:460–464, 1996.

5. P Piccinali, CR Stampanoni Koeferli. Sensory analysis in flavor development. Food Marketing Technol (2)1996.

6. BM King. Sensory profiling of vanilla ice cream: Flavor and base interactions. Lebensm-Wiss u-Technol 27:450–456, 1994.

7. IJ Jeon. Undesirable flavors in dairy products. In: MJ Saxby, ed. Food Taints and Off-Flavours. London: Chapman & Hall, 1993.

8. H Stone, JL Sidel. Sensory Evaluation Practice. San Diego: Academic Press, ·1993, p 212.

9. M Chevalier, Y Prat, P Navellier. Caractérisation par chromatographie en couche mince des constituants aromatiques de la vanille dans les glaces et crèmes glacées. Ann Fals Exp Chim 697:12–16, 1972.

10. D Ehlers. HPLC Untersuchung von Handelsprodukten mit Vanille und Vanillearoma. Deutsche Lebensmittel-Rundschau 95:464–468, 1999.

11. K Kempe, M Kohnen. Deterioration of natural vanilla flavours in dairy products during processing. Adv Food Sci (CMTL) 21:48–53, 1999.

12. E Anklam, S Gaglione, A Müller. Oxidation behaviour of vanillin in dairy products. Food Chem 60:43–51, 1997.

13. J Baumgartner, H Neukom. Enzymatische Oxidation von Vanillin. Chimia 26:366–368, 1972.

14. BJ Demott, OA Praepanitchai. Influence of storage, heat and homogenisation upon xanthine oxidase activity of milk. J Dairy Sci 61:164–167, 1977.

15. NY Farkye. Other enzymes. In: Fox PF, ed. Food Enzymology, Vol 1. London: Elsevier Applied Science, 1991, pp 107–130.

16. A Ravati, B Ahlemeyer, A Becker, J Krieglstein. Preconditioning induced neuroprotection is mediated by reactive oxygen species. Brain Res 866:23–32, 2000.

17. M Dixon. In: Enzymes, 3rd ed. London: Longman Group, 1979, p 281.

18. JM Gebicki, GHJ Bielski. Comparison of the capacities of the perhydroxyl and the superoxide radicals to initiate chain oxidation of linoleic acid. J Am Chem Soc 103:7020–7022, 1981.

19. JF Griveau, E Dumont, JP Callegari, D Le Lannou. Reactive oxygen species, lipid peroxidation and enzymatic defence systems in human spermatozoa. J Reprod Fertil 103:17–23, 1995.

18

Challenges in Analyzing Difficult Flavors

Willi Grab
Givaudan Singapore Pte Ltd, Singapore

I. INTRODUCTION

The analysis of flavors is challenging for various reasons: flavors are complex mixtures of volatile molecules with sensory properties. They occur only in small amounts in a wide range of natural products with different matrix compositions (including fat, protein, carbohydrates, water, minerals, and active enzymes). Flavors and flavoring molecules are prone to chemical and physical alterations such as evaporation, contamination, oxidation, degradation, and enzymatic and intermolecular reactions. All volatile molecules do not contribute equally to the overall flavor profile. An analytical flavor chemist therefore needs extensive knowledge, skills, and tools to isolate the traces, separate the mixture, identify the molecules, and validate the sensory properties. Developments in the last 30 years have made this task manageable and a matter of routine [1–4]. Methods and microtechniques are available to isolate flavors from small samples such as single fruits during a ripening process without destroying the fruit [5]. High-resolution capillary gas chromatography combined with sensitive, fast mass spectrometers allow efficient separation and identification of volatile molecules out of complex mixtures, based on reliable, complete mass spectral databases. Suitable equipment is commercially available. Quanti-

The content of this chapter was presented at the 224th ACS National Meeting in Boston, August 2002.

fication is simplified by the use of stable isotope derivatives of the key components [6]. Training in sensory evaluation has been drastically improved, using standardized descriptive languages and quantitative flavor profiling [7]. Gas chromatography (GC) sniffing is widely used to recognize and quantify key impact molecules [8–10]. Yet, there are still skills and tools to improve:

> Identify the critical steps in analyzing flavors.
> Link analytical and sensory results.
> Handle rare and complex samples.

II. CRITICAL STEPS IN ANALYZING FLAVORS

It is obvious: the analytical sample must represent the target flavor or at least the flavor note to detect/identify. The GIGO principle (garbage in/ garbage out) plays a role. Do we really know what we are looking for? We must realize that no sample preparation method is perfect. We therefore must constantly verify the sensory properties of the sample fractions, eventually by recombination experiments to check the loss of critical molecules. We have to distinguish contaminants and artifacts from sample components. We have to quantify only meaningful and relevant data. Area percentage of an extract is easy to determine and in some cases is also helpful. But is it meaningful in quantifying concentrations of molecules in a product? The direct use of suitable internal standards throughout the analytical process is essential to compensate for all losses. Stable isotope derivatives of the critical molecules have proved to be simple, efficient, and correct [8]. Finally we have to verify the identifications by reconstitution of the flavor with synthetic molecules and validate its sensory profile. Only this last experiment proves whether our results are relevant and correct.

III. LINK ANALYTICAL AND SENSORY RESULTS—THE KLAUS PROFILE

Gas chromatography, (GC) sniffing, aroma extract dilution analysis (AEDA), and CHARM analysis are well accepted tools to identify the sensory properties of individual molecules in complex mixtures [9–10]. In fact, they are the fastest tool to identify typical character impact molecules and odor off-notes. They require well trained people who are able to keep pace with the fast emerging GC peaks: imagine that a small gas balloon of ca. 0.1 mL leaving the sniff port within 3 sec has to be transferred to the olfactory receptors in the upper part of the nose without dilution. Normal breathing just washes it down the lung! Multiple sniffing of a sample at different concentrations leads directly to a quantitative expression of the contribution of individual molecules to the profile. These results are now more and more often published. They add a lot of understanding to the importance of molecules to the overall flavor profile. Klaus Gassenmeier [11] created a very simple method to visualize this flavor profile, which resembles like a sensory profile:

> Arrange the GC-sniff report according to the dilution steps.
> Place similar odor impressions together.
> To compare two products, arrange the report by the difference in dilution steps.

Instead of the dilution steps you may use the log of the flavor dilution (FD) factor. This corresponds to Stevens's law, that sensory impression follows the logarithmic scale of the concentration. The result is a profile that can be read as a sensory profile, and people with a trained memory of the odor impressions of molecules will be able to "smell" the profile. It is, to my understanding, the first direct objective link between analytical results and a sensory flavor profile, linking the contribution of character impact molecules to the perceived sensory profile.

The following figures illustrate the Klaus profile: Fig. 1 represents the difference of the headspace of two butter biscuits: The freshly baked biscuit

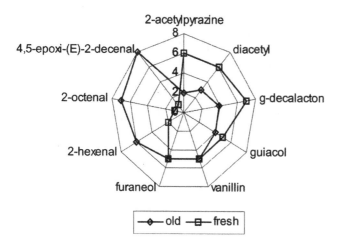

Figure 1 Klaus profile of two butter biscuits: freshly baked and stored. The scale is given in number of dilution steps.

has a nice baked, roasted, buttery, creamy odor. The stored, old biscuit shows rancid, metallic, fatty off-notes. The Klaus profile is based on GC-sniffing experiments of diluted headspace samples of these biscuits. The AEDA results are arranged by the difference of the two samples. Only the main differences are shown. The fresh sample is dominated by the roasted note of 2-acetylpyrazine, the buttery note of diacetyl, and the creamy-buttery note of gamma-decalactone. In the old sample we find a strong metallic note from 2.4-epoxidec-2-enal, a fatty note from 2-octenal, and a green note from 2-hexenal.

Figure 2 is a direct translation of a table of an analysis of two mint oils, *Mentha piperita* and *Mentha arvensis*, published by Benn [12]. The dilution steps are arranged by retention time on the GC column. The two oils are mirrored to get a direct comparison. This is the typical presentation of such results.

In Fig. 3 the same results are presented according to the difference of the two oils, leaving out the middle section, which mainly contains the minty impressions, which are similar in both oils. It directly shows the sensory difference: *Mentha piperita* is dominated by the hay-caramelic note of menthofuran; fruity, pineapple, juniper notes of unidentified molecules; fruity, estery notes of ethyl-2-methylbutyrate and ethylisovalerate; cocoa

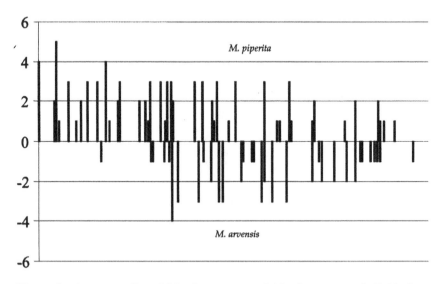

Figure 2 Aroma profiles of *Mentha piperita* and *Mentha arvensis* oil. Table from Ref. 12 translated into an aromagram. Peaks ordered by retention time, intensity in dilution steps.

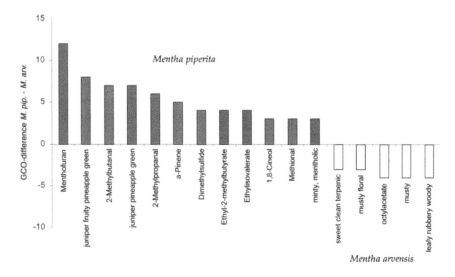

Figure 3 Comparison of *Mentha piperita* and *Mentha arvensis*. Table from Ref. 13 translated into a Klaus Profile.

notes from 2-methylbutanal and 2-methylpropanal, and sulfurous notes from dimethylsulfide and methional. The *Mentha arvensis* is dominated by unidentified leafy, rubbery, woody, musty notes and the oily, mushroom note of octylacetate. Both examples show the power of the Klaus Profile to visualize sensory properties, which are based on objective analytical data.

IV. RARE AND COMPLEX SAMPLES

The flavor industry has a strong interest in creating new, innovative types of flavors. Nature and exotic menus are the source of inspiration. The common fruits are already well known, but there are still unknown resources somewhere in the biosphere. The highest biodiversity is concentrated in the tropical rain forests and within the rain forests on top of the trees, the so-called canopy. Special equipment and long travel are necessary to reach these places. Givaudan has sponsored several scientific missions to French Guyana, Gabon, and Madagascar organized by Pro-Natura International [13–15]. The main tool during these missions was the largest hot air balloon with an engine to control its movements exactly. It allowed individuals to approach individual trees to collect unusual, rare flowers and fruits for sensorial and analytical investigation. Another approach to unusual flavors is the investigation of local food and menus in remote areas in the world. Givaudan explores the food of specialized restaurants and kitchens under the title "TasteTrek." In both locations, the rain forest and the kitchen, sensory evaluation has to be done on the spot, and often only small and unique samples can be recovered. We normally use a dynamic headspace method with portable battery-operated vacuum pumps and solvent desorption or microextraction in a Pasteur pipette. The collected sample is only a few (up to 50 microliters μL) and of uncertain concentration. In some cases we have only a "single-shot" sample, concentrated down to 2 μL (Fig. 4), just enough for a single GC injection. We compromise in GC resolution by combining GC, mass spectrometry (MS), and sniffing in the same run. As a result we often get highly complex MS results with many unresolved peaks. The classical approach to isolate an MS, by subtracting neighbor peaks, is extremely time-consuming. The Automated Mass Spectral Deconvolution and Identification System (AMDIS) of W. Gary Mallard and J. Reed [16] is a very helpful tool for running the entire deconvolution process within a short time. A short survey showed that this program is not yet well known and often used in analytical flavor labs. It is able to uncover hidden, small traces under a larger peak (Fig. 5). This program helped us to identify more than 600 molecules in a Chinese tallow sample (Fig. 6), separated on a 30-m capillary column within a retention time of 35 min (20 unresolved peaks per minute).

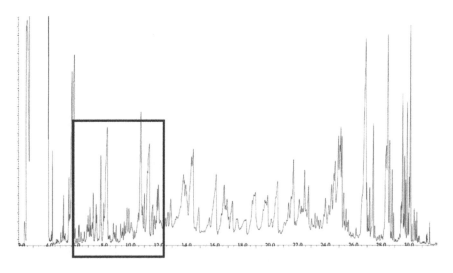

Figure 4   Chinese tallow, headspace extract, an extremely complex mixture. The part in the frame is enlarged in Fig 5. Retention time is indicated in minutes.

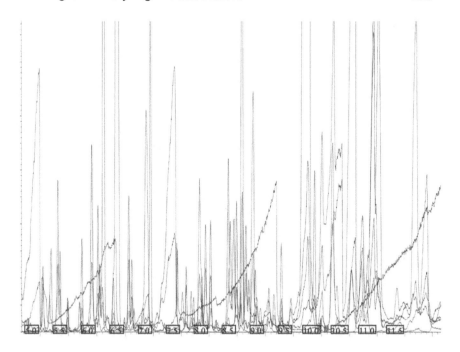

Figure 5 Chinese tallow, enlarged portion with several single-ion traces (from 6 to 12 min).

V. CONCLUSIONS

Flavors are complex mixtures and analytical samples may contain hundreds of molecules. In addition, the analysis of fresh tropical fruit flavors is challenging for different reasons: The fruits are normally not available in the freshest, ripest form in the analytical laboratory; artificial ripening does not in general lead to the genuine flavor. The analytical chemist does not exactly know the quality of the ripe flavor. Sensory evaluation of analytical samples is crucial to validate analytical results. The Klaus Profile links objective analytical results with a sensory profile by linking the contribution of character impact molecules with sensory impressions. Tropical fruits may contain new, unknown molecular structures. This chapter demonstrates our method to handle the freshest samples from unusual fruits from the rain forests. These samples may be very small, unique, and unable to be reproduced a second time. In general the analytical chemist has to preseparate a sample by fractionation (distillation, chromatography) into manageable complexity. Many methods and tools have been developed to

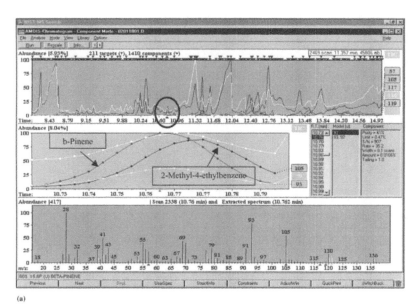

(a)

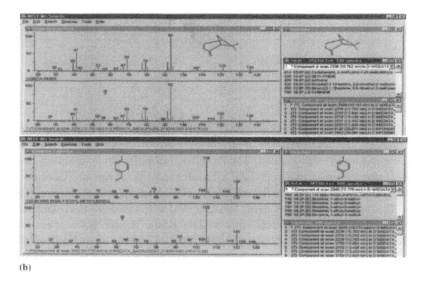

(b)

Figure 6 (a) Chinese tallow, AMDIS [16] deconvolution of complex mixed peaks. Upper part, part of the chromatogram with some single ion traces; middle part, deconvolution of two mixed peaks; lower part, mass spectrum of the mixed peak. (b) Deconvoluted isolated mass spectra of β-pinene and 4-ethyl-methylbenzene from (a) with reference spectra from the libraries.

handle the delicate flavor samples, and microtechniques are available to the skilled chemist. In some cases the sample size does not allow prefractionation, either because of its instability or because of the very small amounts isolated, e.g., by a headspace technique. If the sample is too complex, even good separation by a high-resolution capillary column does not separate all molecules into individual peaks. Time-consuming mass spectrum interpretation may indicate the structure in the mixed peak. Automatic deconvolution of the mass spectra in a peak is a great help to improve the manual interpretation. This chapter demonstrates the power of such programs to identify >600 molecules in small one-shot samples.

REFERENCES

1. ST, Likens, GB, Nickerson, Detection of certain hop oil constituents in brewing products, Am Soc Brew Chem Proc, 5–13 (1964).
2. K, Grob, G, Grob, J Chromatogr, 62(1), 1–13 (1971).
3. P, Schieberle, KH, Engel, Ed. Frontiers of Flavor Science, Proceedings of the Ninth Weurman Flavor Research Symposium, Fresing 1999, Publ. Deutsche Forschungsanstalt für Lebensmittel (2000).
4. F, Etzweiler, N, Neuner-Jehle, E, Senn, Seifen, Oele, Fette, Wachse, 106(15), 419–427 (1980).
5. W, Grab, In: Symposium on Flavors of Fruits and Fruit Juices, Bern 1978, International Federation of Fruit Juice Producers, Juris Druck & Verlag, Zurich (1978).
6. W, Grosch, Trends Food Sci Technol, 4, (3), 68 (1993).
7. CR, Stampanoni, J Sens Stud 9, 383–400 (1994).
8. R, Kerscher, W, Grosch, Z Lebensm Unters Forsch A, 204, 3–6 (1997).
9. P Schieberle, In: Characterization of Foods: Emerging Methods (Gaonkar, A.G., ed.), Elsevier Science B.V., pp. 403–431 (1995).
10. T, Acree, J, Barnard, D, Cunningham, Food Chem, 14, 273 (1984).
11. Gassenmeier, K. Givaudan Dübendorf, Switzerland, private communication.
12. S, Benn, Perfumer Flavorist, 23, 5 (1998).
13. F, Halle, ed. Biologie d'une Canopee de Foret Equatoriale IV, Pro-Natura International & Operation Canopee (1999).
14. R, Kaiser, Chimia, 54(6), 346–363, (2000).
15. K, Gassenmeier, X, Yang, W, Grab, J, Peppet, R, Eilerman, Chimia, 55, 435–440 (2001).
16. Mallard, W.G., Reed, J., US Department of Commerce, National Institute of Standards and Technology NIST: download: http://chemdata.nist.gov/mass-spc/amdis/

19

Nose to Text: Voice Recognition Software for Gas Chromatography Olfactometry

Philippe Mottay
Brechbühler, Inc., Spring, Texas, U.S.A.

I. INTRODUCTION

Gas chromatography olfactometry (GCO) is a valuable technique [1–4]. Combining the human perception of odor and chromatographic separation of compounds offers great possibilities. Applications include correlating sensory responses with volatile chemicals, resolving off-flavor problems, and assessing olfactory acuity of individuals.

One of the main challenges of this technique is the identification of the odors. It requires the panelist to remain focused on the odor throughout the run. Taking notes is one of the challenges. The panelist has to record both the impression of the odors and the time at which they were made i.e., retention time. A simple solution requires an assistant to take notes and record the retention time. This solution is quite inefficient and can be a distraction to the panelist. Other solutions have been developed; however, most of them require actions from the panelist, hence diverting him or her from the sniffing. These include the panelist's selecting from a list of descriptors using a mouse, touch screen, or joy stick while the retention time is internally monitored (Charmware^{TM} from DATU, Inc., Geneva, NY, U.S.A., Aroma Trax® software by Microanalytics, Austin, TX, U.S.A., OSME from Oregon State University, Corvallis, OR, U.S.A.) [5]. We have developed GCO dedicated software based on voice recognition (Brechbühler AG, Schlieren, Switzerland).

II. SOFTWARE DEVELOPMENT

A. Voice Recognition

The power of personal computers is such today that voice recognition is a viable tool and is commercially available. Voice recognition in itself is a complex program. The computer converts an electrical signal into words. To achieve high recognition accuracy, voice recognition software relies on several sources of information:

- An acoustic model, a mathematical model of the sound patterns used by the speaker's language
- A vocabulary, a list of words that the program can recognize; each word in the vocabulary has a text representation and a pronunciation
- A language model, which is statistical information associated with a vocabulary that describes the likelihood of the occurrence of words and sequences of words in the user's speech

The first challenge in speech recognition is to identify what is speech and what is noise. A second challenge is to recognize speech from more than one speaker. Speech recognition software works best when the computer adjusts to each new speaker. The process of teaching the computer to recognize the voice of a speaker is called *training*.

Another challenge is distinguishing between two or more phrases that sound alike. Speech recognition programs do not understand what words mean, so they cannot use common sense as people can. Instead, they keep track of how frequently words occur by themselves and in the context of other words. This information helps the computer choose the most likely word or sentence from several possibilities.

The software must be customized to the voice of the speaker. This is the training part, which consists of the user's reading a passage from a prepared text. The program then adds the data to the information it already knows about the sounds in the language. When training a user, it starts with a standard set of models and then customizes them for the way the user speaks (acoustic model) and the way he or she uses words (vocabulary and associated language model).

B. Voice Recognition Applied to Gas Chromatography Olfactometry

1. Goals

When developing *Nose to Text* we kept in mind its main purpose: to free the user of any unwanted distractions while recording the comments and the retention time during the GCO run.

2. The Challenges

Developing the software presented several challenges. The software had to support multiple users. The user interaction with the software had to be kept to a minimum. Finally the data obtained from the GCO run had to be in an electronic format reporting function. The program also offered several ways to simplify data processing.

3. The Program

The interface (Fig. 1) was designed for easy and minimal interaction. Each user must train the voice recognition engine to be able to work accurately with the program. The user selection is done from a drop-down menu. Voice commands are used to start and end a run. The preparation and start of the run consist of two voice commands: *Ready*, then *Inject* at the injection of the sample. The *Stop* command ends the run. During the run, all of the user's comments are added in a text box along with the retention time. These comments can be edited at the end of the run.

C. Data Processing

The GCO data consist of a list of retention times and comments. Unlike written notes, the data can be used directly to generate reports. The identification of the compounds by their odors is a challenge. The reporting function is based on a text file in tab, comma, or space separated format.

1. Reports for Gas Chromatography Users

The GC reports consist of a list of retention times and intensities (or areas). The GCO comments can be merged with the GC report to give a concise report. The algorithm used is simple (Fig. 2): comments made are associated with the previous peak. Not all the peaks detected by the flame ionization detector (FID) are responsible for an odor. So correlating GC reports and comments made requires an external trace generated by the panelist, as he or she smelled the peaks. This odor intensity trace (odorogram) is used to

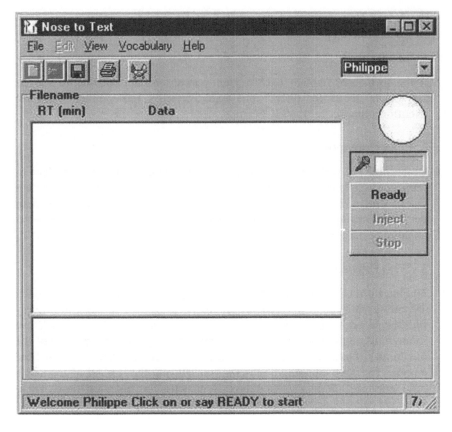

Figure 1 *Nose to Text* interface with user selection and voice-activated commands "Ready, Inject, Stop."

generate a GC-type report that is in turn used to generate the GCO report (Fig. 3).

2. Report for Non–Gas Chromatography Personnel

Non-GC persons may have no interest in retention time. They need a list of the odors present and eventual comments associated with the odors. Often they also need to compare two products. The type of reporting offered is flexible enough to accommodate the different needs of the users. In view of these considerations the program offers a reporting function ignoring the retention time and any other GC-related consideration.

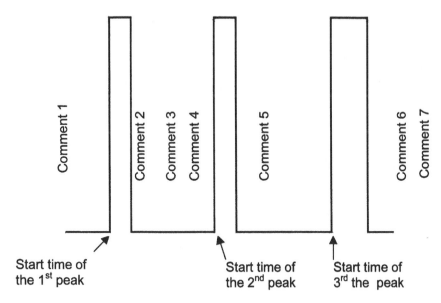

Figure 2 Algorithm for gas chromatography olfactometry reporting.

Chrom-Card Report - Modified by Nose to Text

```
Method Name     :
Method File     : C:\...\Thermo Finnigan\Chrom-Card 32 bit for TRACE\data\
Chromatogram    :
GC Method       :
Operator ID     :                        Company  Name : C.E. Instruments
Analysed        : 25.06.1991  10:22      Printed      : 26.07.2001  18:19
Sample ID       :            (# 1)       Channel      : Channel A
Analysis Type   : Calibration (Area)     Calc.  Method : External STD

Calib. method   : using Response Factors
```

Component Name	Ret.Time	Area	Resp. Fact.	Rel.Ret.Time	GC-O Results
1	0.29	341716	0.0000	0.5062	herbaceous
2	0.80	68727	0.0000	0.5551	citrus
3	1.05	80736	0.0000	0.5795	orange almond
Peak aaa	1.75	901622	901622.0000	0.7419	fruity musky
Peak bbb	2.44	1856328	185633E+07	1.0000	coumarin dill
Peak ccc	3.16	2914266	.291427E+07	0.7190	
7	3.29	54130	0.0000	0.8364	Strawberry pineapple
Peak ddd	3.78	3575910	.357591E+07	0.8632	blackcurrent forest
Peak eee	4.28	3654779	.365478E+07	1.0000	
10	4.36	80449	0.0000	1.0591	nice smell woody coffee
11	4.93	189235	0.0000	1.0903	cigarette tobacco fresh
Totals		20402010			

Figure 3 The gas chromatography report: the comments are added in the last column.

The report consists of the odor and associated comments (i.e., intensity, panelist feelings, etc.). A report using a template can be developed with the user's defining how the comments are organized and what needs to be reported. For example, a panelist describes the odor and its intensity for each peak smelled. The comments made during the run must be consistent with the template. An option allows for the comparison of samples and a brief description for each sample. Another option allows the user to target specific odors and report only on these odors. From a list of predefined odors, the program searches the results for these odors. If found, they are listed along with the comments made. If they are not found, the program marks them as "not detected."

3. Assisting the Identification of the Compounds by Odor

The flavor chemists face another challenge: identifying a compound responsible for an odor. Among the tools available to GC users for identification of compounds are retention indexes (Kovats and/or Ester). Compared to GC, GCO gives us an additional dimension, the odor descriptor. An odor library is available, and combining the two is a valuable tool. Dr. Acree from Cornell University maintains a library combining both the odor and the retention indexes [6]. This library, called the FlavorNet, is, to the best of our knowledge, the only one offering such features to the general public on the World Wide Web (www.nysaes.cornell.edu/flavornet/). Since flavor and fragrance companies often develop their own library, we developed a flexible library file format.

The library consists of a binary file. It consists of a header defining the number and type of custom field as well as the column description for the retention indexes. The data for each compound are recorded in an indexed sequential record according to a proprietary format. Each record includes the parameter pertinent to the compounds:

> The compound name
> The associated odor
> Six user-defined fields (such as MW, formula, CAS#)
> Retention indexes for up to 10 columns

The results of the GCO run can be searched against this library, giving the user an additional tool for identification (Fig. 4). Any of the custom parameters can be searched. The retention indexes are used to narrow the search. Search within results helps narrow the number of hits.

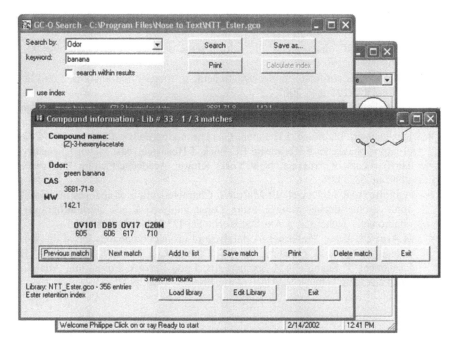

Figure 4 Single-match report listing all the custom fields.

III. CONCLUSIONS

Gas chromatography olfactometry is a valuable technique for food and flavor chemists. It presents numerous challenges. The *Nose to Text* software presented addresses the recording of comments made during such runs and proposed additional tools for data reporting. It was designed to let the panelist focus on the odor in order to increase the accuracy and efficiency of the run. The software was designed to be open and flexible.

ACKNOWLEDGMENT

The author wishes to thank Brechbühler AG for sponsoring the development of the software, all the scientists who gave their inputs, and Dr. Terry Acree for the right to use the FlavorNet GCO library.

REFERENCES

1. TE Acree. GC/olfactometry: GC with a sense of smell. Anal Chem 69:170A–175A, 1997.
2. I Blank. Gas chromatography-olfactometry in food aroma analysis. Food Sci Technol 115:297–331, 2002.
3. JV Leland, P Schieberle, A Buettner, TE Acree. Gas chromatography-olfactometry: The state of the art. Washington, DC: ACS, 2001.
4. KD Deibler, TE Acree, EH Lavin. Gas chromatography-olfactometry (GC/O) of vapor phases. In R Teranishi, EL Wick, I Hornstein, eds. Flavor Chemistry: Thirty Years of Progress. New York: Kluwer Academic/Plenum Publishers, 1999, pp 387–395.
5. A Plotto, MR McDaniel, JP Mattheis. Characterization of changes in "Gala" apple aroma during storage using Osme analysis, a gas chromatography-olfactometry technique. J Am Soc Hortic Sci 125:714–722, 2000.
6. H Arn, TE Acree. Flavornet: A database of aroma compounds based on odor potency in natural products. In ET Contis, C-T Ho, CJ Mussinan, TH Parliament, F Shahidi, AM Spanier, eds. Food Flavors: Formation, Analysis and Packaging Influences. Amsterdam: Elsevier, 1998, p 27.

20

Headspace Sampling: A Critical Part in Evaluating True Gas Phase Concentrations

Gerhard N. Zehentbauer, Cindy L. Eddy, Pete A. Rodriguez, Christa S. Pelfrey, and Jianjun Li
The Procter & Gamble Company, Cincinnati, Ohio, U.S.A.

I. INTRODUCTION

The analysis of volatile compounds, which are perceived by the human nose in the air above a food, has always been a goal in flavor research. Two different techniques, static and dynamic headspace sampling, have been proposed to solve this problem. Although dynamic headspace sampling usually provides sufficient amounts of odorants to obtain a signal, it has a strong dependence of the yield of the odorants on the velocity of carrier gas [1] and the selectivity of the adsorption and desorption processes [2,3]. As it is very difficult to control these parameters accurately, reproducibility of measurements may be poor. On the other hand, as a result of its usually good reproducibility, cheap and rapid analysis, simplicity of the procedure, and availability of automated systems, static headspace sampling became very popular over the last few years. Here, generally two techniques, injection of gas samples by gastight syringes usually combined with cryofocusing or use of an adsorptive fiber such as solid-phase microextraction (SPME) [4–7] or headspace sorptive extraction (HSSE) [8] are used to analyze key odorants. Although both techniques can show good reproducibility, which allows comparison of samples relative to each other, they also suffer from analyte losses and gas phase distortion. Hence, care must be taken if elucidation of the true gas phase composition is the goal. A comparison of both techniques and their specific disadvantages is presented.

II. MATERIALS AND METHODS

A. Chemicals

Carbon-14-labeled *n*-hydrocarbons and 1-butanol, tagged in the one-position, were used. Decanal (7.5 mCi/mmol) and dodecane (5.7 mCi/mmol) were purchased from ICN (Irvine, CA). Tetradecane (5.7 mCi/mmol) and 1-butanol (94.5 mCi/mmol) were purchased from Pathfinder Laboratories Inc. (St. Louis, MO). Hexadecane (53.6 mCi/mmol) was purchased from Amersham (Amersham, Great Britain). All other chemicals were purchased from Sigma-Aldrich (St. Louis, MO).

B. Gas Phase Standards

Radiolabeled hydrocarbon standards were prepared by delivering a pentane solution of the radiolabeled material to a 50-mL Hypo-vial® (Pierce Chemical Co., Rockford, IL) capped with a Miniert® valve (Pierce Chemical Co.) The gas phase concentrations ranged from 0.7 ng/mL (*n*-C_{16}) to 38 ng/mL (*n*-C_{10}).

Butanol standards were prepared after an approximate 1:10 dilution of the labeled material with cold 1-butanol. The diluted material was delivered to a Hypo-vial containing silica gel. The butanol/silica gel ratio was used to establish the desired gas phase concentration (ca. 300 ng/mL).

The radiopurity of the standards was established by means of a gas chromatograph (Hewlett Packard 5890) equipped with an injector/trap [9] and a radioactivity detector [10]. Other radioactivity measurements were performed with a Packard tri-Carb Scintillation spectrometer (model 2001, Packard, Downers Grove, IL) by using an ethanolamine-based mixture.

C. Procedures for Sampling, Establishing of Recovery, and Distribution of Radioactivity

Experiments to measure distribution and recovery of *n*-hydrocarbons were performed with a 10-mL gastight syringe (Hamilton Company, Reno, NV). The syringe was equipped with a fixed, epoxy-cemented 304 stainless steel needle [0.028 in outer diameter (o.d.), 0.016 in inner diameter (i.d), point style 2]. Sampling and delivery were done with an infusion pump (Sage Instruments, Cambridge, MA). The sampling and delivery rates were fixed at 5 mL/min. The corresponding linear velocity in the needle was ca. 60 cm/sec.

D. Recovery of *n*-Hydrocarbons

The gaseous sample was withdrawn to the 10-mL mark of the syringe, the drive was stopped, and the syringe content infused into liquid scintillation mixture (results in Table 1). After infusion, the plunger was withdrawn from the syringe. The syringe walls and needle were washed with pentane (1 mL) and acetone (0.5 mL). The washes were combined in a vial containing scintillation mixture. The plunger tip was rinsed with pentane (1 mL), and the wash transferred to a different vial containing scintillation mixture. The syringe parts were dried in a vacuum oven at room temperature for a minimum of 10 min before assembly. The evacuation operation was performed before any sampling.

Distribution of *n*-hydrocarbons (results in Table 2) was obtained by withdrawing the gaseous sample to the 10-mL mark. Laboratory air was then withdrawn at a rate of 2.5 mL/min, until the plunger was removed from the syringe barrel. The inside of the needle was rinsed with pentane (1 mL) delivered with a 1-mL syringe equipped with a 4-in-long needle. The needle washings were collected in a scintillation mixture. The walls and plunger tip of the 10-mL syringe were rinsed with pentane (1.5 mL) and the washes were collected in a different scintillation mixture. The syringe assembly was dried as indicated previously.

Recovery of 1-butanol (results in Table 3) was obtained by using two 1-mL syringes (Hamilton Company, Reno, NV) equipped with a fixed,

Table 1   Recovery of Gaseous ^{14}C-*n*-Hydrocarbon Standards from a 10-mL Gastight Syringe

	Radioactivity, dpm ($\times 10^4$), percentage of total				Recovered by
n-HC	I (Infused)[a]	II (Wall + needle)[b]	III (Plunger)[b]	IV (Total)	infusion, percentage I/IV
C$_{10}$	43.7	0.21 [<1]	0.13 [<1]	44	99
C$_{12}$	14.2	0.84 [5.5]	0.30 [2.0]	15.3	92
C$_{14}$	4.7	0.80 [14.0]	0.21 [3.7]	5.7	82
C$_{16}$	3.4	0.09 [2.6]	0.05 [1.4]	3.5	96

[a] Radioactivity recovered by infusion of syringe contents into a scintillation mixture.
[b] Residual radioactivity in syringe components after infusion.

Table 2   Distribution of [14]C-*n*-hydrocarbons Within a 10-mL Gastight Syringe Immediately After Sampling Gas Phase Standards

		Radioactivity, dpm ($\times 10^4$), percentage of total			
n-HC	BP, °C	I (Needle)[a]	II (Wall + plunger)[b]	III (Air)[c]	IV (Total)
C_{10}	174	0.03 [<0.1]	1.2[3]	42.8 [97]	44
C_{12}	214	0.14 [1]	0.92 [6]	14.2 [93]	15.3
C_{14}	254	2.4 [42]	1.8 [32]	1.5 [26]	5.7
C_{16}	287	2.4 [69]	0.3 [8]	0.8 [23]	3.5

[a] Radioactivity washed from needle.
[b] Residual radioactivity in syringe components.
[c] Calculated by subtraction of I and II from IV.

epoxy-cemented 304 stainless steel needle [0.028 in o.d., 0.016 in i.d.; point style 5 (side hole)]. The gaseous butanol standard was equilibrated in a constant-temperature water bath, kept at 25.0°C ± 0.1°C. The gaseous butanol was withdrawn manually to the 1-mL mark at a rate of approximately 2–3 mL/min. This sample was then infused into a scintillation mixture. The syringe needle was washed by withdrawing 100 μL of hexane into the syringe and expelling the hexane into a vial containing scintillation mixture. Two additional 100-μL hexane washes of the needle were done and combined with the previous wash. Finally, 1 mL of hexane was withdrawn into the syringe and delivered into a different vial containing scintillation mixture. The syringe assembly was dried as indicated.

Table 3   Recovery of Gaseous ^{14}C-1-butanol Standard from Two 1-mL Gastight Syringes A and B Immediately After Sampling Gas Phase Standards

	Radioactivity, dpm, percentage of total			
Syringe	I (Infused)[a]	II (Needle)[b]	III (Wall + plunger)[b]	IV (Total)
A (*n*=6)	1200 [29.7]	2798 [69]	53 [1.3]	4046
B (*n*=4)	668 [16.6]	3269 [81.4]	78 [1.9]	4015

[a] Radioactivity recovered by infusion of syringe contents into a scintillation mixture.
[b] Residual radioactivity in syringe components after infusion.

E. Solid-Phase Microextraction

For gas phase standard preparation, n-hydrocarbons (C_{10}, 0.6 mg; C_{12}, 5.99 mg; C_{14}, 50 mg; C_{16}, 490 mg; C_{18}, 4997 mg) were dissolved in purified mineral oil (50 g). Aliquots of 10 g were placed into headspace vials (volume 180 mL) and the mixtures were allowed to equilibrate overnight at room temperature. The resulting composition of the gas phase was determined by using large-volume injection system GC [11]. Analysis was performed by using poly dimethyl siloxane (PDMS) fibers (30 μm and 100 μm), which were purchased from Supelco (Bellefonte, PA). The sampling time was from 30 sec to 2-hr at room temperature, followed by desorption for 3 min at 250°C. The GC conditions were described previously [11].

III. RESULTS AND DISCUSSION

A. Static Headspace Sampling Using Gastight Syringes

While performing GC analyses of gas mixtures of n-hydrocarbons ranging from C_{10} to C_{16}, using a gastight syringe as a sampling device, we found that the reproducibility and accuracy of the results became progressively worse as the molecular weight of the hydrocarbons increased. The relative standard deviation of C_{10}-C_{12} analyses was better than 3% but was 30–35% in the case of C_{16}. Because the instrumentation had been shown to produce reproducible and accurate results when injecting liquid standards of the identical radiolabeled materials [10], we suspected the gastight syringes were responsible for the poor results obtained with the gasphase standards. To establish the limitations on accuracy imposed by using gastight syringes as sampling tools for gaseous samples, we performed a mass balance. The results shown in Table 1 indicate a high recovery rate (>90%) for all materials except C_{14}. The high recovery rate of C_{16} was surprising in the light of reproducibility and accuracy problems we had experienced. Furthermore, in the case of C_{14}, and to a lesser extent in the case of C_{12}, a significant fraction of the material remained on the syringe walls, needle, and plunger.

The reason for these results and for the poor reproducibility and accuracy of the GC analysis reported resides in the distribution of materials within the syringe immediately after sampling. During sampling, both C_{14} and C_{16} concentrate in the needle, as shown in Table 2. Therefore, any gas flow caused by expansion or expulsion of excess gas after sampling would result in a disproportionate loss of those two hydrocarbons. Note also that because of the higher fraction of C_{16}, compared to C_{14}, retained in the needle, the fraction that can be adsorbed to the walls and plunger is smaller for C_{16}. The lower losses of C_{16} to the wall and plunger, compared to those of C_{14}, contribute to the higher recovery reported in Table 1. Therefore, we expect the accuracy of analysis to be good (95%) when nonpolar compounds, with boiling points less than 200°C, are sampled with gastight syringes. The accuracy should drop off as the boiling point increases and then should improve as the boiling point is high enough to favor concentration of the compounds within the needle. However, as noted, the accuracy and reproducibility of measurements involving compounds with high boiling points (BP > 250°C) are highly dependent on the technique. Withdrawal of sample must be to the desired volume, and the syringe must be handled carefully to prevent changes in temperature before injection. Repeated flushings of the syringe should be prevented.

Accuracy and reproducibility problems are likely to become more severe as the polarity of the compounds increases. For example, infusion of 1-butanol into a scintillation mixture with a 1-mL syringe resulted in the transfer of only 15%–30% of the 1-butanol (BP 118°C) contained in the sample. Furthermore, most of the material left behind remained within the needle of the syringe, as shown in Table 3. For 1-butanol, the fraction retained in the needle was of no consequence when the analysis was performed with a heated injection port but may be the key factor in determining accuracy and reproducibility of non-GC procedures.

In addition, concentration within the needle precludes the use of valve-equipped syringes for the long-term storage of some volatile materials. When a syringe is to be used for storage and delivery of those materials, it is probably better to purchase a syringe equipped with a needle having a side port. After sampling, the side port is covered with a tightly fitting sleeve, made from a section of Teflon tubing, to prevent losses. This approach worked well with samples of *n*-butanol, with losses of less than 5% observed after 4-hr storage.

B. Static Headspace Sampling Using Solid-Phase Microextraction

Similar to those of sampling headspace with gastight syringes, the results obtained by SPME were highly dependent on absorption conditions as well as type of fiber used. Because of the large variety of fibers available (e.g., PDMS, DVB, Carboxen, including fiber combinations) as well as the development of new and improved techniques [e.g., headspace sorptive extraction (HSSE) [8]], a detailed description would be too extensive for this chapter. However, an excellent review [12] as well as theoretical aspects [4] and applications of SPME in food, flavor, and fragrance have been published [5–7].

After optimization of sampling technique and polarity of fiber, usually a high reproducibility (<10%) can be obtained. Therefore, SPME is very useful for comparing samples to each other. However, determining actual gas phase concentrations from SPME is more complicated. The SPME chromatogram not only depends on the actual gas phase composition but also is affected by sampling time, partition between gas phase and fiber, and other surface effects. As shown in Table 4, SPME GC analysis of a gas phase standard consisting of n-hydrocarbons (C_{10}–C_{18}) in approximately equal amounts in gas phase concentrations revealed a varying degree of distortion of the gas phase profile at different sampling times. At short sampling times, from 30 sec to 5 m, the distortion was relatively insignificant for C_{14} and C_{16}. This is in general agreement with the observations of Roberts and associates [13] and Obretenov and colleagues [14]. However, the distortion was significant even at 30 sec for the high-boiling-point compound n-C_{18}. The peak area ratio obtained for C_{18} to C_{12} was 1.88 when the sampling time was 30 sec. The ratio for C_{18} to C_{12} changed very little when the sampling time changed to 1 and 5 min, whereas the ratio for C_{10} to C_{12} decreased from 0.7 to 0.19. When the sampling time was 120 min, the ratio for C_{10} was further decreased to 0.05, whereas the ratio for C_{18} was increased to 12.84. The explanation of these data can be given through the partition coefficients between gas and fiber and gas and liquid. When the sampling time was short, there was not a great disturbance of the equilibrium between the gas phase and the liquid phase, so the gas phase concentration seen by the SPME fiber should not deviate much from the true gas phase concentration. The fact that C_{10} is underestimated and C_{18} is

Table 4 Effect of Solid-Phase Microextraction Sampling Time on the Changes of Measured Gas Phase Concentrations[a] Using a 10-μm Polydimethyl Siloxane Fiber

n-HC[b]	HS[c]	0.5 Min[d]	1 Min[d]	5 Min[d]	30 Min[d]	60 Min[d]	120 Min[d]
n-C_{10}	0.87	0.70	0.44	0.19	0.09	0.09	0.05
n-C_{12}	1.00	1.00	1.00	1.00	1.00	1.00	1.00
n-C_{14}	0.99	0.69	0.97	1.14	2.95	4.63	5.56
n-C_{16}	1.03	1.03	1.10	1.20	3.74	6.92	8.70
n-C_{18}	1.05	1.88	1.72	1.73	4.42	7.98	12.84

[a] Gas phase concentrations are expressed as ratios based on the measured peak areas of the n-hydrocarbons to that of n-C_{12}.
[b] HC, hydrocarbon.
[c] HS, headspace measured by gas chromatography flame ionization detector (GC-FID) using a large-volume injection system (Ref. 11). This is regarded as the true gas phase concentration.
[d] Solid-phase microextraction (SPME) sampling time.

Table 5 Effect of Solid-Phase Microextraction Sampling Time on the Changes of Measured Gas Phase Concentrations[a] Using a 30-μm Polydimethyl Siloxane Fiber

n-HC[b]	HS[c]	0.5 Min[d]	1 Min[d]	5 Min[d]	30 Min[d]	60 Min[d]
n-C$_{10}$	0.87	0.17	0.10	0.05	0.06	0.08
n-C$_{12}$	1.00	1.00	1.00	1.00	1.00	1.00
n-C$_{14}$	0.99	1.25	1.08	2.15	7.37	8.83
n-C$_{16}$	1.03	1.34	1.65	2.86	14.62	28.80
n-C$_{18}$	1.05	2.94	2.73	4.23	18.97	38.95

[a] Gas phase concentrations are expressed as ratios based on the measured peak areas of the n-hydrocarbons to that of n-C$_{12}$.
[b] HC, hydrocarbon.
[c] HS, headspace measured by gas chromatography flame ionization detector (GC-FID) using a large-volume injection system (Ref. 11). This is regarded as the true gas phase concentration.
[d] Solid-phase microextraction (SPME) sampling time.

overestimated even at sampling time 30 sec showed that gas fiber diffusion was fast and thermodynamics was starting to have an effect. As the sampling time was increased, the gas fiber partitioning took over and the profile seen by SPME was largely determined by the gas-fiber partitioning coefficient.

The effect of the volume of the PDMS phase on the gas phase concentrations "seen" by the fiber is shown in Table 5. Here using the same 30-sec sampling time, great distortion was seen when compared with using a 100-μm SPME fiber. This is because the gas fiber diffusion was faster with the 30-μm fiber so thermodynamics (gas-fiber partitioning) had a greater effect. By the same token, using an even thicker fiber SPME and short sampling time should give an even closer approximation to capture the true gas phase composition. However, although short sampling times give the least distortion, longer sampling times are often necessary to obtain a signal of odorants only present at trace levels.

IV. CONCLUSIONS

The elucidation of the composition of odorants, which are responsible for a certain aroma experience of headspace, using either gastight syringes or SPME as sampling technique, remains a challenge. After careful optimization of the technique, high reproducibility can be obtained for both methods, to allow comparison of samples to each other. However, in cases in which the true gas phase composition of a sample is the goal, the use of labeled internal standards [14,15] or of gas phase models to compensate for gas phase sampling distortion, caused by losses within syringe components or enrichment of high-molecular-weight material on the fiber, is recommended.

REFERENCES

1. P Werkhoff, W Bretschneider, HJ Herrmann, K Schreiber. Labor Praxis 426–430, 1989.
2. W Jennings, M Filsoof. Comparison of sample preparation techniques for gas chromatographic analysis. J Agric Food Chem 25:440–445, 1977.
3. J Schaefer. Comparison of adsorbents in headspace sampling. In: P Schreier, ed. Flavour '81. Berlin: Walter de Gruyter, 1981, pp 301–313.
4. J Pawliszyn. Quantitative aspects of SPME. In: J Pawliszyn, ed. Applications of Solid Phase Microextraction. Cambridge: The Royal Society of Chemistry, 1999, pp 3–21.
5. X Yang, T Peppard. Solid-phase microextraction for flavor analysis. J Agric Food Chem 42:1925–1930, 1994.
6. AD Harmon. Solid-phase microextraction for the analysis of flavors. In: R Marsili, ed. Techniques for Analyzing Food Aroma. New York: Marcel Dekker, 1997, pp 81–112.
7. AJ Matich. Analysis of Food and Plant Volatiles. In: J Pawliszyn, ed. Applications of Solid Phase Microextraction. Cambridge: The Royal Society of Chemistry, 1999, pp 349–363.
8. C Bicchi, C Iori, P Rubiolo, P Sandra. Headspace sorptive extraction (HSSE), stir bar sorptive extraction (SBSE), and solid phase micro extraction (SPME) applied to the analysis of roasted Arabica coffee and coffee brew. J Agric Food Chem 50:449–459, 2002.
9. PA Rodriguez, CL Eddy, GM Ridder, CR Culbertson. Automated quartz injector/trap for fused-silica capillary columns. J Chromatogr 236:39–49, 1982.
10. PA Rodriguez, CL Eddy, GM Ridder, CR Culbertson. Improved radioactivity detector for fused-silica capillary columns. J Chromatogr 264:393–404, 1983.
11. S Maeno, PA Rodriguez. Simple and versatile injection system for capillary gas chromatography columns performance evaluation of a system including mass spectrometry and light-pipe Fourier-transform infrared detection. J Chromatogr 731:201–215, 1996.
12. L Pillonel, JO Bosset. Rapid preconcentration and enrichment techniques for the analysis of food volatile: A review. Lebensm-Wiss u-Technol 35:1–14, 2002.
13. DD Roberts, P Pollien, C Milo. Solid-phase microextraction method development for headspace analysis of volatile flavor compounds. J Agric Food Chem 48:2430–2437, 2000.

14. C Obretenov, J Demyttenaere, KA Tehrani, A Adams, M Kersiene, ND Kimpe. Flavor release in the presence of melanoidins prepared from L-(+)-ascorbic acid and amino acids. J Agric Food Chem 50:4244–4250, 2002.

15. G Zehentbauer, W Grosch. Apparatus for quantitative headspace analysis of the characteristic odorants of baguettes. Z Lebensm Unters Forsch 205:262–267, 1997.

21

Meat Aroma Analysis: Problems and Solutions

J. Stephen Elmore, Donald S. Mottram, and Andrew T. Dodson
The University of Reading, Reading, England

I. INTRODUCTION

A. Sample Preparation and Aroma Extraction

The analysis of cooked meat aroma has been reported many times. Even so, there are a large number of factors that need to be considered when carrying out such an analysis.

1. Before Cooking

Meat has a heterogeneous structure. Adipose tissue surrounds muscle, which contains muscle fibers, marbling fat, and connective tissue. If volatiles derived from the Maillard reaction are considered to be important, then the meat can be cooked with the adipose tissue removed. More reproducible results might be obtained if a muscle is minced and turned into a burger or if a regular shape is cut from the muscle, e.g., a circular slab. Sample size should be considered before cooking. Although the longissimus muscle of a rib steak from a steer may weigh around 120 g after grilling and removal of fat, the equivalent from a lamb may only weigh around 25 g. This may not be a large enough sample for the extraction technique that is being used.

2. The Cooking Process

Grilling, boiling, pressure-cooking, roasting, and frying are some of the cooking processes that can be used. The sample can be cooked for a constant time or to a constant internal temperature, or some visual aspect of the cooked meat can be used as a marker. It is important to employ reproducible methods when cooking the sample. This may be difficult when grilling or frying. The rate of cooking may affect the aroma of the cooked meat. Pressure-cooking, although not as commonly used in the kitchen as some of the other methods, has an advantage in that there is no sample loss and temperature control is straightforward.

3. The Extraction Method

Three commonly used extraction techniques for the analysis of aroma are simultaneous distillation/extraction (SDE), dynamic headspace adsorption on Tenax TA^{TM} (Buchem N.V., Apeldoorn, The Netherlands), and solid-phase microextraction (SPME) [1,2]. All of these techniques have positive aspects and drawbacks, and these are described. In SDE the sample is boiled for 1 to 2 hr and so precooking may not be necessary, although the meat is usually minced to maximize surface area for the extraction process. The other techniques can be used to examine either a chopped or a whole piece of cooked meat.

When a cooked sample is to be analyzed, it is important to extract the sample as soon after cooking as possible. Otherwise generation of aroma volatiles, known as *warmed-over flavor*, occurs when the meat is reheated [3]. However, there is no reason why the extraction cannot be carried out at room temperature.

B. Benefits and Drawbacks of Three Common Aroma Extraction Techniques

Simultaneous distillation/extraction, headspace adsorption on Tenax TA, and SPME have all been widely described. The discussion here is confined to the merits and drawbacks of these techniques.

1. Simultaneous Distillation/Extraction

The benefits of SDE include the following:

1. Efficient stripping of volatiles from foods allows quantitative recoveries to be achieved [4].

2. The aroma extract is obtained in a solvent; therefore, many injections can be performed from one extraction. Hence, one sample

provides material for gas chromatography (GC), GC mass spectrometry (GCMS), and quantitative GC olfactometry (GCO) techniques, such as CharmAnalysis^{TM} (Datu Inc., Geneva, NY, U.S.A.) and aroma extract dilution analysis [5].

3. Fractionation of the extract by liquid column chromatography can be carried out, resulting in increased separation of the components in the extract, facilitating the identification of minor components of the extract.

The drawbacks of SDE include the following:

1. Low-boiling volatiles can be lost when the extract is concentrated, by distilling off the solvent. These include compounds present at high levels in headspace extracts of cooked meat, such as 2-butanone, 2-pentanone, 2- and 3-methylbutanal, diacetyl, 1-propanol, and 1-penten-3-ol.

2. Artifacts can be formed [6].

3. Volatiles can be generated when samples are overcooked during extraction, e.g., enhanced lipid oxidation.

4. The sample is cooked and extracted as a slurry in water, not a typical cooking procedure.

2. Headspace Adsorption on Tenax TA

The benefits of headspace adsorption on Tenax TA include the following:

1. Artifact formation is minimal. Extraction is carried out under nitrogen flow.

2. Sensitivity is high, as the total contents of the trap can be injected onto the GC column.

3. It is suitable for low-boiling compounds.

4. A wide range of volatile compounds (from 3 up to 20 carbon atoms in the molecule) can be analyzed.

The drawbacks of headspace adsorption on Tenax TA include the following:

1. The dedicated injection system may be expensive, especially if automated operation is desired.

2. Accurate quantification is not straightforward as the amounts of different aroma compounds extracted are related to their partition ratio between the food and its headspace. Stable isotope dilution analysis using a labeled internal standard can be used [7].

3. Normally only one GC analysis is obtained from each extraction. Hence, it is not directly suitable for "quantitative" GCO techniques

3. Solid-Phase Microextraction

The benefits of SPME include the following:

1. Artifact formation is minimal.

2. Sensitivity is moderate. The volume of adsorbent material is less than that of the Tenax TA trap.

3. A relatively wide range of volatile compounds (depending on type of fiber) can be analyzed.

4. The technique is relatively sensitive toward polar aroma compounds [8].

5. The method uses a conventional splitless injector.

6. It is simple to use.

The drawbacks of SPME include the following:

1. It is not suitable for accurate quantification.

2. Only one GC analysis is obtained from each extraction.

II. DETERMINATION OF (Z)-4-HEPTENAL IN LAMB, BEEF, AND PORK USING SIMULTANEOUS DISTILLATION/ EXTRACTION

Work by us and by others [10,11] suggests that (Z)-4-heptenal may be an important contributor to cooked lamb flavor, in particular the "pastoral" flavor found in the meat of animals fed diets high in α-linolenic acid, such as grass and linseed. Levels of (Z)-4-heptenal in both the muscle and the adipose tissue of beef, pork, and lamb were determined by using SDE followed by GCMS, to determine whether levels of (Z)-4-heptenal were higher in lamb than in beef or pork.

A. Materials and Methods

Beef, lamb, and pork muscles (m. longissimus dorsi) with attached adipose tissue were purchased from a local retailer. Adipose tissue was separated from muscle, and both were analyzed by SDE. Each sample was minced, and 100 g of either muscle or adipose tissue was placed in a 2-L round-bottomed flask, to which 750 mL of water was added. The samples were then extracted for 2 hr, using 30 mL of pentane/ether (9:1) as solvent. Two drops of silicone antifoaming agent were added to the meat samples. After the first extraction, each sample was extracted a second time, to determine whether any remained in the sample. The extracts were stored in a freezer overnight to remove water and concentrated to 0.2 mL. Methyl decanoate (500 ng) was added as an internal standard. Two replicates were performed for each extraction. A standard of pure (Z)-4-heptenal was also injected to determine its retention time and mass spectrum.

The extracts were analyzed by GCMS, using a CP-Sil 8 CB low-bleed/ MS fused silica capillary column (60 m × 0.25 mm i.d., 0.25-μm film thickness; Varian Chrompack International B.V., Middleburg, The Nether-lands). Quantification of (Z)-4-heptenal in the extracts was performed by using multiple ion monitoring, by preparing a calibration curve using the peak area of the m/z 84 ion for (Z)-4-heptenal and the m/z 74 ion for methyl decanoate internal standard.

B. Results and Discussion

The mean concentrations of (Z)-4-heptenal are shown in Table 1. When comparing the first set of extractions for all of the meats, (Z)-4-heptenal was found at higher levels in adipose tissue than in muscle. The level of (Z)-4-heptenal in lamb adipose tissue was highest, whereas in muscle the level of (Z)-4-heptenal was highest in beef. The suitability of the method for measuring the levels of (Z)-4-heptenal in meat is questionable, however, as the extraction does not appear to be exhaustive. The high levels of (Z)-4-heptenal in the second set of extractions suggest that it is possibly being formed during the extraction process.

Josephson and Lindsay [12] suggested that (Z)-4-heptenal is formed from the decomposition of α-linolenic acid. Enser and associates [13] showed that lamb contains the highest levels of α-linolenic acid in the muscle of the three meats. However, they also showed that pork adipose tissue contains the highest levels of α-linolenic acid; more than 10 times higher than the levels of α-linolenic acid in lamb muscle and nearly 3 times higher than the levels of α-linolenic acid in lamb adipose tissue. As the level of (Z)-4-heptenal in lamb adipose tissue was substantially higher than in pork adipose tissue; this finding suggests that another pathway may also contribute to the formation of (Z)-4-heptenal in cooked meat.

Table 1 Concentrations of (Z)-4-heptenal in Beef, Lamb, and Pork Muscle and Adipose Tissue Determined by Simultaneous Distillation/Extraction

	Mean concentration, ng/100 g			
	First extraction		Second extraction[a]	
Meat	Muscle	Adipose tissue	Muscle	Adipose tissue
Lamb	210	1230	60	540
Pork	110	300	40	110
Beef	330	500	50	140

[a] Second extraction carried out on the same sample.

One way to reduce the formation of (Z)-4-heptenal during the extraction process would be to perform SDE under a static vacuum [4]. Hence cooking conditions would be less severe.

III. COMPARISON OF THE AROMA COMPOSITIONS OF GRILLED AND PRESSURE-COOKED LAMB MUSCLE, USING HEADSPACE ADSORPTION ON TENAX TA

Few researchers have examined the effects of different cooking methods on the aroma of cooked meat. MacLeod and Coppock [14] compared the aromas of boiled and roasted beef using SDE. They suggested that carbonyl compounds, sulfides, pyrroles, and pyridines were associated with roasted aroma, whereas benzenoids and furans may be associated with the desirable qualities of well-cooked boiled beef. In the experiment we describe here, headspace adsorption on Tenax TA was used to compare the aroma profiles of pressure-cooked and grilled lamb muscle.

A. Materials and Methods

The meat from five lambs was studied. Two loin chops from each animal were cooked. The first chop was cooked by grilling to a core temperature of 70°C, measured by using a small thermocouple inserted into the center of the chop. Cooking to a fixed temperature compensated for any variations in the thickness between and within the samples. Samples were turned over every 2 min during cooking. After cooking, the major muscle (m. longissimus dorsi) was separated from the rest of the chop and all visible fat was removed. The second loin chop was pressure-cooked. The major muscle from the loin chop was removed from the bone, trimmed of fat, and cooked in a 100-mL bottle fitted with an airtight polytetrafluoroethylene-lined screw top, at 140°C in an autoclave for 30 min.

Immediately after cooking, the meat was chopped in a blender and samples were immediately taken for volatile analysis. Headspace entrainment on Tenax TA was performed on 10 g of the chopped muscle at 60°C for 1 hr, using a nitrogen flow rate of 40 mL/min [15]. A standard (100 ng 1,2-dichlorobenzene in 1 μL methanol) was added to the trap at the end of the collection, and excess solvent and any water retained on the trap were removed by purging the trap with nitrogen at 40 mL/min for 10 min.

A CHIS injection port (SGE Pty Ltd, Ringwood, Australia), was used to desorb the volatiles thermally from the Tenax TA trap onto a CP-Sil 8 CB low-bleed/MS fused silica capillary column (60 m × 0.25 mm i.d., 0.25-μm film thickness; Varian Chrompack International B.V., Middelburg, The

Netherlands). The volatiles were then analyzed by GCMS, under the conditions reported by Elmore and colleagues [11]. A series of *n*-alkanes (C_5–C_{25}) in diethyl ether was analyzed, under the same conditions, to obtain linear retention index (LRI) values for the lamb aroma components.

Compounds were identified by first comparing their mass spectra with those contained in the NIST/EPA/NIH Mass Spectral Database or in previously published literature. Wherever possible, identities were confirmed by comparison of LRI values, with either those of authentic standards or published values. Quantities of the volatile compounds were approximated by comparison of their peak areas with that of the 1,2-dichlorobenzene internal standard, obtained from the total ion chromatograms, using a response factor of 1. A Student *t*-test was carried out on the quantitative data for each compound identified in the GCMS analyses.

B. Results and Discussion

Ninety compounds were quantified in the lamb. Each of these compounds was present at a level of at least 50 ng per 100 g of sample in the headspace extract of at least one of the samples (Table 2). Of those compounds that were quantified in both sets of samples, 35 were significantly affected by cooking treatment. Of those 35 compounds only the alkyl-substituted benzenes toluene, ethylbenzene, and styrene were at higher levels in the grilled lamb muscle, compounds not usually considered to be major contributors to cooked meat aroma. Heterocyclic compounds were found at much higher levels in the pressure-cooked lamb than in the grilled, many of which, such as thiophene, pyrazine, 2-furfural, and 2-acetylthiazole, could not be detected in the grilled lamb. Maillard reaction products, such as pyrazines, were found in only trace quantities in the grilled lamb, and the only Maillard-derived compounds present above 50 ng in the headspace of 100 g of sample were simple sugar degradation products, such as 3-hydroxy-2-butanone and 2,3-butanedione, and the Strecker aldehydes, 2-methylpropanal, 2-methylbutanal, and 3-methylbutanal.

Many lipid-derived compounds were also present at higher levels in the pressure-cooked meat than in the grilled. In particular, alkylfurans, such as 2-ethylfuran, 2-pentylfuran, and 2-(2-pentenyl) furan, were at levels more than 10 times higher in the pressure-cooked lamb than in the grilled lamb.

Although the surface temperature of grilled meat could exceed 100°C, the core temperature could never reach that temperature until all the water was lost. This explains the far higher levels of volatile compounds formed in pressure-cooked meat, where the product attained 140°C. Although subjected to a relatively intense cooking process, the pressure-cooked meat possessed a desirable meaty taste, but it was difficult to determine the

Table 2 Comparison of the Volatile Aroma Compounds in Grilled and Pressure-Cooked Lamb Muscle Extracted by Headspace Adsorption on Tenax TA

| Compound | Mean concentration in headspace, ng/100 g[a] | | | Method of identification[c] | LRI[d] |
	Grilled	Pressure-cooked	P[b]		
Acetone	66	1058	*	MS + LRI	501
1-Propanol	1251	3522	NS	MS + LRI	558
2-Methylpropanal	82	334	*	MS + LRI	558
2,3-Butanedione	274	712	**	MS + LRI	596
2-Butanone	422	3948	***	MS + LRI	602
2-Methylfuran	—	325		MS + LRI	604
3-Methylbutanal	112	3849	*	MS + LRI	655
Benzene	54	11	NS	MS + LRI	660
1-Butanol	33	111	**	MS + LRI	664
2-Methylbutanal	80	1988	**	MS + LRI	665
Thiophene	—	132		MS + LRI	670
1-Hydroxy-2-propanone	—	336		MS + LRI	680
1-Penten-3-ol	383	721	NS	MS + LRI	686
2-Pentanone	84	522	**	MS + LRI	688
Heptane	123	95	NS	MS + LRI	700
3-Pentanone	23	125	**	MS + LRI	701
2,3-Pentanedione	—	890		MS + LRI	701
Pentanal	441	749	NS	MS + LRI	702
2-Ethylfuran	26	456	**	MS + LRI	703
3-Hydroxy-2-butanone	614	857	NS	MS + LRI	721
Pyrazine	—	517		MS + LRI	735
Thiazole	tr	370		MS + LRI	736
2-Methyl-2-butenal	tr	95		MS + LRI	743
Dimethyl disulfide	tr	130		MS + LRI	745
Pyridine	5	128	***	MS + LRI	746
Pyrrole	8	394	***	MS + LRI	752
Toluene	177	104	*	MS + LRI	768
1-Pentanol	311	675	*	MS + LRI	769
2-Methylthiophene	tr	1099		MS + LRI	773
Cyclopentanone	7	238	***	MS + LRI	796
Octane	193	136	NS	MS + LRI	800
Hexanal	1748	1529	NS	MS + LRI	804
Dihydro-2-methyl-3(2*H*)-furanone	—	782		MS + LRI	811
Methylpyrazine	—	1361		MS + LRI	828
2-Furfural	—	561		MS + LRI	836
2-Furanmethanol	—	285		MS + LRI	859

Table 2 Continued.

| Compound | Mean concentration in headspace, ng/100 g[a] | | | Method of identification[c] | LRI[d] |
	Grilled	Pressure-cooked	P[b]		
Ethylbenzene	126	17	**	MS + LRI	863
1-Hexanol	157	455	*	MS + LRI	871
2-Heptanone	31	145	***	MS + LRI	891
Styrene	54	15	*	MS + LRI	895
(Z)-4-Heptenal	119	281	NS	MS + LRI	902
Heptanal	852	1812	NS	MS + LRI	904
2-Furanmethanethiol	—	302		MS + LRI	914
2,5 (and 2,6)-Dimethylpyrazine	tr	557		MS + LRI	917
Ethylpyrazine	—	184		MS + LRI	920
2,3-Dimethylpyrazine	15	67	**	MS + LRI	922
Dihydro-3(2H)-thiophenone	—	413		MS + LRI	961
Benzaldehyde	331	2738	***	MS + LRI	969
1-Heptanol	91	438	*	MS + LRI	972
Dimethyl trisulfide	—	182		MS + LRI	976
1-Octen-3-ol	151	431	**	MS + LRI	982
2,3-Octanedione	32	65	NS	ms + lri	986
6-Methyl-5-hepten-2-one	37	82	**	MS + LRI	987
2-Octanone	14	90	***	MS + LRI	992
2-Pentylfuran	48	988	***	MS + LRI	992
Dihydro-5-methyl-3(2H)-thiophenone	—	902		ms(19)	993
Dihydro-2-methyl-3(2H)-thiophenone	—	265		MS + LRI	997
3-Formylthiophene	—	71		MS + LRI	1001
(E) or (Z)-2-(2-Pentenyl)furan	5	536	***	ms(20)	1001
2-Ethyl-6-methylpyrazine	—	228		MS + LRI	1002
Octanal	542	1645	*	MS + LRI	1006
Trimethylpyrazine	—	119		MS + LRI	1006
2-Formylthiophene	—	301		MS + LRI	1009
2-Ethyl-(2H)-thiapyran	—	73		MS + LRI	1016
2-Acetylthiazole	—	349		MS + LRI	1025
Limonene	254	84	NS	MS + LRI	1034
5-Ethyl-l-formylcyclopentene	9	64	**	ms(21)	1037
Benzenacetaldehyde	—	89		MS + LRI	1051
(E)-2-Octenal	19	183	*	MS + LRI	1062
1-Octanol	142	372	NS	MS + LRI	1073
3-Ethyl-2,5-dimethylpyrazine	—	102		MS + LRI	1080

Table 2 Continued.

| Compound | Mean concentration in headspace, ng/100 g[a] | | | | |
	Grilled	Pressure-cooked	P[b]	Method of identification[c]	LRI[d]
3-Formyl-2-methylthiophene	—	387		MS + LRI	1091
2-Hexylfuran	6	175	**	MS + LRI	1092
2-Nonanone	11	88	***	MS + LRI	1092
2-Acetylthiophene	—	93		MS + LRI	1097
Nonanal	1100	1878	NS	MS + LRI	1107
(E)-2-Nonenal	42	541	NS	MS + LRI	1164
1-Nonanol	26	77	***	MS + LRI	1173
2-Decanone	6	50	***	MS + LRI	1193
2,3-Dihydro-6-methythieno-2,3c-furan	—	267		MS + LRI	1198
2-[(Methyldithio)methyl]furan	—	61		MS + LRI	1222
(E)-2-Decenal	24	518	NS	MS + LRI	1267
2-Undecanone	8	256	**	MS + LRI	1294
2-Octylfuran	tr	91		MS + LRI	1295
(E,E)-2,4-Decadienal	tr	113		MS + LRI	1325
(E)-2-Undecenal	17	422	NS	MS + LRI	1369
Methyldihydrothieno-thiophene	—	183		ms(19)	1380
2-Tridecanone	tr	185		MS + LRI	1496
2-Pentadecanone	—	139		MS + LRI	1699
1-Phytene	31	82	NS	ms + lri(22)	1782

[a] Means are from five replicate samples; tr, less than 2 ng; -; less than 1 ng in the headspace of 100 g of cooked lamb.

[b] Probability that there is a difference between means; NS, no significant difference between means ($P > 0.05$); *, significant at the 5% level; **, significant at the 1% level; ***, significant at the 0.1% level.

[c] MS + LRI, mass spectrum and LRI agree with those of authentic compound; ms + lri, mass spectrum identified using NIST/EPA/NIH Mass Spectral Database and LRI agrees with literature value (23); ms, mass spectrum agrees with spectrum in NIST/EPA/NIH Mass Spectral Database or with other literature spectrum.

[d] Linear retention index on a CP-Sil 8 CB low-bleed/MS column.

species characteristics. The surface of the grilled meat could become dry at temperatures greater than 100°C. Furthermore, hydrogen sulfide and ammonia, two important intermediates for flavor formation in the Maillard reaction, would be lost from the surface of the meat during grilling, whereas in the closed system of the pressure-cooked bottle, these compounds would be available at relatively high concentrations, allowing formation of heterocyclic compounds to occur. These included compounds formed from the reactions between lipid-derived volatile compounds and those volatile compounds derived from the Maillard reaction. Such compounds, identified in the pressure-cooked sample, include 2-ethyl-(2H)-thiapyran and 3-formylthiophene [16].

IV. TWO-FIBER SOLID-PHASE MICROEXTRACTION FOR THE ANALYSIS OF VOLATILE AROMA COMPOUNDS IN COOKED PORK

Two SPME fibers, a 75-μm Carboxen™/polydimethylsiloxane (PDMS) fiber and a 50/30-μm divinylbenzene (DVB)/Carboxen on PDMS fiber, were compared for the extraction of the aroma volatiles from cooked pork muscle. The GCMS showed that the two fiber coatings gave very different gas chromatographic profiles. Low-boiling aroma compounds dominate in the chromatogram of the aroma compounds desorbed from the 75-μm Carboxen/PDMS fiber, whereas the chromatogram of the aroma compounds desorbed from the fiber coated with 50/30-μm DVB/Carboxen on PDMS contained relatively high levels of higher-boiling aroma compounds.

A procedure was developed so that volatile compounds could be extracted using both fibers at the same time [17]. The contents of these fibers were then desorbed sequentially in the injection port of the GCMS, to give one chromatogram, containing both low-boiling and high-boiling volatile compounds.

A. Materials and Methods

Pork chops were purchased from a local retailer. The major muscle in the chop (m. longissimus dorsi) was trimmed of all visible fat. The muscle was cut into two rectangular pieces with a combined weight of 30 ± 0.1 g. The pieces were cooked in an autoclave as in the previous experiment. After cooling to ambient temperature, the cooked meat was analyzed immediately, without being chopped.

Two SPME devices were used: one contained a 1-cm fused silica fiber coated with a 75-μm layer of Carboxen/PDMS; the second contained a 1-cm Stable-flex fiber coated with 50/30-μm DVB/Carboxen on PDMS. Both fibers were conditioned before use by heating them in a gas chromatograph injection port at 250°C for 30 min.

For each SPME analysis, the screw top used during the cooking of the pork sample was replaced with a similar top containing one or two drilled holes of 3-mm radius, depending on whether one or two fibers were being used in the analysis (Fig. 1). Extractions were carried out with either fiber, or both fibers simultaneously, at 60°C for 30 min. Four pork samples were analyzed using each configuration. After extraction, the SPME device was removed from the sample bottle and inserted into the injection port of the GC/MS system.

GC/MS conditions are reported elsewhere [17]. The volatile compounds on each SPME fiber were desorbed for 3 min. When two fibers were injected, both were desorbed for 3 min and the second fiber was desorbed immediately after the first. Therefore, the splitter was closed for 6 min in this case.

Analysis of variance (ANOVA) was carried out on the quantitative data for each compound identified in the GCMS analyses. For those compounds exhibiting significant difference in the ANOVA, Fisher's least significant difference test was applied to determine which sample means differed significantly ($p < 0.05$).

B. Results and Discussion

Ninety-six compounds were present in the headspace at levels above 5 ng per 100 g of sample in at least one of the extracts [17]. A selection of these compounds showing the full range of responses to the two fibers is shown in Table 3. Low-boiling aroma compounds were present at relatively high levels on the Carboxen/PDMS fiber, whereas high-boiling aroma compounds were present at relatively high levels on the Carboxen/PDMS fiber, whereas high-boiling aroma compounds were present at relatively high levels on the DVB/Carboxen on PDMS fiber.

Using the two fibers together resulted in a chromatogram with only one of the 96 peaks absent [dihydro-2-methyl-3(2H)-furanone, which was also absent in the extract from the DVB/Carboxen on PDMS fiber], i.e., below 0.2 ng per 100 g of sample. All of the peaks found at trace levels, i.e., below 1 ng per 100 g of sample, on the single fibers were found at levels above trace detection, when using the two fibers together. Five compounds were at trace levels or undetected using the Carboxen/PDMS fiber [C_{13} to C_{15} alkanals, 2-(2-ethoxyethoxy)ethanol and an unknown] and 15 compounds were at trace levels or undetected using the DVB/Carboxen on PDMS fiber [including seven carbonyl-containing (C_8 or less), five

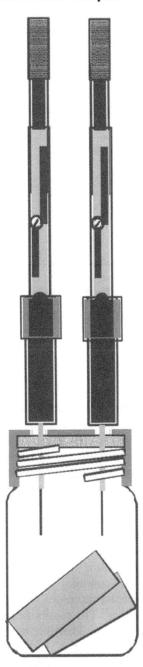

Figure 1 Experimental setup for two-fiber solid-phase microextraction.

Table 3 Volatile Compounds Present in the Aroma Extracts of Boiled Pork, Using Solid-Phase Microextraction with Two Different Stationary Phases, Separately and Combined

Compound	Carboxen PDMS	DVB/Carboxen on PDMS	Both fibers	Method of identification[b]	LRI[c]
	Concentration in headspace, ng/100g[a]				
3-Methylbutanal	65	2	15	MS + LRI	656
2-Methylthiophene	80	2	29	MS + LRI	776
Hexanal	368	18	227	MS + LRI	804
Methylpyrazine	129	15	74	MS + LRI	829
2-Pentylfuran	959	100	766	MS + LRI	994
(E,E),2,4-Decadienal	11	21	12	MS + LRI	1327
Tetradecanal	—	12	14	MS + LRI	1619
Hexadecanal	31	361	245	ms + lri	1825

[a] Means are from four replicate samples; —, less than 0.2 ng in the headspace of 100 g of boiled pork. PDMS, polydimethylsiloxane; DVB, divinylbenzene; LRI, linear retention index.

[b] MS + LRI, mass spectrum and LRI agree with those of authentic compound; ms + lri, mass spectrum identified using NIST/EPA/NIH Mass Spectral Database and LRI agrees with literature value [20].

[c] Linear retention index on a CP-Sil 8 CB low-bleed/MS column.

hydrocarbons (C_{11} or less), two alkylfurans (C_9 or less), and unsubstituted thiophene].

This work demonstrates that the use of the two fibers together should give more information on the aroma of boiled pork than use of either separately. However, for most compounds, the peak area obtained when using the two fibers together was not as great as the peak area obtained with one of the fibers used individually [17]. It is not easy to explain why this happened. These results suggest that using the two fibers together altered the attainment of equilibrium for the sample, the headspace, and the fiber.

Shirey [18] showed that Carboxen was a more effective coating than DVB for low-molecular-weight compounds. DVB contains relatively few micropores (2–20 Å diameter), whereas Carboxen contains relatively similar volumes of micro-, meso- (20–500 Å), and macropores (>500 Å), allowing adsorption of a wider molecular weight range. The results from this work confirmed this finding. High-molecular-weight aldehydes were present at higher levels on the DVB/Carboxen on PDMS fiber, whereas, in general, other volatiles were present at higher levels on the Carboxen/ PDMS fiber.

REFERENCES

1. R Teranishi, S Kint. Sample preparation. In TE Acree, R Teranishi, eds. Flavor Science: Sensible Principles and Techniques. Washington, DC: American Chemical Society, 1993, pp 137–167.
2. J Pawliszyn. Solid phase microextraction. In: RL Rouseff, KR Cadwallader, eds. Headspace Analysis of Foods and Flavors: Theory and Practice. New York: Kluwer Academic/Plenum Publishers, 2001, pp 73–87.
3. AM Pearson, JI Gray. Mechanism responsible for warmed-over flavor in cooked meat. In: GR Waller, MS Feather, eds. The Maillard Reaction in Foods and Nutrition. Washington, DC: American Chemical Society, 1983, pp 287–300.
4. A Chaintreau. Simultaneous distillation-extraction: From birth to maturity—review. Flavour Fragrance J 16:136–148, 2001.
5. TE Acree. Bioassays for flavor. In TE Acree, R Teranishi, eds. Flavor Science: Sensible Principles and Techniques. Washington, DC: American Chemical Society, 1993, pp 1–20.
6. DS Mottram, DJ Puckey. Artefact formation during the extraction of bacon volatiles in a Likens-Nickerson apparatus. Chem Ind 1978:385–386, 1978.
7. C Milo, I Blank. Quantification of impact odorants in food by isotope dilution assay: Strengths and limitations. In CJ Mussinan, MJ Morello, eds. Flavor Analysis: Developments in Isolation and Characterization. Washington, DC: American Chemical Society, 1998, pp 69–77.
8. JS Elmore, E Papantoniou, DS Mottram. A comparison of headspace entrainment on Tenax with solid-phase microextraction for the analysis of the aroma volatiles of cooked beef. In: RL Rouseff, KR Cadwallader, eds. Headspace Analysis of Foods and Flavors: Theory and Practice. New York: Kluwer Academic/Plenum Publishers, 2001, pp 125–132.
9. Reference deleted.
10. OA Young, TJ Braggins, J West, GA Lane. Animal production origins of some meat color and flavor attributes. In YL Xiong, C-T Ho, F Shahidi, eds. Quality Attributes of Muscle Foods. New York: Kluwer Academic/Plenum Publishers, 1999, pp 11–28.
11. JS Elmore, DS Mottram, M Enser, JD Wood. The effects of diet and breed on the major volatiles present in lamb aroma. Meat Sci 55:149–159, 2000.
12. DB Josephson, RC Lindsay. c4-Heptenal: An influential volatile compound in boiled potato flavor. J Food Sci 52:328–331, 1987.
13. M Enser, KG Hallett, B Hewitt, GAJ Fursey, JD Wood. Fatty acid content and composition of English beef, lamb and pork at retail. Meat Sci 42:443–456, 1996.
14. G MacLeod, BM Coppock. A comparison of the chemical composition of boiled and roast aromas of heated beef. J Agric Food Chem 25:113–117, 1977.
15. JS Elmore, DS Mottram. Extraction of novel thiazoles and 3-thiazolines from cooked beef by headspace trapping onto Tenax. In: CJ Mussinan, MJ Morello,

eds. Flavor Analysis: Developments in Isolation and Characterization. Washington, DC: American Chemical Society, 1998, pp 69–77.

16. JS Elmore, MM Campo, DS Mottram, M Enser. Effect of lipid composition on meat-like model systems containing cysteine, ribose, and polyunsaturated fatty acids. J Agric Food Chem 50:1126–1132, 2002.

17. JS Elmore, DS Mottram, E Hierro. Two-fibre solid-phase microextraction combined with gas chromatography-mass spectrometry for the analysis of volatile aroma compounds in cooked pork. J Chromatogr A 905:233–240, 2001.

18. RE Shirey. Optimization of extraction conditions for low-molecular-weight analytes using solid-phase microextraction. J Chromatogr Sci 38:109–116, 2000.

19. LJ Farmer, DS Mottram, FB Whitfield. Volatile compounds produced in Maillard reactions involving cysteine, ribose and phospholipid. J Sci Food Agric 49:347–368, 1989.

20. MS Smagula, SS Chang, C-T Ho. The synthesis of 2-(2-pentenyl) furans and their effect on the reversion flavor of soy bean oil. J Am Oil Chem Soc 56:516–519, 1979.

21. P Werkhoff, J Brüning, R Emberger, M Güntert, R Hopp. Flavor chemistry of meat volatiles: New results on flavor components from beef, pork and chicken. In: R Hopp, K Mori, eds. Recent Developments in Flavour and Fragrance Chemistry. Weinheim, Germany: VCH, 1993, pp 183–213.

22. G Urbach, W Stark. The C-20 hydrocarbons of butter fat. J Agric Food Chem 23:20–24, 1975.

23. N Kondjoyan, J-L Berdagué. A Compilation of Relative Retention Indices for the Analysis of Aromatic Compounds. Saint Genes Champanelle, France: INRA de Theix, 1996.

22

Analysis of Flavor Compounds from Microwave Popcorn Using Supercritical Fluid CO_2 Followed by Dynamic/Static Headspace Techniques

Ramachandran Rengarajan*
Quest International, Hoffman Estates, Illinois, U.S.A.

Larry M. Seitz
Grain Marketing and Production Research Center, Agricultural Research Service, U.S. Department of Agriculture, Manhattan, Kansas, U.S.A.

I. INTRODUCTION

There are a number of similarities in volatile compounds found in cereal-based foods. Bread, crackers, cooked basmati rice, and pearl millet often elicit a response similar to a roasty or popcornlike aroma due to presence of the same type or most often the same compound. There are not many reports on volatile compounds from freshly popped popcorn; the report by

---

* U.S. Cooperative investigations, U.S. Department of Agriculture, Agricultural Research Service (USDA-ARS), and the Department of Grain Science and Industry, Kansas State University. Contribution 03-213-J, Department of Grain Science and Industry, Kansas State Agricultural Experiment Station, Manhattan, KS 66506. Mention of firm names or trade products does not constitute endorsement by the U.S. Department of Agriculture over others not mentioned. The U.S. Department of Agriculture, Agricultural Research Service, Northern Plains Area, is an equal opportunity/affirmative action employer and all agency services are available without discrimination.

Walradt and associates [1] was the first. No other studies were reported until Schieberle [2] used aroma extract dilution analysis (AEDA) to identify the potent odorants in popcorn. The author used an 8-hr Soxhlet extraction method utilizing 600 g of freshly prepared popcorn and concentrated the extract obtained on a Vigreux column to about 150 mL, frozen in liquid nitrogen and subjected to a vacuum sublimation procedure similar to that of Sen and colleagues [3]. The author found 23 odorants, among which 2-acetyl-1-pyrroline (roasty, popcornlike), (E,E)-2,4-decadienal (fatty), 2-furfurylthiol (coffeelike), and 4-vinylguaiacol (spicy) were the most potent compounds with high flavor dilution (FD) factors. In addition, 2-acetyltetrahydropyridine and 2-propionyl-1-pyrroline had a strong roasty odor. The author also found that nona- and decadienals, several pyrazines, 4,5-epoxy-(E)-2-decenal, (E)-β-damascenone, and vanillin had an odor impact. Buttery and associates [4] used high-flow dynamic headspace sampling with Tenax trapping to identify flavor volatiles of some sweet corn products. The authors blended 100 g of cooked corn with 50 mL of water placed in a 1-L round-bottomed flask with an appropriate fitting, before it was purged with purified air at 3 L/min. The volatiles were extracted from Tenax with 50 mL of diethyl ether and concentrated by using a micro-Vigreux column. Dimethylsulfide, 1-hydroxy-2-propanone, 2-hydroxy-3-butanone, and 2,3-butanediol were the major volatiles; pyridine, pyrazine, alkylpyrazines, and 2-acetylthiazole were the additional compounds reported. They concluded that dimethylsulfide, 2-acetyl-1-pyrroline, 2-ethyl-3,6-dimethylpyrazine, acetaldehyde, 3-methylbutanal, 4-vinylguaiacol, and 2-acetylthiazole were the most important compounds for corn aroma. Schieberle [5] used stable isotope dilution analysis to quantify the four roast-smelling odorants in hot-air popped popcorn. The levels of 2-acetyltetrahydropyridine, 2-acetyl-1-pyrroline, 2-propionyl-1-pyrroline, and 2-acetylpyrazine obtained from simultaneous distillation extraction were 1207, 57, 21, and 7 μg/kg, respectively. A one-third decrease in the levels of these compounds was reported after 7 days of storage in a polyethylene bag. Their model studies revealed that a proline/fructose mixture was efficient in generating acetyltetrahydropyridine, whereas 1-pyrroline/2-oxopropanal was most effective for 2-acetyl-1-pyrroline formation. Buttery and co-workers [6] reexamined the popcorn aroma using dynamic headspace analysis with and without addition of water. In addition to the previously reported compounds, several new compounds, such as dimethylsulfide, dimethyl di- and trisulfides, 3-methylindole(skatole), alpha- and beta-ionones, 2-methyl-3-hydroxypyran-4-one (maltol), 2,3-dihydro-3,5-dihydroxy-6-methyl-4H-pyran-4-one (DDMP), 5-methyl-4-hydroxy-3(2H)-furanone (norfuraneol), and geranyl acetone, were identified in popcorn for the first time. The authors also found hydrogen sulfide as a major component of volatiles emitted during popping. Hofmann and associates [7] identified 5-acetyl-2,3-dihydro-1,4-thiazine, which had an intense popcornlike aroma, from the reaction of cysteine and ribose. The propionyl and butanoyl homologs also elicited a roasty popcornlike odor. Hofmann and Schieberle [8] identified the formation of the roast-smelling odorants 2-propionyl-1-pyrroline and 2-propionyltetrahydropyridine in Maillard-type reactions from two different model mixtures of proline and glucose reacted under aqueous and dry-heating conditions. The latter compound and the thiazine, mentioned previously, were not directly observed in the studies using popcorn samples. The objectives of this work were (a) to combine a sample extraction technique utilizing supercritical CO_2 with other headspace techniques using commercially available microwavable popcorn products and (b) to evaluate such a technique for rapid isolation of volatiles during routine flavor analysis.

II. EXPERIMENTAL PROCEDURES

A. Popcorn

Freshly popped popcorn was obtained by placing a bag of gourmet microwave popcorn (94% fat-free, artificially butter flavored) in a commercial grade (Amana Radarrange, Amana, IA, U.S.A., Model RFS8MP) microwave oven for 3 min at 70% medium-high strength. A portion of the popped corn was blended at room temperature using a commercial grade blender (Waring, Torrington, CT, U.S.A., Model No. 34BL97) at speed 1 for 30 sec. Immediately after blending, about 1.3 g of the coarse powder was placed into a 3/4-in sparger tube and 1 μL of 50 ng d_{10}-ethylbenzene internal standard was added to the sparger before it was connected to the purge and trap concentrator. The sample was subjected to a direct helium purge (DHP) procedure, as discussed later. Another 1.3 g of the blended sample was placed into a stainless steel supercritical fluid extraction (SFE) sample tube. Two microliters of 50 ng d_8-naphthalene was added to the tube before it was placed into the SFE extractor.

B. Direct Helium Purge

The Hewlett Packard 7695 (Palo Alto, CA, U.S.A.) purge and trap concentrator (equivalent to Tekmar, Model 3000 purge and trap instrument) was used with helium as the purge gas at a flow rate of 30 mL/min. The purged volatile compounds were trapped in Tenax, kept cold at 25 °C by a Turbocool™ system. The concentrator normally had nickel tubing for all internal plumbing connections. The system was modified with

Silicosteel^{TM} tubing (Restek, Bellefonte, PA, U.S.A.), which has an inert coating that minimizes thermal degradation of compounds. Also, the original moisture control system in the instrument was bypassed by using a short Silicosteel^{TM} line. The trapped compounds were desorbed and cryofocused before injection. A HP 5890 Series II gas chromatograph was used with an SGE DBX-5 column (SGE, Austin, TX, U.S.A.) (50 m, 0.32-mm inner diameter [i.d.], and 0.25-μm film thickness). The oven parameters were 40 °C initial temperature for 8 min, 2 °C/min up to 250 °C. The effluents from the column were detected by using a Hewlett Packard 5971 mass selective detector operating in an electron impact mode set at 70 eV. Further, the peak mass spectrum was matched with database spectra in the Hewlett Packard 59943B Wiley PBM MS database. The conditions for the purge and trap concentrator are listed in Table 1. One microliter of ethylbenzene-d_{10} internal standard was added to the sample just before the sample purge.

C. Supercritical Fluid Extraction

The SFE method used carbon dioxide to extract the volatile compounds from the food matrix. The operating temperature and pressure were manipulated to adjust the solvating properties of the fluid. There were several advantages to this technique, including no need to use large amounts of organic solvents, no solvent disposal fees, easy sample preparation, and low cost. This method used only small sample quantities, typically 1 g for popcorn experiments. The sample was placed into a stainless steel extraction vessel, which was then placed into an ISCO SFX220 (Lincoln, NE, U.S.A.)

Table 1 The Conditions for the Purge and Trap Concentrator

Parameter	Setting
Purge	10 min
Dry purge	10 min
Sample temperature	80 °C
Mount temperature	80 °C
Oven	200 °C
Transfer line	180 °C
Cryounion	200 °C
Cryocool	− 140 °C
Desorb	6 min At 220 °C
Inject time	0.85 min
Bake time	15 min At 225 °C

extractor attached to a high-pressure pump. The surface of the pump was kept at $-4\,°C$ by using a water circulating cooler. The SFE extraction conditions were 3000 psi and sample chamber temperature of 50 °C; 30 mL of supercritical fluid (SF) CO_2 was swept through the sample at an average flow rate of 0.7 mL/min with a 5-min initial equilibration. The flow of SF-CO_2 containing the extracted compounds was directed to a cooled collection container through a heated restrictor that was maintained at 145 °C. Depending on the experiment, the volatile compounds emerging out of the restrictor were collected differently, as discussed later. Unflavored microwave popcorn was microwaved for 3.5 min at 70% power level. After popping, about one-third of the bag was transferred to a blender and blended for 30 sec. About 1.3 g of the sample was immediately transferred into the stainless steel SFE sample tube. Two microliters of d_8-napthalene standard solution was added to the tube before it was placed into the extractor.

1. Supercritical Fluid Extraction

The restrictor was inserted into a purge and trap ∪-shaped sparger tube kept at $-78\,°C$ by using dry ice in a dewar flask. Just after the SFE run was completed, the sparger tube was removed from the dewar and was attached to the purge and trap concentrator. Then the sparger tube was subjected to purge conditions similar to those mentioned in the DHP method. One microliter of ethylbenzene-d_{10} internal standard was added to the sample just before the sample purge. These method parameters were very similar to those in previously published work [9].

2. Supercritical Fluid Extraction Solid-Phase Microextraction

In the SFE SPME experiment the restrictor was inserted into a purge and trap sparger tube, similarly to the SFE DHP procedure. The SFE isolation was repeated five times, each time with a new fresh sample of 1.3 g per run. The volatiles were collected in the same sparger tube kept at $-78\,°C$ by inserting the restrictor inside the sparger. After five subsequent SFE runs, the restrictor was removed from the sparger and the sparger was removed from the dewar flask. A freshly activated polydimethyl siloxane (PDMS) (100 µm, Supelco, Bellefonte, PA) fiber was then exposed by inserting the fiber into the sparger headspace while the sparger tube was slowly warming up to room temperature. After 30 min the fiber was retracted and exposed again in the gas chromatography (GC) injector for 2 min at 250 °C. The top of the column just below the injector was kept cold at $-140\,°C$ by using a microcryotrap (Model 981LN2, Scientific Instrument Services, Rigoes, NJ,

U.S.A.). Immediately after the 2-min fiber exposure, the GC run was started by rapidly heating the top of the column.

3. Supercritical Fluid Extraction Direct Injection

In the SFE DI experiment the restrictor was inserted into an SFE sample collection tube, which was kept cold by immersion into dry ice. These experiments were also repeated five times, each time with a fresh sample, and the volatiles were trapped/combined in the same tube. After the SFE isolations, about 1 mL of methylene chloride was added to dissolve the residue sticking on the side walls of the sample collection tube. The extract was then transferred into a small tapered vial. The vial was kept under dry ice and concentrated with a gentle stream of nitrogen to 0.5 mL, and 1 µL of this extract was manually injected with the GC in splitless injection mode. For experiments 2 and 3 an SFE chamber pressure of 4000 psi was used. These two experiments were also performed on a different gas chromatography mass spectrometry (GCMS) system, the Hewlett Packard (Palo Alto, CA, U.S.A.) 5890 Series II GC coupled with a 5970 mass selective detector.

All the experiments were performed in duplicate and the average area between two runs was used for the calculations in Table 2. A 4.8% error limit was calculated for the SFE experiment, using the d_8-napthalene internal standard. The added internal standard was not detected in the SFE SPME experiments; hence the results are reported as total ion chromatogram (TIC) peak area.

III. RESULTS AND DISCUSSION

A. Comparison of Dynamic Headspace Purge and Supercritical Fluid Extraction Dynamic Headspace Purge

Table 2 shows the combined data from DHP and SFE DHP analyses of artificially butter flavored popcorn. In both DHP and SFE DHP methods, tetramethylpyrazine was most abundant in the chromatogram. In the case of DHP, 3-hydroxy-2-butanone, acetic acid, 2,5-dimethylpyrazine, 2-furancarboxaldehyde, and 2-methylbutanal were most abundant. In contrast, acetic acid, 3-hydroxy-2-butanone, nonanal, and 2,5-dimethylpyrazine were most abundant in the SFE DHP experiment. Figure 1A shows the relative amounts of normal aldehydes from C_6 to C_{11}. Clearly the SFE DHP method extracted a higher quantity of aldehydes than the DHP. Nonanal was present at a higher level in SFE DHP. The level of hexanal was similar for SFE DHP and DHP. It would seem the amount of hexanal should be higher

Table 2 Comparison of Volatile Compounds from Artificially Butter Flavored Commercial Microwave Popcorn Using Direct Helium Purge and Supercritical Fluid Extraction Followed by Dynamic Headspace Purge[a]

Peak No.	Time, min	Compound	DHP	SFE DHP
1	3.28	Ethanol	0.11	4.73
2	3.81	Propanal, 2-methyl-	25.71	0.88
3	3.92	2-Propenal, 2-methyl-	0.54	0.00
4	4.11	2,3-Butanedione	15.50	4.56
5	4.17	2-Butanone	4.48	2.52
6	4.59	Acetic acid	55.98	138.60
7	5.17	Butanal, 3-methyl-	14.72	1.50
8	5.36	Butanal, 2-methyl-	27.65	2.37
9	5.79	2-Propanone, 1-hydroxy-	24.13	42.00
10	5.88	2-Pentanone	1.32	0.90
11	6.21	2,3-Pentanedione	11.96	3.58
12	7.03	2-Butanone, 3-hydroxy-	100.25	106.34
13	7.21	Propanoic acid	0.00	5.33
14	7.71	1-Butanol, 3-methyl-	0.34	0.29
15	7.73	Pyrazine	1.15	0.57
16	8.01	Dimethyldisulfide	2.75	0.66
17	8.41	Pyridine	0.00	2.77
18	8.46	1,2-Propanediol	145.77	62.60
19	9.04	Benzene, methyl-	61.13	60.36
20	9.16	Unknown (m/z 43,45,75,86)	1.50	0.92
21	9.28	1-Pentanol	2.18	4.02
22	9.63	2-Butanone, 1-hydroxy-	0.89	1.52
23	10.30	2,3-Hexanedione	0.39	0.00
24	10.57	2,3-Butanediol + Butanoic acid	2.96	3.99
25	11.00	Butanoic acid, ethyl ester	6.14	2.30
26	11.11	Hexanal	11.96	11.94
27	11.39	2,3-Butanediol	0.00	6.55
28	11.87	3(2H)-Furanone, dihydro-2-methyl-	2.83	3.35
29	12.37	Butanoic acid	3.09	27.15
30	12.95	Pyrazine, 2-methyl	13.23	12.35
31	13.23	Cyclohexane, 1,1,3-trimethyl-	5.43	1.12
32	13.58	2-Furancarboxaldehyde	32.69	43.82
33	14.33	Oxazole, 2,4,5- trimethyl-	5.11	7.71
34	14.70	Ethylbenzene-d_{10} (internal standard)	50.00	50.00
35	15.11	2-Furanmethanol	14.76	0.00
36	15.58	Unknown (m/z 43,72,98)	trace	1.41
37	15.77	Benzene, 1,4-dimethyl-	1.86	3.09
38	16.08	1-Hexanol	1.11	3.38
39	16.27	2-Propanone, 1-(acetyloxy)- (acetol acetate)	0.50	3.27

Table 2 Continued.

Peak No.	Time, min	Compound	DHP	SFE DHP
40	17.10	1-Acetoxy-2-propanol (propylene glycol acetate)	1.38	0.00
41	17.13	3-Heptanone	0.00	1.55
42	17.30	Butanoic acid, 3-methyl	0.00	2.41
43	17.35	Protoanemonine (2-cyclopentene-1,4,-dione)	1.19	1.20
44	17.49	2-Heptanone + Styrene	2.63	3.70
45	17.72	Nonane	0.60	2.79
46	18.47	Heptanal	4.84	13.94
47	18.55	Ethanol, 2-butoxy-	21.47	0.00
48	19.02	(E)-1-Buten-1-ol acetate	0.74	0.98
49	19.29	Ethanone, 1-(2-furanyl)- (2-acetyl furan)	1.61	2.47
50	19.46	Pyrazine, 2,5-dimethyl-	33.34	61.07
51	19.66	Pyrazine, ethyl-	1.25	0.00
52	19.80	Pyrazine, 2,3-dimethyl-	7.21	16.16
53	19.95	2(3H)-Furanone, dihydro-	trace	4.43
54	20.24	2-Acetyl-1-pyrroline	0.80	1.80
55	20.26	Propylcyclohexane	0.00	0.92
56	20.73	2-Buten-2-ol acetate	0.46	1.16
57	20.80	Pentanoic acid	0.48	3.69
58	21.82	2(3H)-Furanone, 5-methyl-	trace	0.18
59	22.32	2(3H)-Furanone, 2,5-dimethyl-	0.30	0.76
60	22.52	2-Heptanone, 6-methyl-	0.00	0.74
61	22.60	2-Heptenal, (E)-	0.19	0.23
62	22.81	3,3-Diethoxy-2-butanone	0.46	0.65
63	22.88	Benzene, 1-ethyl-2-methyl-	0.51	0.55
64	22.99	2-Heptenal, (E)-	0.81	2.63
65	23.03	2-Furancarboxylic acid	1.11	5.80
66	23.48	2-Furancarboxaldehyde, 5-methyl	1.23	2.66
67	23.57	Benzaldehyde	1.79	3.86
68	23.83	Dimethyltrisulfide	0.84	1.38
69	23.95	1-Heptanol	0.65	2.87
70	24.46	1-Octen-3-one	0.49	0.98
71	24.64	1-Octen-3-ol	1.33	2.21
72	24.73	2(5H)-Furanone, 3-methyl	0.00	1.06
73	25.08	6-Methyl-5-hepten-2-one	0.65	2.66
74	25.22	Furan, 2-pentyl-	2.61	5.04
75	25.44	2-Octanone	0.61	1.66
76	25.71	Decane	0.41	1.30
77	26.12	Pyrazine, 2-ethyl-5-methyl-	1.68	3.34
78	26.46	Pyrazine, 2,3,5-trimethyl-	16.87	42.16
79	26.63	Octanal	8.80	44.27

Table 2 Continued.

Peak No.	Time, min	Compound	DHP	SFE DHP
80	26.92	2-Thiophenecarboxaldehyde	1.28	5.42
81	26.95	2-Formyl-1-methylpyrrole	1.37	5.85
82	27.50	Phenol	0.00	3.23
83	27.74	Pyrazine, 2-methyl-6-vinyl-	0.53	1.96
84	28.19	Unknown (m/z 39,43,53,80,123)	12.74	44.05
85	28.34	1H-pyrrole-2-carboxaldehyde	0.75	6.91
86	28.68	Hexanoic acid	0.46	1.95
87	28.85	Benzene, 1-propenyl	8.81	7.83
88	28.97	1-Hexanol, 2-ethyl	1.47	1.22
89	29.07	Pyridine, 3-acetyl	0.70	1.37
90	29.20	Cyclotene	0.00	2.08
91	29.37	3-Octen-2-one	0.23	0.61
92	29.54	2,6-Dihydro-2H-pyran-2-one	0.00	3.70
93	29.76	2-Acetyl-1,4,5,6-tetrahydropyridine	0.00	2.18
94	29.92	Unknown (*m/z* 39,43,53,67,78,94,137,140)	0.92	2.52
95	30.14	2-Acetyl-3,4,5,6-tetrahydropyridine	5.26	30.49
96	30.18	2-Phenylacetaldehyde	2.74	15.38
97	30.99	2-Octenal, (E)-	1.28	4.74
98	31.49	2,3-Octanedione	0.68	1.15
99	31.94	Pyrrole, 2-Acetyl	0.46	0.69
100	32.23	Pyrazine, 3-ethyl-2,5-dimethyl-	4.21	12.78
101	32.40	Benzene, 2-methyl-1-propenyl	0.75	0.80
102	32.65	1-Octanol	0.00	8.75
103	33.40	Pyrazine, tetramethyl-	193.28	616.87
104	33.50	2-Nonanone	11.10	22.70
105	33.78	Unknown	0.00	5.11
106	34.10	Dimethyl-2-vinylpyrazine	1.45	4.68
107	34.62	Nonanal	11.54	75.73
108	35.03	Unknown (*m/z* 80,139)	1.19	4.59
109	35.19	Unknown (*m/z* 39,43,55,68 + 119,126,136)	0.00	1.89
110	35.26	2-Phenylethanol	1.42	2.18
111	35.79	Unknown (m/z 80,137)	1.11	3.71
112	35.92	2-Acetyl-3-methyl pyrazine	0.80	2.80
113	36.98	1H-Indene, 2,3-dihydro-5-methyl-	1.48	2.20
114	37.29	2-Methyl-5H-6,7-dihydrocyclopentapyrazine	0.26	1.15
115	37.65	2-Acetyl-3,4,5,6-tetrahydropyridine	4.84	11.65
116	37.69	1H-Indene, 2,3-dihydro-5-methyl-	2.64	2.30
117	38.21	2,3,5-Trimethyl-6-ethylpyrazine	1.26	4.81
118	38.69	Nonenal	3.56	3.96
119	39.02	2-Allyl-6-methylpyrazine	0.00	3.44

Table 2 Continued.

Peak No.	Time, min	Compound	DHP	SFE DHP
120	39.38	1-Nonanol	0.47	3.03
121	40.15	1H-Pyrrole, 1-(2-furanylmethyl)-	2.14	5.88
122	40.41	Naphthalene-d_8 (internal standard)	0.00	15.10
123	40.62	Naphthalene	3.68	3.17
124	40.81	2-Decanone	0.63	0.99
125	40.97	Dodecane	0.97	1.08
126	41.07	Octanoic acid, ethyl ester	2.41	3.41
127	41.38	2-Methyl-5H-6,7-dihydrocyclopentapyrazine	0.41	2.34
128	41.89	2-Tert-butylquinone	0.00	20.49
129	41.91	Decanal	2.47	33.48
130	42.82	2,4-Nonadienal	0.07	0.74
131	43.64	4-Ethylindan	1.25	0.82
132	45.91	2-Decenal, (Z)-	0.75	2.13
133	47.84	2-Undecanone	0.46	0.50
134	47.93	Tridecane	1.42	1.26
135	48.23	2,4-Decadienal, (E,Z)-	0.14	0.92
136	48.35	Naphthalene, 1 -methyl-	1.44	1.30
137	48.79	Undecanal	0.43	2.92
138	49.12	4-Vinyl-2-methoxy-phenol	4.25	26.59
139	49.34	2,4-Decadienal, (E,E)-	0.83	3.41
140	50.59	2-Dodecenal	0.70	1.49
141	50.68	Propanoic acid, 2-me, 2-et-3-oh-hexyl ester	0.53	1.26
142	51.14	Decanoic acid, ethyl ester	1.99	2.98
143	51.32	5-Acetyl-2,3-dihydro-1 H-pyrrolizine	0.84	2.07
144	51.46	Dodecanal	0.29	0.11
145	51.53	Napthalene, 2,7-dimethyl	0.37	0.11
146	51.67	1-Furfuryl-2-formyl pyrrole	0.34	0.92
147	51.85	Naphthalene, 1,3-dimethyl-	0.20	0.23
148	52.80	Delta-decalactone	0.00	2.07

[a] DHP, direct helium purge; SFE DHP, supercritical fluid extraction direct helium purge.

in the SFE DHP method, similarly to other higher aldehydes; however, this could be due to the higher volatility of this compound. The 20-min SFE isolation, which used dry ice for trapping the volatiles and no adsorbents inside the sparger tube, affected the amount of low-boiling, high-volatile compounds, as evident from the data summarized in Table 2. The higher boiling compounds, which were relatively nonvolatile, were least extracted

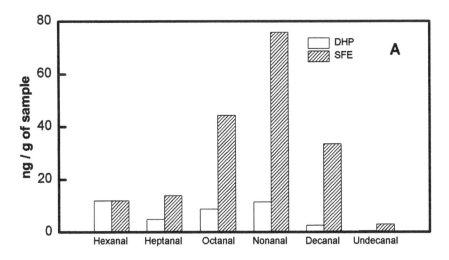

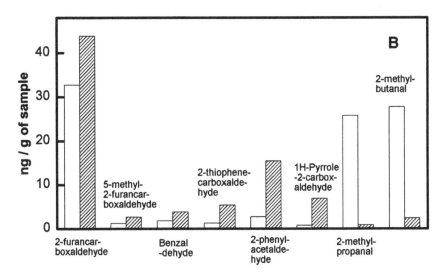

Figure 1 Relative amounts of aldehydes extracted by DHP and SFE DHP procedures. DHP, dynamic headspace purge; SFE DHP, supercritical fluid extraction DHP.

by the DHP method. Similarly, the branched aldehydes (Fig. 1B), such as 2-methyl-1-propanal and 2-methyl-1-butanal, were very volatile and not efficiently trapped during the SFE experiments.

Among the straight-chain ketones, 2-nonanone was most abundant in both the SFE DHP and DHP methods, even though the SFE DHP method showed two times greater extraction than the DHP method for this compound. High-volatile compounds, such as dimethyldisulfide, ethyl butanoate, and 2-methyl-1-propanal, were more abundant in the DHP experiment. As expected, 2,3-butanedione, which elicits a butterlike aroma, was observed three times more often in DHP because of its volatile nature. Other diketones, such as pentanedione, hexanedione, and octanedione, were all present in smaller amounts when compared to that of 2,3-butanedione. Organic acids from C_2 to C_6, along with 2-furancarboxylic acid and 3-methylbutanoic acid, were also present. In both methods acetic acid was most abundant. Propanoic acid and 3-methylbutanoic acid were present only in trace levels in the DHP method. Hexanoic acid, an important compound having a cheesy note, was also present at a lower level. Heptenal, octenal, nonenal, and decenal were the only enals found. Nonadienal and decadienal were also present in smaller amounts. It is evident from Table 2 that the SFE DHP method extracted the enals and di-enals better than the DHP method. Figure 2 shows the comparison of levels of the previously reported high-flavor-dilution (high-FD) compounds [10] detected by SFE DHP and DHP that were observed in this study. In both methods many of the character impact compounds were observed even with a small amount of sample. 2-Acetyl-1-pyrroline, a potent aroma compound with popcornlike aroma, was observed in this study from a 1.3-g sample of artificially butter flavored popcorn. Buttery and colleagues [4] reported that the heated zones in the purge and trap equipment destroyed many of the heat-labile compounds. Most of the compounds that were previously reported by several authors were observed. This suggested that modern purge and trap instruments were capable of handling reasonably heat-labile compounds, provided that proper modifications were made to make the system inert. Other potent aroma compounds, such as 2-acetyl-tetrahydropyridine, 2-vinylguaiacol, and 2,4-decadienal, were also observed in this study. When compared to DHP, in the SFE DHP method, three different isomers of 2-acetyltetrahydropyridine were found at elevated levels. This suggested that the combined SFE DHP method is more sensitive and has higher extraction efficiency than the DHP method by itself. 2-Methyl-5H-6,7-dihydrocyclopentapyrazine and cyclotene, both potent aroma compounds with a nutty note, were also found with the SFE DHP method. 5-Acetyl-2,3-dihydropyrrolizine was reported as a proline-specific Maillard reaction product [11]. This compound, which elicits a "smoky, haylike" character, has also been found in bread (crust) made by proline-enriched dough [12]. We observed this compound in both methods.

Furaneol[TM] (Firmenich), 4-hydroxy-2,5-dimethyl-3(2H)-dihydrofuranone, a compound with intense "cotton-candy like" aroma, has been

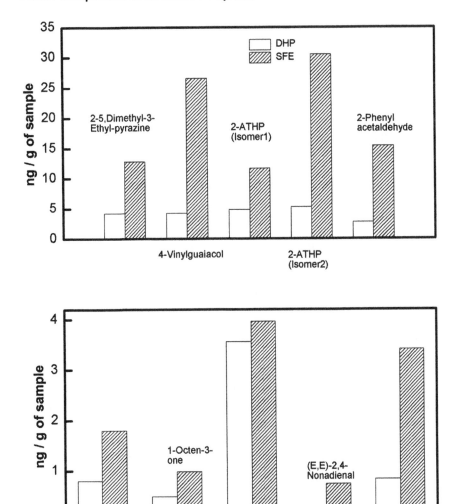

Figure 2 Relative amounts of selected compounds (high-FD) extracted by DHP and SFE DHP procedures. FD, flavor dilution; DHP, dynamic headspace purge; SFE DHP, supercritical fluid extraction DHP.

reported as a potent aroma compound in popcorn, bread, and many other food systems. This compound was not detected in popcorn by either the SFE DHP or the DHP method. On the other hand, both direct SPME and SFE SPME methods showed a good amount of Furaneol, a result that suggested that purge and trap was a less sensitive technique for observing this compound. Also, 2-furfurylthiol and β-damascenone were not observed in popcorn by either method. In this case, sample size and the reported [2,5] occurrence at low levels prevented detection of these compounds.

The observed n-alcohols were from C_5 to C_9 (Fig. 3B); 1-nonanol was the most abundant. This was consistent with nonanal and 2-nonanone's being the most abundant compounds among alcohols and ketones, respectively. Figure 3A shows differences in levels of some miscellaneous compounds detected by the two methods. It was evident that dimethyl disulfide level was highest in the DHP method, whereas dimethyl trisulfide level was highest in the SFE DHP method. Other compounds, such as 2-phenylethanol, delta-decalactone, and cyclotene, were present at higher levels in the SFE DHP method than in the DHP method. Only a trace of delta-decalactone was observed with the DHP method.

Figure 4 shows the relative amounts of pyrazines observed in both methods. The SFE DHP method was better in extracting all the observed 15 pyrazines except 2-methylpyrazine, which is fairly volatile. Tetramethylpyrazine (peak 103, Table 2), 2,5-dimethylpyrazine (peak 50, Table 2), and trimethylpyrazine (peak 78, Table 2) were the most abundant pyrazines observed with both methods. The presence of these pyrazines at such high levels compared to the previously reported levels suggests that these were added to enhance the flavor artificially.

In Fig. 5, which shows partial total ion chromatograms of extracts obtained by the DHP and SFE DHP methods, the differences in levels for some of the marked compounds are clearly evident. An unknown compound similar to the one previously reported [13] was also observed in popcorn for the first time. The MS and the infrared (IR) spectra for this compound and both spectra were consistent with those previously reported for 2-acetyl-2,3-dihydropyridine [13]. This compound was observed with both methods.

B. Results from Supercritical Fluid Extraction and Supercritical Fluid Extraction Direct Injection

The experiments were carried out on two different days using the same batch of popcorn, but freshly popped each time just before the experiment. The direct injection method results should represent the true nature of the sample as extracted by supercritical CO_2. The aims of this experiment were to apply the use of the SFE SPME and SFE DI methods in volatile compound analysis for the first time and to evaluate the feasibility of using this technique for flavor analysis from difficult solid matrices using smaller sample quantities.

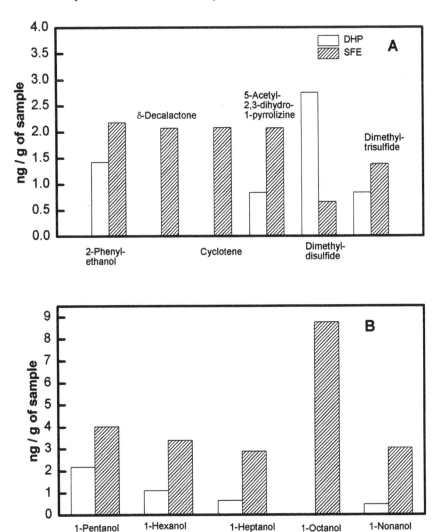

Figure 3 Relative amounts of selected compounds extracted by DHP and SFE procedures. DHP, dynamic headspace purge; SFE, supercritical fluid extraction.

Table 3 shows the list of compounds observed by both methods. Apparently, the most volatile compounds, pyrazines, aldehydes, etc., were observed more with the SFE SPME method than with the SFE DI method. Acetic acid, 2,5-dimethylpyrazine, 4-vinylguaiacol, and nonanal were the most abundant compounds observed with the SFE SPME method. On the other hand, palmitic acid, maltoxazine, nonanal, 2,3-dihydro-3,5-dimethyl-

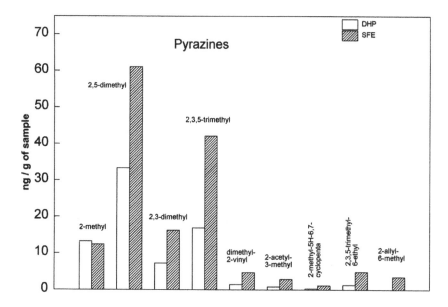

Figure 4 Relative amounts of pyrazine compounds extracted by DHP and SFE procedures. Tetramethyl pyrazine is not shown because of its higher abundance. DHP, dynamic headspace purge; SFE, supercritical fluid extraction.

4(H)-pyran-4-one (DDMP), and 4-vinylguaiacol were the most abundant compounds observed with the SFE DI method. In general, the SFE DI method showed more response for all the observed compounds except the highly volatile pyrazines and aldehydes. The level of cyclotene was similar in both methods. 2-Cyclohexen-1-one, a compound with sweet-brown, boiled odor, did not elicit any response with the SFE SPME method. One odor-impact compound, 2-acetyltetrahydropyridine, which has been previously reported in the literature [10], showed a slight decrease in level with the SFE DI method. The amount of acetic acid for the SFE DI method is underestimated because of the solvent delay used in the experiment. Several acids were present, including butanoic, hexanoic, octanoic, nonanoic, and undecanoic acids. Higher fatty acids such as palmitic and lauric were found only by the SFE DI experiment. Enals from C_6 to C_{11} were also present. More volatile enals such as hexenal and heptenal were not found by the SFE SPME method, whereas the higher enals were observed with SFE SPME method. Decenal was the most abundant among all of the observed enals with the SFE SPME method. Figure 6 compares all the observed pyrazines for both methods. It is apparent from the graph that the pyrazines had the highest response in the SFE SPME method. 2,3-Dimethyl and

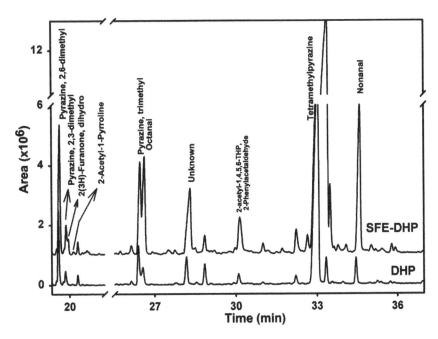

Figure 5 A section of total ion chromatogram from freshly popped popcorn (butter flavored) using SFE and SFE DHP methods. Note the improvement in 2-acetyl-1-pyrroline, a character-impact compound, by the SFE method. SFE, supercritical fluid extraction; SFE DHP, SFE dynamic headspace purge.

2-ethylpyrazines were found only with the SFE SPME method. 2,5-Dimethylpyrazine and 2,5-dimethyl-3-ethylpyrazine showed the greatest difference in these two methods. Dihydro-2(3H)-furanone, 2-ethyl-1-hexanol, tetrahydroquinolene, and 2,3,6,7-tetrahydrocyclopenta-azepine-8(1H)-one (THCA) had slightly higher abundance by the SFE SPME method than by the SFE DI method. It is interesting to note that THCA was observed about two times more often by SPME than by DI. This compound was reported [12] among other azepinones to have a bitter taste and was considered important in bread flavor. The authors did not specify the odor associated with this compound. 5-Acetyl-2,3-dihydro-1H-pyrrolizine, among several other pyrrolizines, were reported by the same authors as an important new bread crust flavor compound having a smoky, haylike odor. In the present experiment, this compound followed the same trend as THCA: i.e., the SFE SPME experiment showed more of this compound than the SFE DI method. Even though these compounds have relatively

Table 3 Comparison of Volatile Compounds from Unflavored Commercial Microwave Popcorn Using Supercritical Fluid Extraction Direct Injection and Supercritical Fluid Extraction Solid-Phase Microextraction[a]

Peak No.	Time	Compound	TIC Area	
			SFE DI	SFE SPME
1	3.28	Acetic acid	1648	5181
2	4.21	2,3-Pentanedione	0	87
3	4.60	Pentanal	118	0
4	4.64	Isopropanol	1029	0
5	6.61	Toluene	255	0
6	7.78	Octane	2004	0
7	8.36	Hexanal	2226	0
8	10.09	Pyrazine, methyl-	104	580
9	10.53	Furfural	545	394
10	11.61	E-2-Hexenal	900	0
11	11.99	2-Furanmethanol	264	315
12	12.74	1-Pentanol, 4-methyl-	0	22
13	13.11	6-Undecanone	0	32
14	13.84	Unkown	1723	0
15	13.94	Butanoic acid	0	389
16	14.63	Heptanal	878	0
17	15.44	2-Acetyl furan	31	69
18	15.50	2,5-Dimethyl pyrazine	1120	3386
19	15.75	Ethylpyrazine	0	255
20	15.88	2,3-Dimethyl pyrazine	0	169
21	15.93	2(3H)-Furanone, dihydro-	0	1217
22	16.87	2-Cyclohexen-1-one	1540	0
23	18.20	5-Methylfurfuryl alcohol	111	170
24	18.33	2-Heptenal, (E)-	270	0
25	18.83	5-Methyl furfural	142	223
26	19.19	1-Heptanol	0	141
27	19.67	Unknown (m/z 144)	178	48
28	20.31	Hexanoic acid	1942	1208
29	20.91	Pyrazine, 2-ethyl-6-methyl	100	502
30	21.11	Octyl aldehyde + trimethylpyrazine	1733	322
31	22.14	2-Methyl-6-vinyl pyrazine	43	121
32	22.30	2-Formyl pyrrole	174	199
33	22.42	2-Acetyl-dihydropyridine	170	261
34	22.63	2-Ethyl-1-hexanol	86	477
35	22.67	Cyclotene	6	7
36	22.83	Cyclohexanol, 4-chloro-, trans-	2560	163
37	23.63	2-Acetyl-1,4,5,6-tetrahydropyridine	86	97
38	23.86	2-Phenylacetaldehyde	232	98

Table 3 Continued.

Peak No.	Time	Compound	TIC Area SFE DI	SFE SPME
39	24.48	E-2-Octenal	381	327
40	24.84	Furaneol	685	224
41	25.15	2-Acetyl pyrrole	309	469
42	25.42	2,5-Dimethyl-3-ethyl pyrazine	244	1965
43	25.83	3,5-Dimethyl-2-ethyl pyrazine	0	188
44	26.00	2,3-Dimethyl-5-ethylpyrazine	0	196
45	26.23	Nonane	86	41
46	26.27	Dihydromaltol	136	126
47	26.42	4-Heptenal	309	47
48	26.74	Dimethyl-2-vinylpyrazine	0	264
49	27.00	Nonanal	5499	2638
50	27.73	Maltol (2-hydroxy-3-ethyl-4H-pyran-4-one)	119	78
51	27.78	Ethyl-*n*-propyldisulfide	0	140
52	28.07	Unknown (m/z 80, 137)	74	47
53	28.16	2-Acetyl-3-methyl pyrazine	93	160
54	28.79	2-Formyl-1-methylpyrrole	52	68
55	28.94	1-Tetradecanol	0	41
56	29.19	5-Methyl-(5H)-6,7-dihydrocyclopentapyrazine	75	245
57	29.47	2-Methyl-3,5-dihydroxy-5,6-dihydro-pyran-4-one (DDMP)	5413	0
58	29.71	Pyrazine; 3,5-diethyl; 2-methyl-	0	177
59	30.11	2-Nonenal	445	747
60	30.82	Octanoic acid	783	219
61	31.23	*N*-furfuryl pyrrole	58	93
62	31.69	Dodecane	156	242
63	32.19	2-Methyl-(5H)-6,7-dihydrocyclopentapyrazine	126	343
64	32.41	Decanal	1352	1625
65	33.52	4-Vinyl phenol	1300	333
66	33.94	Quinolene; 1,2,3,4-tetrahydro-	52	105
67	34.09	5-Hydroxymethylfurfural	1311	269
68	35.22	Tridecane	0	77
69	35.34	E-2-decenal	831	915
70	35.78	Nonanoic acid	2950	186
71	36.09	Unknown (m/z 67,94,109,122,137)	2257	200
72	36.41	Sulfurol (5-thiazoleethanol,4-methyl)	600	73
73	36.66	Tridecane	374	339
74	37.00	(E,E)-2,4-Decadienal	327	210
75	37.46	Tridecanal	312	381

Table 3 Continued.

Peak No.	Time	Compound	TIC Area	
			SFE DI	SFE SPME
76	37.92	4-Vinyl guaiacol	5400	3414
77	38.20	(E,Z)-2,4-Decadienal	699	559
78	39.26	Triacetin	6191	642
79	39.60	Butyl butyryl lactate	426	275
80	40.20	2-Undecenal	891	465
81	40.47	2,2,4- Trimethyl-1-(3-methyl-butoxy)	273	296
82	41.19	E-4,5-Epoxy-2-decenal	177	60
83	41.33	Tetradecane	360	131
84	41.90	5-Acetyl-2,3-dihydro- 1H-pyrrolizine	365	497
85	42.01	2,4,7,9-Tetramethyl-5-dicyne-4,7-diol	1567	12
86	42.16	Vanillin (3-methoxy-4-hydroxy benzaldehyde)	1321	185
87	42.74	1-Furfuryl-2-formyl pyrrole	158	46
88	43.50	1-Butanone, 1-phenyl	268	0
89	45.02	Decanol	290	0
90	47.45	2,3,6,7-Tetrahydrocyclopent(b)azepine-8(1H)-one	456	1041
91	48.75	Undecanoic acid	1317	0
92	52.77	Maltoxazine	5773	0
93	53.60	delta-Dodecalactone	29	0
94	54.48	Phenyl benzoate	1439	0
95	55.39	Isopropyl benzoic ester	1467	0
96	56.47	Lauric acid	1464	0
97	63.32	Palmic acid	14897	0

[a] TIC, total ion chromatogram; SFE DI, supercritical fluid extraction direct injection; SFE SPME, supercritical fluid extraction solid-phase microextraction.

high molecular weights, they showed a tendency similar to that of other observed pyrazines of being strongly attracted to the SPME fiber. This makes the SFE SPME method an excellent technique for observing nitrogen-containing polar compounds from the SFE extract.

Figures 7A and 7B show the relative differences of a few selected compounds from Table 3. 4-Vinylguaiacol and 4-vinylphenol were observed best with the SFE DI method. The compound 4-vinylguaiacol was considered a potent aroma compound for the popcorn aroma. Buttery

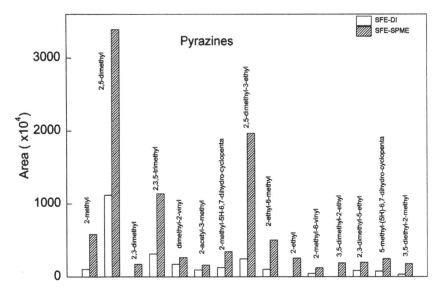

Figure 6 Relative amounts of pyrazine compounds extracted from 5.5 g of freshly popped nonflavored popcorn by SFE DI and SFE SPME procedures. Tetramethyl-pyrazine was not found. SFE DI, supercritical fluid extraction direct injection; SFE SPME, SFE solid-phase microextraction.

and coworkers [6] observed higher levels of 4-vinylguaiacol and 4-vinylphenol in their wet method than in their dry purge and trap method. Furaneol, among other cyclic alcohols, was most abundantly observed by the SFE DI method. This was anticipated because Furaneol appears to be difficult to detect by headspace methods [14]. This was verified in this study by analyzing a fresh sample of Furaneol. Vanillin was present in about seven times higher amounts in the direct injection method, as was expected because of its low-volatile nature [14]. 2-Phenylacetaldehyde and 2,4-decadienals, potent aroma compounds, were also observed better by the SFE DI method; DDMP was observed more often in the SFE DI method. This compound was reported by Cutzach and colleagues [15] to have a roasty, brown aroma. The compound sulfurol, which has a meaty aroma, was also observed more by the SFE DI method. Similarly, maltoxazine was found at very high levels with the SFE DI method. The spectrum of maltoxazine was similar to the one reported [6]. The odor impact of this compound has not been very clearly determined.

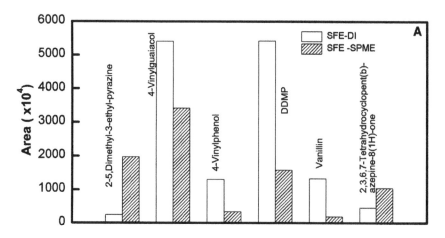

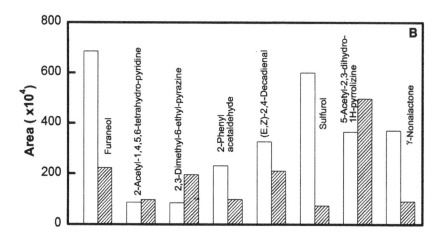

Figure 7 Relative amounts of selected compounds from 5.5 g of freshly popped popcorn extracted by SFE DI and SFE SPME methods. B, Sulfurol (4-methyl-5-thiazolethanol), a compound with meaty aroma, was most abundantly observed by the SFE DI method. SFE DI, supercritical fluid extraction direct injection; SFE SPME, SFE solid-phase microextraction; DDMP, 2,3-dihydro 3,5-dimethyl-4(H)-pyran-4-one.

IV. CONCLUSIONS

The combination of SFE and DHP has the following advantages: (a) no extensive sample preparation involving solvents was used; (b) smaller sample quantities were used than in conventional methods; (c) the extracted lipids (fats) and other nonvolatile components stayed in the sparger tube and did not transfer to the GC; (d) the method had enhanced sensitivity when compared to DHP alone. For highly volatile compounds, acetaldehyde, dimethylsulfide, etc., SFE DHP is not a choice because of inefficient trapping. Relative to SFE DI, SFE SPME proved more sensitive for nitrogen-containing compounds and other more volatile compounds. Further work is needed to evaluate methodologies utilizing an internal standard with SFE SPME. The SFE DI method was more sensitive for nonvolatile compounds.

REFERENCES

1. JP Walradt, RC Lindsay, LM Libbey. Popcorn flavor: Identification of volatile compounds. J Agric Food Chem 18:926–928, 1970.
2. P Schieberle. Primary odorants of popcorn. J Agric Food Chem 39:1141–1144, 1991.
3. A Sen, G Laskawy, P Schieberle, W Grosch. Quantitative determination of β-damascenone in foods using a stable isotope dilution analysis. J Agric Food Chem 39:757–759, 1991.
4. RG Buttery, DJ Stearn, LC Ling. Studies on flavor volatiles of some sweet corn products. J Agric Food Chem 42:791–795, 1994.
5. P Schieberle. Quantification of important roast-smelling compounds in popcorn by stable isotope dilution assays and model studies on flavor formation during popping. J Agric Food Chem 43:2442–2448, 1995.
6. RG Buttery, LC Ling, DJ Stearn. Studies on popcorn aroma and flavor volatiles. J Agric Food Chem 45:837–843, 1997.
7. T Hofmann, R Hassner, P Schieberle. Determination of the chemical structure of the intense roasty, popcorn-like odorant 5-acetyl-2,3-dihydro-1,4-thiazine. J Agric Food Chem 43:2195–2198, 1995.
8. T Hofmann, P Schieberle. Flavor contribution and formation of the intense roast-smelling odorants 2-propionyl-1-pyrroline and 2-propionyltetrahydropyridine in Maillard-type reactions. J Agric Food Chem 46:2721–2726, 1998.
9. LM Seitz, MS Ram, R Rengarajan. Volatiles obtained from whole and ground grain samples by supercritical carbon dioxide and direct helium purge methods: Observations on 2,3-butanediols and halogenated anisoles. J Agric Food Chem 47:1051–1061, 1999.
10. W Grosch, P Schieberle. Flavor of cereal products—a review. Cereal Chem 74:91–97, 1997.

11. R Tressl, B Helak, H Koppler, D Rewicki. Formation of 2-(1-pyrrolidinyl)-2-cyclopentenones and cyclopent(b)azepin-8(1H)-ones as proline specific Maillard products. J Agric Food Chem 33:1132–1137, 1985.

12. WLP Bredie, M Boesveld, G Budolfsen, L Dybdal. Modification of bread flavor with enzymes. Poster presented at the American Association of Cereal Chemists annual meeting, Minneapolis, 1998.

13. CY Chang, LM Seitz, E Chambers IV. Volatile flavor components of breads made from hard red winter wheat and hard white winter wheat. Cereal Chem 72:237–242, 1995.

14. J Budin. A review of methods for the analysis of oxygen-containing aroma compounds. In: G Reineccius, T Reineccius, ed. Heteroatomic Aroma Compounds. ACS Symposium Series 826, American Chemical Society, Washington, DC, 2002, 192–206.

15. I Cutzach, P Chatonnet, R Henry, D Dubourdieu. Identification of volatile compounds with a "toasty" aroma in heated oak used in barrelmaking. J Agric Food Chem 45:2217–2224, 1997.

23

Representative Sampling of Volatile Flavor Compounds: The Model Mouth Combined with Gas Chromatography and Direct Mass Spectrometry

Saskia M. van Ruth, Michael D. Geary, Katja Buhr, and Conor M. Delahunty
University College Cork, Cork, Ireland

I. INTRODUCTION

In the early stages of flavor research, most emphasis was on the development of methods to establish the chemical identity of constituents found only in trace quantities. Because flavor chemistry was considered a special application of organic chemistry, the methods worked out in flavor chemistry were applied in other fields in which small quantities of organic compounds can have profound biological effects, and vice versa. Examples are found in nutrition, air or water pollution, plant and animal hormones, and insect attractants [1].

The analytical task in flavor chemistry is rather complicated, as a relatively simple flavor may be composed of 50–200 constituents, which produce a combined effect to yield the characteristic flavor of a food. The large number of volatile flavor compounds that make up the total amount of flavor chemicals (>6000 [2]) complicates the analytical task even further. Now that many volatile flavor compounds have been identified, the task remains to determine biological activities as well as compounds' behavior in specific food products.

An instrumental approach to volatile flavor characterization can be regarded as a two-phase arrangement. The first phase comprises representative sampling of the volatile compounds. The importance of this sampling step cannot be overemphasized. Sample size requirements of analytical instruments have become smaller and smaller with developments. However, depending on the method employed, care must be taken so that heat-labile compounds are not destroyed by harsh conditions, highly volatile compounds not lost, and the original proportions of the compounds present in either the food or its headspace are retained. The analysis of the flavor compounds is the second step in volatile flavor analysis. Gas chromatography combined with a large variety of detectors, as well as mass spectrometry techniques, are among those most commonly used.

In the present study a model food and a real food were subjected to flavor analysis to determine the influence of the sampling method and the analysis technique on the volatile flavor profile obtained. The model food consisted of five volatile flavor compounds in sunflower oil. Rehydrated diced red bell peppers were used as a real food example. With regard to sampling method, static headspace sampling, dynamic headspace sampling, and model mouth sampling were compared. The analysis techniques included gas chromatography (GC) combined with flame ionization detection (FID) and mass spectrometry (MS), as well as the direct MS techniques atmospheric pressure chemical ionization–time of flight MS (APcI-TOFMS) and proton-transfer-reaction MS (PTRMS).

II. EXPERIMENTAL PROCEDURES

A. Experimental Design

The experimental design of the model food and the real food are presented in Tables 1 and 2, respectively.

B. Sampling Material

For the model food experiments, cold-pressed sunflower oil (Suma Wholefoods, Dean Clough, Halifax, UK) was used. The volatile flavor compounds 2-butanone, ethyl acetate, hexanal, and 2-heptanone were obtained from Sigma-Aldrich (Steinheim, Germany). Ethyl butyrate was purchased from Merck (Hohenbrunn, Munich, Germany). Solutions containing the five volatile compounds were prepared in triplicate for each measurement. Concentrations varied from 0.001% to 0.1% v/v for each compound, depending on the analytical method used (methods are specified in the following sections).

Table 1 Design of the Experiments Carried Out to Evaluate Sampling and Analysis Methods for Analysis of the Model Food (Five Flavor Compounds in Sunflower Oil)[a]

Analysis method	Sampling method		
	SH	DH	MM
GC	X	X[b]	X[b]
APcI-TOFMS		X[b]	
PTRMS		X[b]	X[c]

[a] SH, static headspace sampling; DH, dynamic headspace sampling; MM, model mouth sampling; GC, gas chromatography; APcI-TOFMS, atmospheric pressure chemical ionization–time of flight mass spectrometry; PTRMS, proton-transfer-reaction mass spectrometry.
[b] Time-average release, release measured over 1 min.
[c] Temporal release profile was determined.

For the real food studies, commercially dried diced red bell peppers from Turkey were supplied by Top Foods b.v. (Elburg, the Netherlands). For rehydration, bell peppers were placed in a glass flask and distilled water was added (1.2:10 w/w). The bell peppers were boiled for 10 min and allowed to cool down for 4 min.

C. Static Headspace Sampling and Analysis

The model food samples were subjected to static headspace sampling. An aliquot of sunflower oil solution (2.5 mL, 0.01% v/v for each individual

Table 2 Design of the Experiments Carried Out to Evaluate Sampling and Analysis Methods for Analysis of the Real Food (Rehydrated Diced Bell Peppers)[a]

Analysis method	Dynamic headspace sampling
GC-FID	X
GCMS	X
APcI-TOFMS	X
PTRMS	X

[a] GC-FID, gas chromatography flame ionization detection; GCMS, gas chromatography mass spectrometry; APcI-TOFMS, atmospheric pressure chemical ionization–time of flight mass spectrometry; PTRMS, proton-transfer-reaction mass spectrometry.

compound) was transferred into a 10-mL headspace vial. The samples were incubated at 37°C and agitated at 750 rpm for 6 min in the automated headspace unit of the GC. Then 2 mL of headspace was injected into the GC and analyzed according to the method described previously [3]. Five concentrations of each of the compounds were analyzed in triplicate for calibration, allowing quantification of the compounds in the air phase. Concentrations used varied with the compounds to ascertain that the experimental concentrations were included. Air phase concentrations (w/v) were divided by the concentrations in the oil (w/v) to determine air/oil partition coefficients.

D. Dynamic Headspace Sampling

For sampling of the model food samples, 10 mL of sunflower oil solution (0.1% v/v for each individual compound for GC analysis, 0.001% v/v for each individual compound for direct MS analysis) was transferred into the dynamic headspace vessel (50 mL). For the real food studies, rehydrated diced bell peppers (1.2 g dry weight) were placed in the vessel. The model food samples were analyzed by GCMS, APcI-TOFMS, and PTRMS. The real food samples were subjected to GC-FID, GCMS, APcI-TOFMS, and PTRMS analysis. Sampling conditions were similar for GC and direct MS analyses. Instrumental and analytical conditions employed were as described previously [4].

For the GCMS analysis, the headspace was flushed with nitrogen gas and after 5 min the effluent was trapped on Tenax TA^{TM} (Supelco, Bellefonte, PA, U.S.A.) for 1 min. Previous studies showed, after the initial minute, linear release for these compounds from sunflower oil over the 12 min [5]. For quantification of the volatile release from the model food samples, seven concentrations of the volatile compounds in pentane were analyzed in triplicate for calibration. For the real food samples, total ion counts were used in GCMS analyses and FID peak areas in GC-FID analysis. For the direct MS analysis, the headspace was drawn by a vacuum pump and led through a heated transfer line into the APcI-TOFMS or PTRMS instrument. A constant reading for at least 5 min was obtained, after an initial purge to account for the dead volume of the system. The outputs of individual pusher cycles of the APcI-TOFMS analysis were combined in an embedded computer to form an average spectrum over the full mass range recorded, at a rate of 1 spectrum/sec. The measurement was repeated twice.

In PTRMS analysis, the mass spectrometric data were collected for m/z 20–180 at a rate of 1 mass/sec for 10 cycles. Headspace concentrations of the compounds were calculated and averaged, as reported by Lindinger and associates [6].

E. Model Mouth Sampling

The model food (10 mL sunflower oil solution, 0.001% v/v for each individual compound) was placed in the sample flask of the latest version of the model mouth [7]. The headspace was flushed and the effluent analyzed by GCMS and PTRMS, as described for the dynamic headspace analysis. Headspace concentrations were calculated according to Ref. 6. For real-time PTRMS analysis one specific ion (2-butanone 73; ethyl acetate 89; hexanal 83; 2-heptanone 115; ethyl butyrate 117) was monitored for the individual compounds in order to obtain a temporal release profile. Data were corrected for fragmentation.

F. Statistical Analysis

Spearman's ranked correlation tests were conducted on the data sets to compare the various sampling and analysis techniques (Spearman's ranked correlation coefficient $= \rho$). A significance level of $p < 0.05$ was used throughout the study.

III. RESULTS AND DISCUSSION

A. The Model Food System

Various sampling and analysis techniques were compared for determining the volatile flavor profile of the model food. The design of the model food experiments is presented in Table 1. The release of the five flavor compounds from sunflower oil under static headspace, dynamic headspace, and model mouth conditions is shown in Tables 3 and 4. Under the experimental conditions used, sampling under model mouth (MM) conditions resulted in larger quantities of volatiles sampled than dynamic conditions (DH, Table 3) with the exception of ethyl acetate. Although a larger quantity does not necessarily coincide with better analysis results, it may be beneficial when samples are analyzed close to the detection limits of the analysis instrument. The relative release of the five flavor compounds from sunflower oil using the three sampling techniques is shown in Table 3. Not only did the sampling techniques result in quantitative differences, they affected the proportions of the compounds released as well. When comparing the static conditions and the dynamic conditions, the proportions of the more hydrophobic compounds, such as hexanal, 2-heptanone, and ethyl butyrate,

Table 3 Volatile Flavor Compounds Sampled Under Static Headspace, Dynamic Headspace, and Model Mouth Conditions Determined by Gas Chromatography and Their Octanol/Water Partition Coefficients (Log P)

Compound	log P^a	SH[b] [K × 1000]	[%][d]	DH[c] [µg]	[%][d]	MM[c] [µg]	[%][d]
2-Butanone	0.23	4.8	39	3.18	41	10.85	22
Ethyl acetate	0.64	5.3	43	1.13	14	23.13	46
Hexanal	1.29	0.6	5	0.84	11	4.21	8
2-Heptanone	1.82	0.5	4	1.16	15	5.42	11
Ethyl butyrate	1.70	1.1	9	1.47	19	6.35	13

[a] Log P data from Ref. 13.
[b] Air/oil partition coefficients. SH, static headspace.
[c] Cumulative release over 1 min. DH, dynamic headspace.
[d] Proportions [%] of volatiles sampled. MM, model mouth.

increased. Under static conditions, thermodynamic factors determined the headspace concentrations only [8]. De Roos [9] showed similarly that hydrophobic compounds were more retained under static than under dynamic conditions. In the present study, a higher release of hexanal, 2-heptanone, and ethyl butyrate was found under model mouth conditions (Table 3, comparison of DH and MM). However, ethyl acetate especially benefited from mastication. Data agree with other studies [8], which reported compound-dependent differences in release under dynamic and model mouth conditions, causing the proportions to change. Ethyl acetate is relatively hydrophilic (log $P = 0.64$) and has a high saturated vapor pressure

Table 4 Spearman's Ranked Correlation Coefficients of Proportions of Five Volatile Flavor Compounds[a]

	SH	DH	MM
log P	− 0.80*	− 0.20	− 0.60
SH		0.20	0.90[b]
DH			0.40

[a] Sampled under static headspace, dynamic headspace, and model mouth conditions and their octanol/water partition coefficients (log P). Raw data in Fig. 1.
[b] Significant correlation ($p < 0.05$).
  SH, static headspace; DH, dynamic headspace; MM, model mouth.

(12266, 25°C [10]), which explains its relatively large proportion under static conditions. According to Voilley and colleagues [10], the compound has a relatively high resistance to mass transfer in oil (27.4) compared to ethyl butyrate (6.7). This may explain the lower proportion under mild dynamic conditions, as well as the beneficial effect of rigorous movements under model mouth conditions.

The two direct MS techniques, APcI-TOFMS and PTRMS, were compared for release of the five volatile flavor compounds from sunflower oil under dynamic conditions (Table 5). The proportions of the compounds determined by the two techniques did not correlate significantly ($\rho = 0.60$, $p < 0.05$). The spectrum obtained with APcI-TOFMS lacked a response for hexanal and resulted in a relatively high reading of ethyl acetate. The ionization in the mixture may have been affected, as can occur if one compound has a greater affinity for protons than another [selective suppression, [11]]. Measurement of different concentrations of the volatile mixture showed a decrease of specific compounds, which may indeed indicate selective suppression [4]. Nonquantitative measurements with APcI-MS may also be related to the way the APcI source is operated [12]. In APcI-TOFMS, which measures all ions simultaneously, the response for an individual compound may not have been optimal as a result of the overall optimization. PTRMS results were more similar to the GC data, with a large fraction of 2-butanone. However, the proportions of 2-heptanone and ethyl butyrate were lower with PTRMS analysis.

The changes in flavor release with time (temporal release) under model mouth conditions were examined by PTRMS (Fig. 1). The maximal intensities of the compounds varied, but the time to reach this intensity did not differ significantly among the compounds. Ethyl acetate showed largest

Table 5 Individual Flavor Compounds' Proportions (Percentage) of the Total Volatile Flavor Released from Sunflower Oil Under Dynamic Headspace Conditions[a]

Compound	APcI-TOFMS	PTRMS
2-Butanone	33.2	68.1
Ethyl acetate	37.9	14.4
Hexanal	0.0	10.6
2-Heptanone	4.7	1.4
Ethyl butyrate	24.3	5.5

[a] Determined by atmospheric pressure chemical ionization-time of flight mass spectrometry (APcI-TOFMS) and proton-transfer-reaction mass spectrometry (PTRMS).

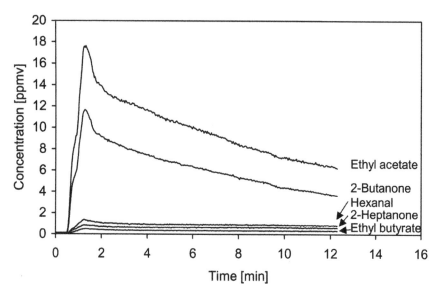

Figure 1 Temporal release profile of five volatile flavor compounds released from sunflower oil under model mouth conditions determined by proton-transfer-reaction mass spectrometry. (PTRMS). Sample was introduced after 0.5 min.

release, followed by 2-butanone, hexanal, 2-heptanone, and ethyl butyrate, respectively. The order in maximal intensities for the various compounds corresponded to the time-average GC data (Table 3), taking into account the variance in the GC analysis. Hexanal, 2-heptanone, and ethyl butyrate did not differ significantly in release in GC analysis. The variance in PTRMS analysis was lower than in the GC analysis of Tenax traps, giving a higher resolution and, therefore, significant differences between the compounds in Fig. 1. The order of hexanal, 2-heptanone, and ethyl butyrate followed the order of their hydrophobicity (Table 3): increased retention with higher hydrophobicity. As indicated, ethyl acetate showed a higher release than expected from its hydrophobicity only. Mass transfer must play a considerable role in the release of ethyl acetate.

B. The Real Food System

Four analysis techniques were compared for the analysis of the volatile flavor of rehydrated bell peppers: GC-FID, GCMS, APcI-TOFMS, and PTRMS (Table 6). Those compounds that demonstrated odor activity in rehydrated bell peppers previously were selected [4]. For all four methods,

Table 6 Individual Flavor Compounds' Proportions (Percentage) of the Total Volatile Flavor Released from Rehydrated Diced Red Bell Peppers Under Dynamic Headspace Conditions[a]

Compound	GC-FID	GCMS	APcI-TOFMS	PTRMS
2-Methylpropanal	9.8	11	18.5	8.8
2-Methylbutanal	24.4	21.6	33.5[b]	31.4
3-Methylbutanal	49.7	32.4	33.5[b]	36.9
Diacetyl	0.2	3.3	0	3.7
1-Penten-3-one	0.2	3.2	5	1.1
Hexanal	12.8	21.3	2.8	16.5
Heptanal	1.2	1.9	1	1.4
1-Octen-3-one + Dimethyl trisulfide	1.4	5	5	0.3
2-Methoxy-3-isobutylpyrazine	0.1	0.3	0.8	0.1

[a] Determined by gas chromatography flame ionization detection (GC-FID), gas chromatography mass spectrometry (GCMS), atmospheric pressure chemical ionization–time of flight mass spectrometry (APcI-TOFMS), and proton-transfer-reaction mass spectrometry (PTRMS).
[b] 2- and 3-Methylbutanal could not be separated.

2-methylpropanal and 2- and 3-methylbutanal were among the major compounds in quantity. Differences in composition were observed for the compounds in the lower concentration range. The GC-FID analysis resulted in a low response for diacetyl and 1-penten-3-one compared with that in the other method, but was still quite similar to the GCMS results ($\rho = 0.95$). In APcI-TOFMS analysis, low proportions of diacetyl and hexanal were found. The APcI-TOFMS correlated less significantly with the other techniques: with GCMS ($\rho = 0.75$), GC-FID ($\rho = 0.78$), and with PTRMS ($\rho = 0.61$). As mentioned, this may be caused by selective ion suppression [11] or suboptimal APcI source operation for these particular compounds [12]. The GC-FID/MS and PTRMS analyses gave fairly similar proportions for the various compounds ($\rho = 0.87$ and $\rho = 0.88$, respectively).

IV. CONCLUSIONS

The present study showed significant differences in volatile flavor profiles of a model food and a real food as a result of sampling method (SH, DH, MM) and analysis technique (GC-FID, GCMS, APcI-TOFMS, PTRMS).

REFERENCES

1. R Teranishi. Challenges in flavour chemistry: An overview. In CJ Mussinan, MJ Morello, eds. Flavour Analysis: Developments in Isolation and Characterization. Washington, DC: American Chemical Society, 1998, pp 1–6.
2. H Maarse, CA Visscher. Volatile Compounds in Foods. Zeist: TNO-Voeding, 1991.
3. SM van Ruth, I Grossmann, M Geary, CM Delahunty. Interactions between artificial saliva and 20 aroma compounds in water and oil model systems. J Agric Food Chem 49:2409–2413, 2001.
4. SM van Ruth, E Boscaini, D Mayr, J Pugh, M Posthumus. Evaluation of three gas chromatography and two direct mass spectrometry techniques for aroma analysis of dried red bell peppers. Int J Mass Spectrom 223–224:55–65, 2003.
5. SM van Ruth, CH O'Connor, CM Delahunty. Relationships between temporal release of aroma compounds in the model mouth system and their physicochemical characteristics. Food Chem 71:393–399, 2000.
6. W Lindinger, A Hansel, A Jordan. Proton-transfer-reaction mass spectrometry (PTR-MS): On-line monitoring of volatile organic compounds at pptv levels. Chem Soc Rev 27:347–354, 1998.
7. SM van Ruth, JP Roozen. Influence of mastication and saliva on aroma release in a model mouth system. Food Chem 71:339–445, 2000.
8. M Harrison. Mathematical models of release and transport of flavors from foods in the mouth to the olfactory epithelium. In DD Roberts, AJ Taylor, eds. Flavor Release. Washington, DC: American Chemical Society, 2000, pp 179–191.
9. KB de Roos. Physicochemical models of flavor release from foods. In DD Roberts, AJ Taylor, eds. Flavor Release. Washington, DC: American Chemical Society, 2000, pp 126–141.
10. A Voilley, MA Espinosa Diaz, C Druaux, P Landy. Flavor release from emulsions and complex media. In DD Roberts, AJ Taylor, eds. Flavor Release. Washington, DC: American Chemical Society, 2000, pp 142–152.
11. J Sunner, G Nicol, P Kebarle. Sensitivity enhancements obtained at high temperatures in atmospheric-pressure ionization mass-spectrometry. Anal Chem 60:1300–1307, 1988.
12. AJ Taylor, RST Linforth, BA Harvey, A Blake. Atmospheric pressure chemical ionization mass spectrometry for in vivo analysis of volatile flavour release. Food Chem 71:327–338, 2000.
13. DR Lide. CRC Handbook of Chemistry and Physics. New York: CRC Press, 1997.

24

Effects of Oral Physiological Characteristics on the Release of Aroma Compounds from Oil-in-Water Emulsions Using Two Mouth Simulator Systems

Michael D. Geary, Saskia M. van Ruth, and Conor M. Delahunty
University College Cork, Cork, Ireland

Edward H. Lavin and Terry E. Acree
Cornell University, Geneva, New York, U.S.A.

I. INTRODUCTION

The release of aroma compounds into the gas phase above a food or beverage is primarily governed by the physicochemical properties of the corresponding food or beverage matrix. Extensive reviews of the effects of lipids [1–3], carbohydrates [4–5], and proteins [6–8] and various related changes in pH [9–11] and viscosity [12–13] on the rate and amount of aroma released have been previously published. Additionally, the relative importance of each of these parameters has been found to vary in relation to the physicochemical properties (functional group, log P, chain length, etc.) of the aroma compounds [14–15].

The influence of specific physicochemical properties of a food matrix on the volatility of an aroma compound can be observed using static headspace analysis and quantified in terms of the air-to-matrix partition coefficient for that compound [16–17]. Partition coefficients describe the thermodynamic component and the extent of aroma release under equilibrium conditions. Such conditions, however, rarely provide the true amount of aroma stimulus presented to the olfactory receptors during consumption via the retronasal pathway [18–19]. In dynamic headspace analysis, sweeping of the aroma compounds from the headspace above the sample with an inert gas to a cold trap, or adsorbents such as Tenax TA^{TM} (Supelco, Bellefonte, PA) or the various fibers used in solid-phase microextraction (SPME), before gas chromatographic injection, not only allows for greater sensitivity [20] but incorporates the influence of the mass transfer coefficients of the aroma compounds. Mass transfer coefficients describe the kinetic component of aroma release and the rate at which equilibrium conditions are achieved under nonequilibrium conditions. The related importance of the rate of release of aroma to changes in the aroma perception of differing solid food matrices was proposed in recent studies [21–22].

During consumption, both the structure and the composition of the food matrix are influenced by the oral physiological parameters of the individual. Some of the significant changes experienced by foods in the mouth include breakdown of the food structure and changes in the viscosity of the liquid phase during chewing, mixing, dilution, and/or hydration of the food matrix components with saliva [23–24]. Additionally, the aroma profiles of many fruits and vegetables contain aroma compounds that are generated through enzymatic reactions as a result of the disruption of the tissue cells during consumption [25–28] and/or the presence of saliva proteins [29]. Furthermore, large variations between people have been measured for both masticatory efficiency [30] and saliva composition and flow rate [31].

In order to incorporate the described effects of the oral physiological parameters on the mechanisms governing the release of aroma compounds from food systems to existing in vitro analysis, a number of devices (mouth simulators) have been developed [12,32–38]. Additionally, these devices were also developed to relate the aroma release more closely to the aroma profile reaching the olfactory receptors in vivo. Both the retronasal aroma simulator (RAS) and the model mouth system have been shown to differ significantly from both purge and trap and conventional headspace analysis [37,39–40]. Furthermore, verification of both the RAS and model mouth system with in vivo measurements has been undertaken [39,41].

The effects of the primary structural and compositional properties of emulsions on the release of aroma have been both systematically investigated [42–43] and mathematically predicated [44–45] in previous studies. The aim of this chapter is to highlight the use of both the RAS and the model mouth system, in two separate studies, in understanding the effects of oral physiological characteristics on the release of aroma as a function of the physicochemical properties of model emulsion systems during consumption.

II. MATERIALS AND METHODS

A. Sample Materials

In both studies, the five aroma compounds used were 2-butanone, hexanal, and 2-heptanone supplied by Sigma-Aldrich (Steinheim, Germany) and 1-butanol and n-butyl acetate supplied by Lancaster (Walkerburn, UK). The artificial saliva was composed of $NaHCO_3$ (5.208 g), K_2HPO_4 (1.369 g), NaCl (0.877 g), KCl (0.477 g), $CaCl_2$ (0.441 g), NaN_3 (0.5 g), mucin (2.1610 g; Sigma-Aldrich) and α-amylase (200,000 units; Fluka Chemie, Buchs, Switzerland). Tween-20 (polyoxyethylene sorbitan monolaurate; Fluka Chemie) was used as an emulsifier in both studies.

B. Model Mouth Study

1. Emulsion Preparation

Tween-20 (1% w/v) was dispersed in distilled water before the addition of the sunflower oil (40% w/v; Mazola, Bestfoods, Esher, Surrey, UK) and further mixing. The solution was then passed through a prototype single-valve homogenizer until the required mean particle size ($d_{32} = 0.85\,\mu m$) was achieved. The mean particle size of the final emulsion was measured by using a Malvern Mastersizer laser diffractometer (model S Ver. 2.15, Malvern Instruments, Malvern, UK) and remained constant throughout the experiment. The five aroma compounds were added at final concentrations of 0.001% v/v per compound to the emulsion. The emulsion was stored at 4°C in the absence of light until used.

2. Static Headspace Gas Chromatography

Static headspace gas chromatography (SHGC) was used to study the effects of saliva volume on the retention of the five aroma compounds by the emulsion. Emulsion to saliva ratios included 100:0, 80:20, 60:40, and 40:60. The 2 mL of the samples was transferred to 10-mL headspace vials, which were then incubated at 37°C and agitated at 750 rpm for 6 min, using an automated headspace unit (Combipal-CTC Analytics; JVA Analytical Ltd., Dublin, Ireland). One milliliter of the headspace was injected and analyzed by using a gas chromatograph (GC; Varian CP-3800; JVA Analytical Ltd.) equipped with a flame ionization detector (FID). The injector and detector temperatures were 225°C and 300°C, respectively. A BPX5 capillary column [60 m × 0.32 mm inner diameter (i.d.), 1.0-μm film thickness; SGE, Kiln Farm, Milton Keynes, UK] was used. The carrier gas was helium at a flow rate of 1.9 mL/min. The chromatographic oven was at an initial temperature of − 30°C for 1 min, increased to 40°C at a rate of 100°C/min and held for 4 min, and subsequently increased to 90°C, 130°C, and finally 270°C at rates of 2°C, 4°C, and 8°C/min, respectively.

Triplicate samples of each solution were analyzed. Calibration graphs for each of the compounds were prepared by GC-FID analysis of five pentane solutions containing known amounts of each compound. Air/liquid partition coefficients ($K_{a/l}$) for each of the five compounds were calculated by dividing the air phase concentrations (w/v) by the liquid phase concentrations (w/v) for each of the five compounds.

3. Analysis of Aroma Release Under Varying Model Mouth Parameters

The model mouth system used in the study is primarily composed of a 70-mL glass flask with connecting parts and incorporating both controllable gas flow and mixing. Two variable-voltage controllers and a variable motor allowed the accurate control of both the up-and-down and circular motions (interdependent) of the Teflon® plunger to simulate chewing and mixing of the sample. Nitrogen gas swept the headspace over the sample (100 mL/min) during isolation of the volatile compounds (1 min), while a continuous water flow through the double-walled sample flask kept the temperature at 37°C. The released volatiles were trapped on a 60/80 mesh of Tenax TA adsorbent. The emulsion was placed into the model mouth and sampled by using the five mixing rates and four emulsion-to-saliva ratios design listed in Table 1. The volatile compounds were thermally desorbed at 220°C over 4 min, using a Tekmar Purge and Trap 3000 concentrator (JVA Analytical Ltd.), transferred via a heated line and cyrofocused at − 120°C using a Tekmar Cryofocusing Module (JVA Analytical Ltd.). The compounds were thermally injected for 2 min at 235°C onto a GC (Varian Star 3400 CX, JVA Analytical Ltd.) and detected by using a mass spectrometer (MS; Varian Saturn 3, JVA Analytical Ltd.). The GC column used was a BPX5 capillary column (60 m × 0.32 mm i.d., 1.0-μm film thickness; SGE) with a helium carrier gas flow of 1.9 mL/min. The GC oven temperature program used for the analysis was identical to that previously employed on the GC-FID for static headspace GC (SHGC). The ionization of the samples was achieved at 70 eV, with the MS scanning from m/z 40 to 400 (3 scans/sec). The compounds were identified and confirmed, using the retention times and mass spectra of single authentic compounds in pentane. As previously

Table 1 Range of Model Simulator Parameters Incorporating Varying Mixing Rates and Emulsion-to-Saliva Ratios

Parameter	Model mouth	Retronasal aroma simulator
Mixing rate	0 Chewing cycle/min	0 rpm (0/sec)
	25 Chewing cycles/min	200 rpm (41.5/sec)
	52 Chewing cycles/min	300 rpm (62.2/sec)
	79 Chewing cycles/min	400 rpm (83/sec)
	107 Chewing cycles/min	500 rpm (103/sec)
Emulsion-to-saliva ratio	100:0	100:0
	80:20	80:20
	60:40	60:40
	40:60	40:60
Total sample volume	10 mL	180 mL

five pentane solutions containing known amounts of each compound were analyzed in triplicate in order to quantify the amount of each compound released in the model mouth.

C. Retronasal Aroma Simulator Study

1. Emulsion Preparation

Starch (1.2% w/v; Sigma-Aldrich) was dispersed in distilled water and heated to 85°C at 2°C/min by using a Buchi Rotovapor-R (Brinkmann Instruments, Westbury, NY). The starch solution was allowed to cool to room temperature. Tween-20 (1% w/v), which was used as an emulsifier, was dispersed in the starch solution by using a stirring bar before the addition of the sunflower oil (40% w/v; Wegmans Food Markets, Rochester, NY) and further mixing for 10 min with a Polytron homogenizer (Brinkmann Instruments). The solution was then passed through a high-pressure homogenizer (Manton-Gaulin Manufacturing Co. Inc., Everett, MA) twice at 4000 psi. The final emulsion had a mean particle size (d_{32}) of 0.85 μm, which was measured by using a laser diffraction particle size analyzer, model LS120 (Coulter Instruments, Hialeah, FL), and which remained constant throughout the experiment. For static headspace analysis, the five aroma compounds were added at final concentrations of 0.01% v/v per compound to the emulsion and at a final concentration of 0.0005% v/v per compound to the emulsion for the dynamic release analysis. The emulsion was stored at 25°C until used.

2. Static Headspace Gas Chromatography

Again SHGC was used to study the effects of saliva volume on the retention of the five aroma compounds. Emulsion-to-saliva ratios included 100:0, 80:20, 60:40, and 40:60. In each case, 2 mL of solution was transferred to 10-mL headspace vials. Analysis of each solution was performed in triplicate. The vials were then incubated at 37°C and agitated at 150 rpm for 25 min, using an orbital shaker incubator. Of the headspace 0.2 mL was injected by using a heated gastight syringe and analyzed by using a GC-FID (Hewlett-Packard 6890 series; Hewlett Packard, Palo Alto, CA). The injector and detector temperatures were 225°C and 300°C, respectively. A HP-5 megabore guard column (20 cm × 0.53 mm i.d., 1.5-μm film thickness; HP) followed by a HP-5 capillary column (60 m × 0.32 mm i.d., 0.25-μm film thickness; HP) was used. A GC-Cryo-Trap (Scientific Instrument Services, Inc., Ringoes, NJ) was attached to the bottom of the GC injection port inside the GC oven and was set at − 70°C during the trapping phase and 220°C during the sample release phase and GC run time. The GC oven was held at 40°C during the injection phase and cryofocusing and for an additional 4 min after the sample was released from the cryotrap. The subsequent GC oven temperature program used for the analysis was identical to that previously described in the model mouth experiments. Calibration graphs for each of the compounds were prepared by analysis of five pentane solutions containing known amounts of each compound. Air/liquid partition coefficients ($K_{a/l}$) for each of the five compounds were calculated as described previously.

3. Analysis of Aroma Release under Varying Retronasal Aroma Simultaneous Parameters

The RAS was primarily composed of a 1-L stainless steel blender incorporating both controlled gas flow and mixing. A voltage controller and variable-speed motor allowed the precise control of the impeller speed to replicate chewing and mixing of the sample. A controllable gas supply (N_2) continuously swept over the sample (1300 mL/min) during collection (1 min) to simulate flushing and renewing of the headspace above the sample. Additionally, a continuous water flow around the blender maintained the temperature at 37°C during sampling. The range of shear rates used (Table 1) was based on the characteristic that chewing and mixing of liquid samples in the mouth correspond to shear rates of 10 to 100 per second experienced by the food sample, depending on its viscosity [46–47].

The rotational speeds (Table 1) of the blender impeller were measured by using a Digital Stroboscope (The Pioneer Electric &Research Cooperation, Forest Park, IL). The corresponding shear rates (γ) (Table 1) were estimated by using the previously calculated value for k, the proportionality constant for the impeller [37]. The emulsion was placed into the RAS and sampled by using the five mixing rates and four emulsion-to-saliva ratio designs listed in Table 1. The released volatile compounds were trapped on a 60/80 mesh of Tenax TA adsorbent (Supelco). The gas flow rate from the RAS to the Tenax trap was split 12:1 using a needle-nose valve. The Tenax was desorbed by using a TD-4 model Short Path Thermal Desorber and Cryotrap Unit (Scientific Instruments Services, Inc). The TD-4 was connected to a Hewlett Packard 6890 Series gas chromatograph equipped with a flame ionization detector, as previously employed for SHGC analysis. The GC injection port was set to 250°C and a 10:1 split ratio was used. As previously, five pentane solutions containing known amounts of each compound were analyzed in triplicate in order to quantify the amount of each compound released in the RAS.

D. Statistical Analysis

For all reported results, triplicate data measurements were analyzed by using SPSS (ver. 10; SPSS Inc., Chicago, IL). Means were calculated and tested by using a one-way analysis of variance (ANOVA), to determine whether a statistical difference existed at $P < 0.05$, with Duncan's multiple comparison tests used to identify statistical separation among the means.

III. RESULTS AND DISCUSSION

The model mouth and the RAS were used in two separate studies to determine the release of aroma from two separate model emulsion systems with known amounts of five aroma compounds, while varying the saliva volumes and mastication rates. In each study, only the effects of mastication and saliva volume on the release of aroma at one emulsion-to-saliva ratio and one mastication rate, respectively, are presented and discussed. Before both studies, static headspace analysis was used to determine the gas-emulsion partition coefficients (K_{ge}) of the five aroma compounds above the emulsions, while varying the emulsion-to-saliva ratio.

A. Static Headspace Analysis

The aroma compounds involved in both studies included 2-butanone, 1-butanol, hexanal, *n*-butyl acetate, and 2-heptanone and were selected to represent a range of both functional group and hydrophobic character. Octanol/water partition coefficients (log *P*) of the compounds included 0.25 (2-butanone), 0.84 (1-butanol), 1.78 (hexanal), 1.82 (*n*-butyl acetate), and 1.98 (2-heptanone), where increasing values for log *P* represent a relative increase in compound hydrophobicity [48]. The initial part of each study investigated the effect of saliva volume on the distribution of the aroma compounds under equilibrium conditions above and within the emulsion (Figs. 1 and 2). Solely in the case of the oil-in-water emulsion, the emulsion can be divided into four phases whereby aroma compounds can distribute themselves on the basis of their molecular structure: the interior of the oil droplets, the continuous water phase, the oil-water interfacial region, and the gas phase above the emulsion [49]. The addition of increasing saliva volumes was found to increase the hydrophilic nature of the emulsion, which, as expected, had no significant effect ($p < 0.05$) on the K_{ge} value of the more relatively hydrophilic compound 2-butanone (Fig. 1). In contrast, decreasing the lipid fraction of the total sample with increasing saliva addition resulted in significant increases in the K_{ge} values of the more hydrophobic compounds, 1-butanol [$F(3, 8) = 37.585$, $P < 0.05$], hexanal

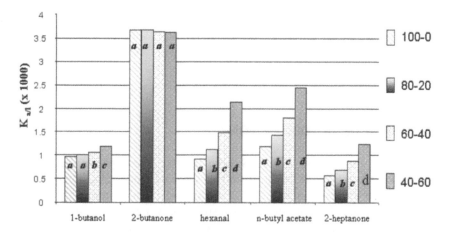

Figure 1   Study 1. Sunflower oil-in-water emulsion. Air/liquid partition coefficients ($\times 1000$) for the five volatile compounds with decreasing emulsion-to-saliva ratios (100–0, etc.). Mean values with different symbols (*a–d*) indicate a significant effect ($P < 0.05$) of saliva volume in emulsion for each compound shown.

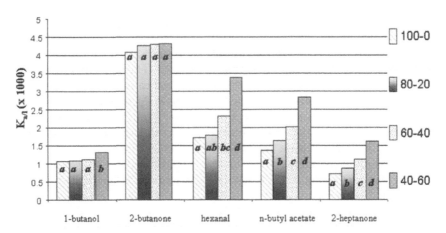

Figure 2 Study 2. Sunflower oil-in-starch emulsion. Air/liquid partition coefficients ($\times 1000$) for the five volatile compounds with decreasing emulsion-to-saliva ratios (100–0, etc.). Mean values with different symbols (a–d) indicate a significant effect ($P < 0.05$) of saliva volume in emulsion for each compound shown.

$[F(3, 8) = 4298.385, \ P < 0.05]$, n-butyl acetate $[F(3, 8) = 2080.566, \ P < 0.05]$, and 2-heptanone $[F(3, 8) = 1911.211, \ P < 0.05]$ being observed (Fig. 1). These effects have previously been reported in a similar study with pure sunflower oil samples [50] and in studies using emulsions with varying lipid concentrations [17,43,51]. In study 2, similar patterns of results were observed with increasing saliva fraction for the oil-in-starch emulsions (Fig. 2). Although the addition of starch affected the K_{ge} values of some of the compounds (results not shown), a decrease in the lipid fraction of the samples for both emulsions resulted in a relative increase in the K_{ge} values of the more hydrophobic aroma compounds, as predicted by Buttery and coworkers [52]; no significant effect was observed for the more hydrophilic compound 2-butanone.

B. Effect of Saliva on the Release of Aroma

Determination of the K_{ge} values for each compound with increasing saliva-to-emulsion ratios at equilibrium highlighted the affinity of each compound for the matrix as a function of the hydrophobicity of each compound. By using both the model mouth and the RAS to mimic conditions observed during eating, the effect of saliva on the release of aroma compounds from an emulsion at nonequilibrium conditions was determined by quantification of the amount of each compound released from emulsions with varying emulsion-to-saliva rations (Table 1). In both studies, using the model mouth and the RAS, the decrease in aroma compound concentration in the sample due to increasing saliva-to-emulsion rations corresponded to a decrease in the release of several compounds (Figs. 3 and 4). This is likely due to the fact that the primary effect of the addition of saliva to aqueous food systems such as emulsions results in dilution of the sample. Increasing the saliva-to-emulsion ratio of the sample reduces both the oil fraction of the sample and the concentration of the aroma compounds in the sample. Proportional decreases in aroma release relative to the decrease in aroma compound concentration in the sample were, however, not observed.

In the case of the RAS (Fig. 4), the amounts of 2-heptanone released from samples containing 20%, 40%, and 60% saliva expressed as a percentage of the amount released from the 100% pure emulsion sample were 111%, 109%, and 93%, respectively. Similar results were observed for 1-butanol, hexanal, and butyl acetate; the more hydrophobic compounds, i.e., those having a greater relative difference in K_{ge} values between samples (Fig. 2), maintained a higher than expected concentration in the headspace relative to the less hydrophobic compounds. Doyen and coworkers, using atmospheric pressure chemical ionization mass spectrometry (APcI-MS), also found that given both a compound's air/water and air/emulsion

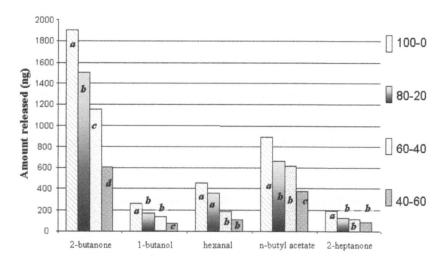

Figure 3 Model mouth. Effect of increasing saliva volume on the release of five volatile compounds at 52 chewing cycles/min from a sunflower oil-in-water emulsion. Mean values with different symbols (*a–d*) indicate a significant effect (*P* < 0.05) of saliva volume.

partition coefficients a higher than expected release of volatiles into breath from an emulsion system upon dilution with saliva during swallowing was observed [51]. Malone and coworkers [53] found a similar result as emulsions with higher lipid concentrations were chewed in the mouth for longer periods. In the case of the more hydrophilic compound 2-butanone, for which no significant differences in K_{ge} were observed for increasing volumes of saliva (Fig. 2), the release of the compound in the RAS was also greater than expected relative to the dilution of the sample with saliva. This result can be related to the decreased resistance to the mass transfer of the compound with decreasing shear viscosity of the sample on dilution with saliva. The importance of the mass transfer coefficient to the release of aroma compounds from emulsions in an open system has been detailed in mathematical models [54–55] and in studies investigating changes in the physicochemical properties of emulsions on aroma release [42].

In the case of the model mouth, a general decrease is also observed in the amount of each compound released from the sample on dilution with saliva (Fig. 3). However, the decreases in amount of each compound released were not proportional to the decreases in the aroma compound concentration of the samples on dilution with increasing volumes of saliva. The amounts of 2-heptanone released from samples containing 20%, 40%, and 60% saliva expressed as a percentage of the amount released from the 100% emulsion sample were 66%, 58%, and 46%, respectively. The decreased effect of the K_{ge} values on the release of the aroma compounds observed in the model mouth may be partially explained by the differences in the operating procedures between the model mouth and the RAS. In the case of the RAS, the initial conditions before dilution of the headspace above the sample with nitrogen and collection of the aroma compounds onto Tenax were very close to those used in static headspace analysis unless initially flushed before sampling. In the model mouth, the sample flask cannot be sealed before collection of the aroma compounds; thus equilibrium conditions in the flask are never achieved before headspace dilution.

C. Effect of Mastication Rate on the Release of Aroma

The effect of mastication rate on the release of aroma from emulsions was determined by using both the model mouth and the RAS at various settings and incorporating various saliva volumes (Table 1). Figures 5 and 6 show the effect of increasing mastication rate on the release of aroma compounds from samples each containing 40% saliva, using the model mouth and the RAS, respectively. Since all samples within each study had the same composition, and mastication rate was the only parameter to vary, only the

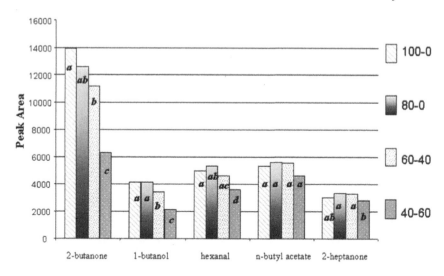

Figure 4 Retronasal aroma simulator. Effect of increasing saliva volume on the release of five volatile compounds at 300 rpm sunflower oil-in-starch emulsion. Mean values with different symbols (*a–d*) indicate a significant effect (*P* < 0.05) of saliva volume.

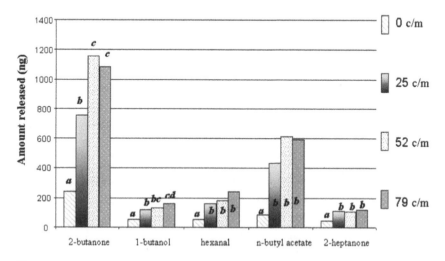

Figure 5 Model mouth. Effect of increasing mastication rate on the release of five volatile compounds at 60:40 emulsion-to-saliva ratio. Mean values with different symbols (*a–d*) indicate a significant effect (*P* < 0.05) of mastication rate.

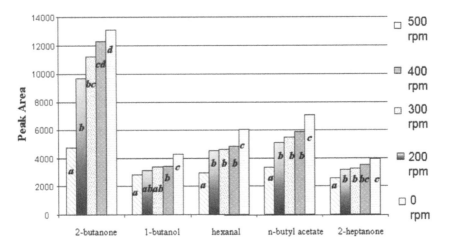

Figure 6 Retronasal aroma simulator. Effect of increasing mastication rate on the release of five volatile compounds at 60:40 emulsion-to-saliva ratio. Mean values with different symbols (*a–d*) indicate a significant effect (*P* < 0.05) of mastication rate.

release rates of the compounds are expected to differ and the K_{ge} values to remain the same.

In the RAS study, shearing was found to have a significant effect (*P* < 0.05) in the amount of each compound released (Fig. 6). Similarly, using the RAS, Roberts and Acree [37] found that shearing of a grape beverage sample significantly increased the volatility rate constants of the aroma compounds. Further increases in the impeller speed also resulted in increases in the amounts released for all compounds (Fig. 6). This can be related to the measured decrease in apparent viscosity of the oil-in-starch emulsion as the shear rate was increased (shear thinning) from 0 to 105 per second (results not shown). The reasons for the occurrence of shear thinning in emulsions are described by McClements [56]. The shear viscosity of the emulsion has been predicted to be the most important physical parameter affecting the diffusion of a compound in an emulsion, with the mass transfer coefficient inversely proportional to the viscosity [44]. Therefore, measured increases in aroma release with increasing impeller speed corroborate the predicted influence of mass transfer on the release of aroma compounds from emulsions. Similarly, Bakker and coworkers [12] found that the release rate of diacetyl from an aqueous solution was proportional to the square root of the stirring rate and inversely proportional to the square root of the viscosity.

In the case of the model mouth, similar results for the effect of increasing chewing cycles per minute on the release of aroma were observed (Fig. 5). A significant effect of mastication ($P < 0.05$) was measured for all compounds over the full range of chewing cycle rates studied; a general increase in release was observed with increasing chewing cycle rate. In addition to the described effect of decreasing viscosity with increasing shear rate experienced by the sample, an increase in the surface area of the sample exposed to the gas phase occurred as a result of the vertical motion of the plunger. The surface area of the gas–emulsion interface has also been predicted to be a primary factor governing the release of aroma compounds from an emulsion [54–55].

IV. CONCLUSIONS

The model mouth and the RAS demonstrated that they could be applied to the study of oral physiological effects on the extent of aroma release. Saliva addition was found to affect both the thermodynamic and kinetic components of aroma release; mastication was shown to affect the kinetic component of release. The effects could be related to specific changes in the physicochemical properties of the food matrix and to the hydrophobicity of the aroma compounds.

REFERENCES

1. LC Hatchwell. Overcoming flavor challenges in low-fat frozen desserts. Food Technol 48:98–102, 1994.
2. K de Roos. How lipids influence food flavor. Food Technol 51:60–62, 1997.
3. H Plug, P Haring. The role of ingredient–flavour interactions in the development of fat-free foods. Trends Food Sci Technol 4:150–152, 1993.
4. I Goubet, J-L Le Quere, AJ Voilley. Retention of aroma compounds by carbohydrates: Influence of their physiological characteristics and of their physical state: A review. J Agric Food Chem 46:1981–1990, 1998.
5. MA Godshall. How carbohydrates influence food flavor. Food Technol 51:63–67, 1997.
6. N Fischer, S Widder. How proteins influence food flavor. Food Technol 51:68–70, 1997.
7. S Lubbers, P Landy, AJ Voilley. Retention and release of aroma compounds in foods containing proteins. Food Technol 52:68–74, 1998.
8. H Kim, DB Min. Interaction of flavor compounds with protein. In: DB Min, TH Smouse, eds. Flavor Chemistry of Lipid Foods. Champaign, IL: American Oil Chemists Society, 1989, pp 404–420.

9. A Hansson, J Andersson, A Leufvén, K Pehrson. Effect of changes in pH on the release of flavour compounds from a soft drink–related model system. Food Chem 74:429–435, 2001.

10. SM van Ruth, E Villeneuve. Influence of β-lactoglobulin, pH and presence of other aroma compounds on the air/liquid partition coefficients of 20 aroma compounds varying in functional group and chain length. Food Chem. 79:157–164, 2002.

11. E Jouenne, J Crouzet. Effect of pH on retention of aroma compounds by beta-lactoglobulin. J Agric Food Chem 48:1273–1277, 2000.

12. J Bakker, N Boudaud, M Harrison. Dynamic release of diacetyl from liquid gelatine in the headspace. J Agric Food Chem 46:2714–2720, 1998.

13. DD Roberts, JS Elmore, KR Langley, J Bakker. Effects of sucrose, guar gum, and carboxymethylcellulose on the release of volatile flavor compounds under dynamic conditions. J Agric Food Chem 44:1321–1326, 1996.

14. SM van Ruth, CH O'Connor, CM Delahunty. Relationships between temporal release of aroma compounds in a model mouth system and their physico-chemical characteristics. Food Chem 71:393–399, 2000.

15. EN Friel, RST Linforth, AJ Taylor. An empirical model to predict the headspace concentration of volatile compounds above solutions containing sucrose. Food Chem 71:309–317, 2000.

16. I Andriot, M Harrison, N Fournier, E Guichard. Interactions between methyl ketones and β-lactoglobulin: Sensory analysis, headspace analysis, and mathematical modeling. J Agric Food Chem 48:4246–4251, 2000.

17. SM van Ruth, G de Vries, MD Geary, P Giannouli, CM Delahunty. Influence of composition and structure of oil–water emulsions on retention of aroma compounds. J Sci Food Agric 82:1028–1035, 2002.

18. C Yven, E Guichard, A Giboreau, DD Roberts. Assessment of interactions between hydrocolloids and flavor compounds by sensory, headspace, and binding methodologies. J Agric Food Chem 46:1510–1514, 1998.

19. F Rousseau, C Castelain, JP Dumont. Oil–water partition of odorant: Discrepancy between sensory and instrumental data. Food Qual Preference 7:299–303, 1996.

20. TG Hartman, J Lech, K Karmas, J Salinas, RT Rosen, C-T Ho. Flavor characterization using adsorbent trapping–thermal desorption or direct thermal desorption–gas chromatography and gas chromatography–mass spectrometry. In: C-T Ho, CH Manley, eds. Flavor Measurement. New York: Marcel Dekker, 1993, pp 37–60.

21. I Baek, RST Linforth, A Blake, AJ Taylor. Sensory perception is related to the rate of change of volatile concentration in-nose during eating of model gels. Chem Senses 24:155–160, 1999.

22. RST Linforth, I Baek, AJ Taylor. Simultaneous instrumental and sensory analysis of volatile release from gelatine and pectin/gelatine gels. Food Chem 65:77–83, 1999.

23. AJ Taylor. Volatile flavor release from foods during eating. Crit Rev Food Sci Nutr 36:765–784, 1996.

24. JR Piggott. Dynamism in flavour science and sensory methodology. Food Res Int 33:191–197, 2000.

25. F Boukobza, PJ Dunphy, AJ Taylor. Measurement of lipid oxidation–derived volatiles in fresh tomatoes. Postharvest Biol Technol 23:117–131, 2001.

26. L Jiang, K Kubota. Formation by mechanical stimulus of the flavor compounds in young leaves of Japanese pepper (*Xanthoxylum pipeeritum* DC). J Agric Food Chem 49:1353–1357, 2001.

27. KE Ingham, RST Linforth, AJ Taylor. The effect of eating on aroma release from strawberries. Food Chem 54:283–288, 1995.

28. C Palma-Harris, RF McFeeters, HP Fleming. Solid-phase microextraction (SPME) technique for measurement of generation of fresh cucumber flavor compounds. J Agric Food Chem 49:4203–4207, 2001.

29. C Hoebler, A Karinthi, M-F Devaux, F Guillon, DJG Gallant, B Bouchet, C Melegari, J-L Barry. Physical and chemical transformations of cereal food during oral digestion in human subjects. Br J Nutr 80:429–436, 1998.

30. C Lassauzay, M-A Peyron, E Albuisson, E Dransfield, A Woda. Variability of the masticatory process during chewing of elastic model foods. Eur J Oral Sci 108:484–492, 2000.

31. A Bardow, D Moe, B Nyvad, B Nauntofte. The buffer capacity and buffer systems of human whole saliva measured without loss of CO_2. Arch Oral Biol 45:1–12, 2000.

32. WE Lee III. A suggested instrumental technique for studying dynamic flavor release from food products. J Food Sci 51:249–250, 1986.

33. K Nassl, F Kropf, H Klostermeyer. A method to mimic and to study the release of flavour compounds from chewed food. Z Lebensm Unters Forsch 201:62–68, 1995.

34. SM van Ruth, JP Roozen. Gas chromatography/sniffing port analysis and sensory evaluation of commercial dried bell peppers (*Capsicum annum*) after rehydration. Food Chem 51:165–170, 1994.

35. H Plug, P Haring. The influence of flavour–ingredient interactions on flavour perception. Food Qual Preference 5:95–102, 1994.

36. JS Elmore, KR Langley. A novel vessel for the measurement of dynamic flavor release in real time from liquid foods. J Agric Food Chem 44:3560–3563, 1996.

37. DD Roberts, TE Acree. Simulation of retronasal aroma using a modified headspace technique: Investigating the effects of saliva, temperature, shearing, and oil on flavor release. J Agric Food Chem 43:2179–2186, 1995.

38. L Margomenou, L Birkmyre, JR Piggott, A Paterson. Optimisation and validation of the Strathclyde simulated mouth for beverage flavour research. J Int Brewing 106:101–105, 2000.

39. SM van Ruth, JP Roozen, JL Cozijnsen. Volatile compounds of rehydrated French beans, peppers and leeks. Part 1. Flavour release in the mouth and in three model systems. Food Chem 53:15–22, 1995.

40. SM van Ruth, JP Roozen, M Posthumus, FJHM Jansen. Volatile composition of sunflower oil-in-water emulsions during initial lipid oxidation: Influence of pH. J Agric Food Chem 47:4365–4369, 1999.

41. KD Deibler, EH Lavin, RST Linforth, AJ Taylor, TE Acree. Verification of a mouth simulator by in vivo measurements. J Agric Food Chem 49:1388–1393, 2001.

42. SM van Ruth, C King, P Giannouli. Influence of lipid fraction, emulsifier fraction, and mean particle diameter of oil-in-water emulsions on the release of 20 aroma compounds. J Agric Food Chem 50:2365–2371, 2002.

43. S-M Miettinen, H Tuorila, V Piironen, K Vehkalahti, L Hyvönen. Effect of emulsion characteristics on the release of aroma as detected by sensory evaluation, static headspace gas chromatography, and electronic nose. J Agric Food Chem 50:4232–4239, 2002.

44. M Harrison, BP Hills, J Bakker, T Clothier. Mathematical models of flavor release from liquid emulsions. J Food Sci 62:653–658, 664, 1997.

45. P Overbosch, WGM Afterof, PGM Haring. Flavor release in the mouth. Food Rev Int 7:137–184, 1991.

46. A Hansson, J Andersson, A Leufvén. The effect of sugars and pectin on flavour release from a soft drink–related model system. Food Chem 72:363–368, 2001.

47. CC Elejalde, JL Kokini. The psychophysics of pouring, spreading and in-mouth viscosity. J Texture Stud 23:315–336, 1992.

48. DR Lide. CRC Handbook of Chemistry and Physcis. Boca Raton, FL: CRC Press, 1997.

49. DJ McClements. Food Emulsions. Boca Raton, FL: CRC Press, 1999, pp 267–294.

50. SM van Ruth, I Grossmann, MD Geary, CM Delahunty. Interactions between artificial saliva and 20 aroma compounds in water and oil model systems. J Agric Food Chem 49:2409–2413, 2001.

51. K Doyen, M Carey, RST Linforth, M Marin, AJ Taylor. Volatile release from an emulsion: Headspace and in-mouth studies. J Agric Food Chem 49:804–810, 2001.

52. RG Buttery, DG Guadagni, LC Ling. Flavor compounds' volatilities in vegetable oil and oil-in-water mixtures: Estimation of odor thresholds. J Agric Food Chem 21:198–201, 1973.

53. ME Malone, AM Appelqvist, TC Goff, JE Homan, JPG Wilkins. A novel approach to the selective control of lipophilic flavor release in low fat foods. In: DD Roberts, AJ Taylor, eds. Flavor Release: Linking Experiments, Theory and Reality. Washington, DC: American Chemical Society, 2000, pp 212–227.

54. M Harrison, BP Hills. Effects of air flow-rate on flavour release from liquid emulsions in the mouth. Int J Food Sci Technol 32:1–9, 1997.

55. M Harrison. Effect of breathing and saliva flow on flavor release from liquid foods. J Agric Food Chem 46:2727–2735, 1998.

56. DJ McClements. Food Emulsions. Boca Raton, FL: CRC Press, 1999, pp 235–266.

25

Identification of Nonvolatile Flavor Compounds by Hydrophilic Interaction Liquid Chromatography– Electrospray Mass Spectrometry

Hedwig Schlichtherle-Cerny, Michael Affolter, and Christoph Cerny
Nestlé Research Center, Lausanne, Switzerland

I. INTRODUCTION

Liquid chromatography coupled to electrospray ionization mass spectrometry (LC-ESI-MS) was introduced in the 1980s [1]. Today it has become a standard method for separation and characterization of nonvolatile compounds. Reversed-phase high-performance liquid chromatography (RP-HPLC) coupled to ESI-MS is the method of choice for peptide and protein analysis, but also used for the characterization of contaminants, therapeutic drugs, and food additives [2–5]. More than 75% of HPLC analyses are run on RP stationary phases, and a wide range of columns are available with various substituents of the silica matrix, base deactivation, endcapping, and column dimensions.

However, RP-HPLC is challenging for the separation of polar compounds that are poorly retained on the hydrophobic stationary phase. This limits the application of RP-HPLC for the analysis of nonvolatile flavor compounds. Many taste-active compounds, especially those with reported savory (*umami* = glutamate-like) taste are of a hydrophilic nature. Hydrophilic di- to tetrapeptides consisting of high molar ratios of acidic and other hydrophilic amino acids such as Glu-Asp, Glu-Glu, Thr-Glu, Asp-

Glu-Ser, and Asp-Asp-Asp-Asp were described as eliciting a lingering umami taste and mouthfeel similar to that of monosodium glutamate (MSG) [6,7]. Other molecules reported to possess umami taste include organic acids such as citric, tartaric, and succinic acids [8–10]. Lactic acid contributes to the glutamate-like taste of bouillon and stewed beef juice [11,12].

The analysis of hydrophilic substances by RP-HPLC has some limitations, which are mainly due to the restricted retention of these analytes on the hydrophobic stationary phase. Ion exchange chromatography of ionic compounds and derivatization of the molecules of interest to render them more hydrophobic are two options to overcome this problem. Not all compounds, however, have ionic character, and derivatization is only specific for a given functional group; consequently, each functional group requires time-consuming and tedious derivatization.

Hydrophilic interaction liquid chromatography (HILIC) was proposed for the separation of diketopiperazines such as cyclo(Ala-Ser), cyclo(Gly-Leu), and cyclo(Ala-Gly), for carbohydrates, and nucleotides by Alpert [13]. Yoshida [14] analyzed peptides on a carbamide derivatized stationary phase. Later Strege [15] used HILIC-ESI-MS for the discovery of polar natural drugs such as vancomycin, terreic acid, and cephalosporin C.

Figure 1 compares the principles of RP-HPLC and HILIC: RP-HPLC employs hydrophobic stationary phases such as alkyl or phenyl derivatized silica, and retention is achieved by interactions of the analyte with the hydrophobic column material and partitioning between the mobile and the stationary phases. Typically, the mobile phase in RP-HPLC contains a low concentration of the organic modifier in the aqueous buffer at the beginning,

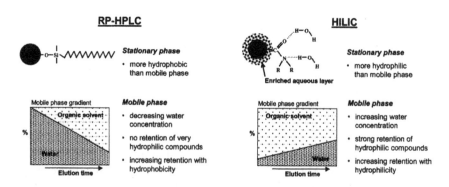

Figure 1 Principles of reversed-phase high-performance liquid chromatography (RP-HPLC) and hydrophilic interaction liquid chromatography (HILIC).

which increases during the HPLC run. In contrast, the stationary phases used with HILIC have polar functional groups such as polyhydroxyethyl aspartamide, carbamoyl-derivatized, cyclodextrin, cyano, amino, or others. The mobile phase consists of a high concentration of organic solvent with increasing water concentrations during the run [13]. A layer of mobile phase enriched with water is partially stagnant on the surface of the stationary phase and the analytes partition between the hydrophobic mobile phase and the enriched aqueous layer. Hydrophilic compounds are strongly retained and elute after the more hydrophobic components. The elution order observed in HILIC is in general opposite to that in RP-HPLC [13,16,17].

The objective of the present study was to evaluate HILIC-ESI-MS for the analysis and characterization of small, polar compounds in a taste-active fraction of an acid-deamidated wheat gluten hydrolysate.

II. MATERIALS AND METHODS

A. Chemicals and Materials

All chemicals used were commercially available and of analytical grade. Solvents were of HPLC gradient grade. The test mixture for HILIC contained amino acids (0.5 mg/mL each) and peptides (0.25 mg/mL each, 0.05 mg for Trp-Gly-Tyr) in acetonitrile (ACN). For the organic acid test solution, lactic acid, malic acid, pyroglutamic acid, succinic acid, and tartaric acid were dissolved in ACN with 40% aqueous ammonium acetate buffer (6.5 mmol/L, pH 7.0).

Amadori compounds were prepared from 30% solutions of equimolar mixtures of the respective reducing sugar (glucose, maltose) and amino acid, which were adjusted to pH 5.5, heated at 95°C for 30 min, vacuum-dried, and used without purification.

Wheat gluten hydrolysate was obtained from partially acid-deamidated wheat gluten by enzymatic hydrolysis using Flavourzyme[TM] L 1000 (Novozymes, Bagsvaerd, Denmark) and fractionated by gel permeation chromatography (GPC). Taste-testing of the obtained fractions (LC-tasting) was performed after freeze-drying and redissolution of the lyophilisate in water, as described by Schlichtherle-Cerny and Amadò [18].

Parmesan cheese extract was obtained from defatted freeze-dried Parmesan cheese by extraction with deionized water and subsequent ultrafiltration of the extract (molecular weight cutoff $M_r > 3000$). The freeze-dried ultrafiltrate was fractionated by GPC on Sephadex G10 with 15% ethanol in water as eluent into four fractions.

B. Reversed-Phase High-Performance Liquid Chromatography

Reversed-phase high-performance liquid chromatography was performed on a Grom-Sil 120 octadecylsilyl (ODS) 4HE column 4.0 mm × 25 cm (Grom, Herrenberg, Germany) using ammonium acetate buffer and methanol as mobile phase. The fractions were evaluated sensorially after freeze-drying and subsequent redissolution in water, as previously described [18].

C. Hydrophilic Interaction Liquid Chromatography– Electrospray Ionization Mass Spectrometry

The HILIC-ESI-MS procedure was performed using a TSK-Gel Amide 80 column (1.5 mm × 25 cm, TosohBioSep, Tokyo, Japan) coupled to a Thermo Finnigan LCQ ion trap mass spectrometer with an ESI source (Thermo Finnigan, San Jose, CA, U.S.A.). Nitrogen was used as sheath gas. The capillary temperature was 200°C and the spray voltage 4 kV in both positive and negative ionization modes. The MS^n spectra were obtained at a relative collision energy of 35% in the data-dependent scan mode with a signal intensity threshold of 2.5×10^5 for positive and 1.0×10^5 for negative ESI. The flow was 100 μL/min and the column effluent was split and a flow of 40 μL/min was diverted into the ESI source. The solvents used in the positive ionization mode were 10% aqueous ammonium acetate buffer (6.5 mmol/L) pH 5.5 in 90% ACN (solvent A) and 40% aqueous ammonium acetate buffer (6.5 mmol/L) pH 5.5 in ACN (solvent B) (15). The gradient was 13% to 40% water in 90 min. The solvents used in the negative ionization mode were 10% aqueous ammonium acetate buffer (6.5 mmol/L) pH 7.0 and 90% ACN (solvent A) and 40% aqueous ammonium acetate buffer (6.5 mmol/L) pH 7.0 in ACN (solvent B). The gradient was 13% to 27% water in 45 min and 27% to 40% water in 5 min.

III. RESULTS AND DISCUSSION

Different methods have been proposed for the analysis of polar peptides, free amino acids, and Amadori compounds. Anion exchange chromatography can be employed for acidic, negatively charged components and cation exchange chromatography for more basic, positively charged molecules such as peptides. Ion exchange chromatography was used to separate hydrophilic glutamyl and aspartyl peptides of hydrolyzed fish protein [7]. Precolumn derivatization of polar peptides and amino acids using 9-fluorenylmethyl chloroformate (FMOC) [19] was reported for polar peptides in cocoa [20] and dairy products [21], and derivatization with phenylisothiocyanate before RP-HPLC was proposed as a sensitive technique for amino acid analysis by Bidlingmeyer and associates [22]. Since the derivatives are more hydrophobic, they can more easily be separated on RP phases. Amadori compounds can be determined by gas chromatography after oxime formation and subsequent trimethylsilylation [23] or by anion exchange chromatography and postcolumn derivatization with triphenyltetrazolium chloride [24]. However, these methods either use high salt concentrations, which are hardly compatible with mass spectrometric detection, or use derivatization methods that are only specific for certain compound classes and difficult for the analysis of complex mixtures. For the present study HILIC was selected because it allows the separation of various hydrophilic compounds and can be directly coupled to MS. The mobile phase consisted of ammonium acetate buffer and acetonitrile, as described by Strege [15]. Ammonium acetate is a weak ion pairing reagent and produces more hydrophilic ion pairs than those formed with other reagents, e.g., trifluoroacetic acid, and consequently permits better separation by HILIC. In addition, ammonium acetate as buffer salt does not impair ionization of the analytes in ESI, neither in positive nor in negative mode. The optimal gradient for efficient analyte separation was found to be 13% to 40% water in 90 min. Sensory evaluation of fractions obtained by the present HILIC method and reconstitution tests may be envisaged after thorough removal of acetonitrile.

A. Test Mixtures

1. Amino Acids and Peptides

A test mixture of free amino acids and small di- and tripeptides was analyzed by RP-HPLC using an octadecylsilyl (ODS) derivatized stationary phase. The chromatogram in Fig. 2 shows that only the hydrophobic tripeptide Trp-Gly-Tyr and leucine were retained on the ODS phase. The other free amino acids, proline, threonine, serine, glutamine, and glutamic acid, as well as the polar peptides, represented the void volume peak, indicating an insufficient retention and separation by the RP-C18 packing. On the other hand, nearly baseline separation of these components was achieved by HILIC, as illustrated in Fig. 3. The more hydrophobic compounds, leucine and Trp-Gly-Tyr, eluted first, followed by proline, threonine, serine, and glutamine, which coeluted, and glutamic acid. The dipeptide Glu-Thr eluted before the other peptides Thr-Glu, Ser-Glu, Gly-Glu-Gly, followed by Glu-Glu, Glu-Lys, Lys-Glu, and Lys-Gly. The

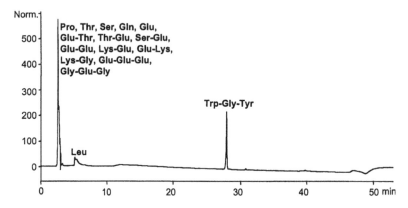

Figure 2 Reversed-phase high-performance liquid chromatography (RP-HPLC) of a test mixture composed of amino acids and hydrophilic peptides.

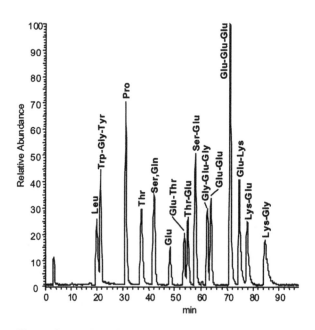

Figure 3 Hydrophilic interaction liquid chromatography (HILIC) of a test mixture composed of amino acids and hydrophilic peptides.

opposite elution order from RP-HPLC described by Alpert [13] and Oyler and colleagues [16] could only be partly confirmed, because leucine and Trp-Gly-Tyr eluted in the same order observed in RP-HPLC.

Interestingly, the dipeptides carrying an N-terminal glutamic acid residue eluted before their sequence analogs with glutamic acid at the C-terminus, indicating a sequence-specific elution of the tested dipeptides. Identification of the components was based on identical retention times and MS spectra obtained with the corresponding reference compounds. Furthermore, unambiguous assignment to a specific peptide identity was confirmed by spiking experiments with reference standards. Figure 4 shows the MS spectra of the dipeptides Glu-Thr and Thr-Glu in the positive ionization mode as an example. The full-scan mass spectra of both isobaric peptides show the predominant pseudomolecular ions $(M + H)^+$ at m/z 249; the dimeric ions $(2M + H)^+$ and the potassium adducts $(2M + K)^+$ at m/z 497 and 535, respectively. In addition, the mass spectrum of Glu-Thr shows the sodium adduct $(M + Na)^+$ m/z 271. For both peptides, the MS/MS spectra of the parent ion m/z 249 show a predominant product ion at m/z

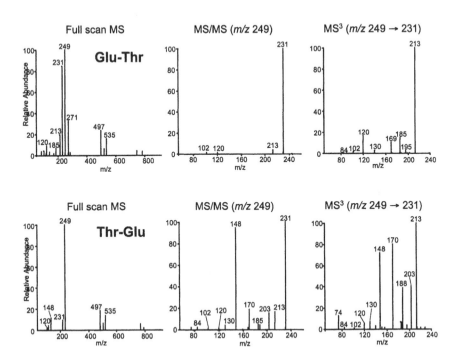

Figure 4 Mass spectra (MS) of Thr-Glu and Glu-Thr.

231, which is due to the loss of one water molecule. In addition to the signal at m/z 231, the MS/MS spectrum of Thr-Glu reveals an intense signal m/z 148 due to the y_1'' ion of the C-terminal glutamic acid [25]. Only the MS^3 spectrum of Glu-Thr gives sufficient information for unambiguous identification based on the presence of the fragments m/z 120 for the y_1'' ion and the b_1 ion m/z 130 corresponding to the C-terminal threonine and the N-terminal glutamic acid residue, respectively. In contrast to the MS^3 spectrum of Thr-Glu, the fragment m/z 148 corresponding to the C-terminal glutamic acid (y_1'' ion) is lacking. The fragments y and b are named according to Roepstorff and Fohlman [26]. Similarly, weaker retention was observed for the dipeptides Glu-Ser and Glu-Asp, which eluted before their sequence analogs Ser-Glu and Asp-Glu, respectively.

2. Organic Acids

Negative ionization mode was employed for the HILIC-ESI-MS analysis of organic acids contributing to the taste of foods. The mobile phase was adjusted to pH 7.0 to ensure negative charge of the organic acids. A higher pH value was not considered because of the limited stability of the silica gel matrix of the column under alkaline conditions. The selected ion mass chromatogram in Fig. 5 shows the separation of lactic, pyroglutamic, succinic, malic, and tartaric acids.

B. Food Samples

1. Wheat Gluten Hydrolysate

Hydrolyzed plant proteins are commonly used as savory ingredients in culinary products. An enzymatic hydrolysate from partially acid-deamidated wheat gluten was fractionated by gel permeation chromatography to locate the taste-active fractions. For the sensory evaluation the collected fractions were freeze-dried to remove water and solvents, then redissolved in water and taste-tested. This stepwise fractionation combined with the sensory evaluation of the fractions after freeze-drying and redissolution is called *LC-tasting*. The low-molecular-weight fraction ($M_r < 700$) containing small peptides and amino acids was found to be the most taste-active one [18]. Subfractionation of this fraction by RP-HPLC and LC-tasting revealed the void volume as the most intensely tasting fraction that could not be further separated by RP-HPLC. Separation and characterization, however, were achieved by HILIC-ESI-MS as the last step of the molecular identification of the taste components. Figure 6 shows the HILIC-ESI-MS base peak chromatogram and indicates the identified compounds. Apart from free amino acids, the glutamic acid–containing peptides Glu-Ala, Glu-

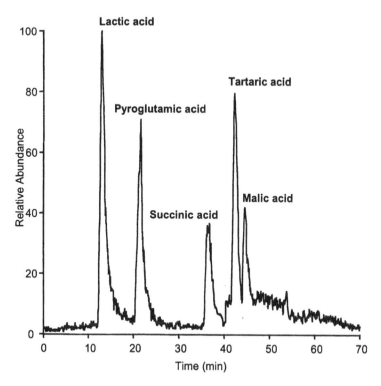

Figure 5   Hydrophilic interaction liquid chromatography (HILIC) of a test mixture of organic acids (selected ion mass chromatograms M-H$^+$).

Thr, Glu-Gly, Glu-Ser, Ala-Glu, Gly-Glu, Gly-Glu-Gly, Glu-Glu, and Glu-Asp were found. These polar peptides were reported to elicit a savory umami taste [7,27–29]. Frerot and coworkers [30] identified small peptides in Parmesan cheese. They observed that tripeptides such as Glu-Leu-Glu or Glu-Asp-Phe consisting of a hydrophobic amino acid residue and at least one acidic and one either acidic or hydrophilic amino acid residue impart mouthfeel and umami taste to foods and could replace MSG to some extent. Van den Oord and van Wassenaar [31], however, discuss glutamyl di- and tripeptides controversially and generally question the existence of glutamate-like-tasting peptides.

Dipeptides with glutamic acid at the N-terminus (Glu-Ala, Glu-Gly) eluted before their sequence analogs. This is in agreement with the observation made for the dipeptides Glu-Thr and Glu-Lys in the analysis of the amino acid/peptide test mixture.

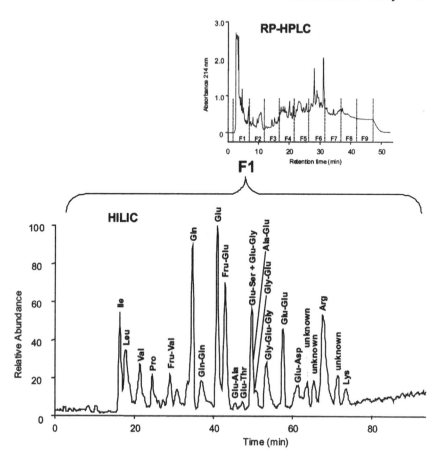

Figure 6 Hydrophilic interaction liquid chromatography (HILIC) separation of the taste-active subfraction F1 of an acid-deamidated wheat gluten hydrolysate that was not retained by reversed-phase high-performance liquid chromatography (RP-HPLC). Fru-Val, *N*-(1-deoxy-D-fructos-1-yl)-valine; Fru-Glu, *N*-(1-deoxy-D-fructos-1-yl)-glutamic acid (cf. Fig. 7).

Several mass spectra could not be assigned to peptides. Mass differences of 162 and 324 between the parent and the daughter ions did not correspond to a protein derived amino acid but indicated the presence of a glycoconjugate between an amino acid and a mono- or disaccharide [32]. The compounds were identified as Amadori rearrangement products on the basis of the comparison of the retention times, MS/MS, and MS^3 spectra with those of reference compounds. Table 1 lists the glycoconjugates

Table 1 Identified Glycoconjugates (Amadori Compounds) in a Taste-Active Fraction of Wheat Gluten Hydrolysate

Retention time, min	Signal [M+H],$^+$ m/z	Amadori compound
29.47	280.1	N-(1-Deoxy-D-fructos-1-yl)-valine
41.24	309.1	N-(1-Deoxy-D-fructos-1-yl)-glutamine
43.49	310.1	N-(1-Deoxy-D-fructos-1-yl)-glutamic acid
52.46	472.1	N-(D-Glucosyl-(1→4)-1-deoxy-D-fructos-1-yl)-glutamic acid
53.10	471.1	N-(D-Glucosyl-(1→4)-1-deoxy-D-fructos-1-yl)-glutamine
59.75	439.2	N-(1-Deoxy-D-fructos-1-yl)-glutamyl glutamate
76.23	309.2	N-(1-Deoxy-D-fructos-1-yl)-lysine
85.68	471.2	N-(D-Glucosyl-(1→4)-1-deoxy-D-fructos-1-yl)-lysine

identified in the polar taste–active subfraction of wheat gluten hydrolysate formed by the reaction of glucose with valine, glutamine, glutamic acid, lysine, and the dipeptide Glu-Glu. N-(Glucosyl-(1 → 4)-1-deoxy-fructos-1-yl)-glutamine (Maltu-Gln), N-(glucosyl-(1 → 4)-1-deoxy-fructos-1-yl)-glutamic acid (Maltu-Glu), and N-(glucosyl-(1 → 4)-1-deoxy-fructos-1-yl)-lysine (Maltu-Lys) were found as maltose-derived Amadori compounds. The structures of these Amadori compounds are depicted in Fig. 7. The MS/MS spectrum measured in negative ionization mode in Fig. 8 shows the fragmentation of the parent ion m/z 308 corresponding to N-(1-deoxy-fructos-1-yl)-glutamic acid (Fru-Glu), the Amadori rearrangement product of glutamic acid and glucose. The prominent signals at m/z 290, m/z 218, and m/z 200 can be explained by the loss of one water molecule, a carbon dioxide molecule, and a carbon monoxide molecule from the parent ion. The product ion at m/z 146 corresponds to glutamic acid. Amadori compounds are early intermediates of the Maillard reaction. They have been identified in various foods, such as malt and dried vegetables [23,24]. Their properties as aroma precursors were reviewed by Yaylayan and Huyghues-Despointes [33]. Little, however, is known on the taste of Amadori rearrangement products. In 2002, N-(glucosyl-(1 → 4)-1-deoxy-fructos-1-yl)-glutamic acid was described as eliciting a pronounced umami taste and as having bouillonlike taste qualities similar to those of MSG [34].

2. Parmesan Cheese

An aqueous extract of defatted 3-year-old Parmiggiano Reggiano cheese was ultrafiltered (molecular weight cutoff $M_r < 3 \,kDa$) and fractionated by

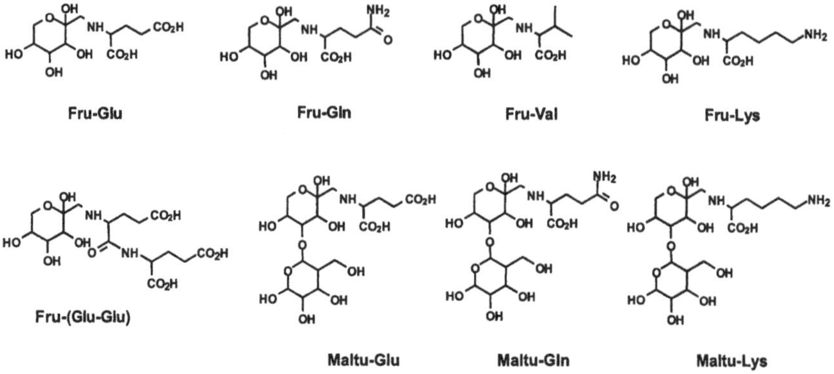

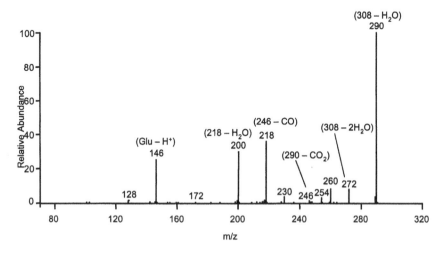

Figure 7 Structures of Amadori compounds identified in wheat gluten hydrolysate.

GPC on Sephadex G10. The fraction containing low-molecular-weight compounds below 700 Da was analyzed by RP-HPLC but could not be separated because of insufficient retention of most of the components on the hydrophobic stationary phase. Separation and characterization were achieved by HILIC-ESI-MS, as shown in Fig. 9. Apart from free amino acids such as lysine and arginine, which showed large peaks, the glutamyl dipeptides Glu-Val, Glu-Gly, Glu-Ser, Glu-His, Glu-Glu, Glu-Lys, Glu-

Figure 8 Mass spectrum (MS/MS) of Fru-Glu.

Arg, and Lys-Glu were identified, among other peptides. Roudot-Algaron and associates [35] analyzed small peptides in cheese by RP-HPLC and mass spectrometry and found several γ-glutamyl peptides. The sensory evaluation of γ-Glu-Tyr revealed a sour and salty taste, whereas γ-Glu-Phe elicited a brothy, slightly sour, salty, and metallic taste. Ferranti and colleagues [36] identified the N-terminal sequences of 38 oligopeptides in the low-molecular-weight fraction ($M_r < 3\,kDa$) of Grana Padano cheese by the combination of fast atom bombardment mass spectrometry (FAB-MS) and Edman degradation. They found oligopeptides with the N-terminal sequences Glu-Val-, Glu-Asp-, Glu-Leu-, Glu-Glu-, Glu-Gln-, Glu-Arg-, Glu-Asn-, and Glu-Ala- originating mainly from α_{S1}-, α_{S2}-, and β-caseine, respectively. The glutamyl dipeptides identified in the present study in Parmiggiano Reggiano cheese by HILIC-ESI-MS are likely to derive from further degradation of these oligopeptides. The dipeptide Glu-Lys eluted before Lys-Glu, confirming again the sequence-specific elution order of glutamyl peptides in HILIC with the carbamoyl derivatized phase employed for this study.

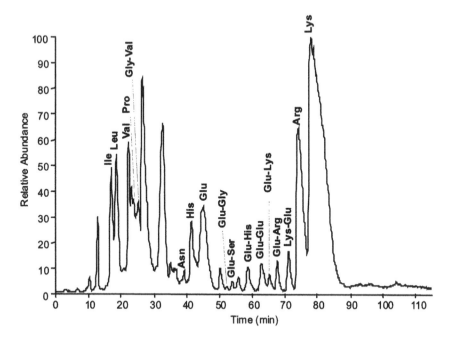

Figure 9 Hydrophilic interaction liquid chromatography–electrospray ionization mass spectrometry (HILIC-ESI-MS) base peak chromatogram of a polar Parmesan cheese extract fraction.

IV. CONCLUSIONS

Polar taste compounds that are not retained on reversed-phase stationary phases were successfully separated without derivatization by HILIC, using a carbamoyl derivatized stationary phase, which allowed the sequence-specific elution of glutamyl dipeptides. Dipeptides with N-terminal glutamic acid residues eluted before their C-terminal reverse sequence analogs. The mobile phase consisting of volatile ammonium acetate buffer and acetonitrile ensured compatibility with ESI-MS in positive as well as negative ionization mode. Interfacing HILIC with ESI-MS^n allowed unambiguous characterization also of coeluting isobaric compounds because of their unique MS^n spectra. The HILIC-ESI-MS procedure offers an alternative for the separation and characterization of polar nonvolatile flavor compounds. It can be used as a complementary tool with almost orthogonal specificity to RP-HPLC in the analysis of taste compounds such as amino acids, peptides, glycoconjugates, and organic acids. However, LC-tasting of HILIC fractions might be considered only after thorough removal of all solvent traces or complete replacement of acetonitrile by other organic solvents, such as ethanol or propanol, which are more compatible with sensory evaluation.

REFERENCES

1. JB Fenn, M Mann, CK Meng, SF Wong, CM Whitehouse. Electrospray ionization for mass spectrometry of large biomolecules. Science 246:64–71, 1989.
2. DR Bobbitt, KW Ng. Chromatographic analysis of antibiotic materials in food. J Chromatogr 624:153–170, 1992.
3. T Herraiz. Sample preparation and reversed phase–high performance liquid chromatography analysis of food-derived peptides. Anal Chim Acta 352:119–139, 1997.
4. M Veerabhadrarao, MS Narayan, O Kapur, CS Sastry. Reverse phase liquid chromatographic determination of some food additives. J Assoc Off Anal Chem 70:578–582, 1987.
5. R Malisch, G Heusinger. Analysis of residues and contaminants. In: R Mattisek, R Wittkowski, eds. High performance liquid chromatography in food control and research. Hamburg: Behr, 1993, pp 299–346.
6. I Matsushita, S Ozaki. Purification and sequence determination of tasty tetrapeptide (Asp-Asp-Asp-Asp) from beer yeast seasoning and its enzymic synthesis. Peptide Chem 32:249–252, 1995.
7. M Noguchi, S Arai, M Yamashita, H Kato, M Fujimaki. Isolation and identification of acidic oligopeptides occurring in a flavor potentiating fraction from a fish protein hydrolysate. J Agric Food Chem 23:49–53, 1975.

8. KH Ney. Flavor enhancing effect of L-glutamate and similar compounds. Z Lebensm Unters Forsch 146:141–143, 1971.
9. J Velisek, J Davidek, V Kubelka, T Tran Thi Bich, J Hajslova. Succinic acid in yeast autolysates and its sensory properties. Nahrung 22:735–743, 1978.
10. R Warmke, HD Belitz, W Grosch. Evaluation of taste compounds of Swiss cheese (Emmentaler). Z Lebensm Unters Forsch 203:230–235, 1996.
11. H Schlichtherle-Cerny, W Grosch. Evaluation of taste compounds of stewed beef juice. Z Lebensm Unters Forsch A 207:369–376, 1998.
12. T Warendorf, HD Belitz. The flavor of bouillon. Part 1. Quantitative analysis of nonvolatiles. Z Lebensm Unters Forsch 195:209–214, 1992.
13. AJ Alpert. Hydrophilic-interaction chromatography for the separation of peptides, nucleic acids and other polar compounds. J Chromatogr 499:177–196, 1990.
14. T Yoshida. Peptide separation in normal phase liquid chromatography. Anal Chem 69:3038–3043, 1997.
15. MA Strege. Hydrophilic interaction chromatography–electrospray mass spectrometry analysis of polar compounds for natural product drug discovery. Anal Chem 70:2439–2445, 1998.
16. AR Oyler, BL Armstrong, JY Cha, MX Zhou, Q Yang, RI Robinson, R Dunphy, DJ Burinsky. Hydrophilic interaction chromatography on amino-silica phases complements reversed-phase high-performance liquid chromatography and capillary electrophoresis for peptide analysis. J Chromatogr 724:378–383, 1996.
17. BY Zhu, CT Mant, RS Hodges. Hydrophilic-interaction chromatography of peptides on hydrophilic and strong cation-exchange columns. J Chromatogr 548:13–24, 1991.
18. H Schlichtherle-Cerny, R Amado. Analysis of taste-active compounds in an enzymatic hydrolysate of deamidated wheat gluten. J Agric Food Chem 50:1515–1522, 2002.
19. S Einarsson, B Josefsson, S Lagerkvist. Determination of amino acids with 9-fluorenylmethyl chloroformate and reversed-phase high-performance liquid chromatography. J Chromatogr 282:609–618, 1983.
20. K Gartenmann, S Kochhar. Short-chain peptide analysis by high-performance liquid chromatography coupled to electrospray ionization mass spectrometer after derivatization with 9-fluorenylmethyl chloroformate. J Agric Food Chem 47:5068–5071, 1999.
21. JM Roturier, D Le Bars, JC Gripon. Separation and identification of hydrophilic peptides in dairy products using FMOC derivatization. J Chromatogr 696:209–217, 1995.
22. BA Bidlingmeyer, SA Cohen, TL Tarvin. Rapid analysis of amino acids using pre-column derivatization. J Chromatogr 336:93–104, 1984.
23. R Wittmann, K Eichner. Detection of Maillard products in malts, beers, and brewing couleurs. Z Lebensm Unters Forsch 188:212–220, 1989 (in German).

24. M Reutter, K Eichner. Separation and determination of Amadori compound by high pressure liquid chromatography and post-column reaction. Z Lebensm Unters Forsch 188:28–35, 1989 (in German).

25. DF Hunt, JR Yates III, J Shabanowitz, S Winston, CR Hauer. Protein sequencing by tandem mass spectrometry. Proc Natl Acad Sci U S A 83:6233–6237, 1986.

26. P Roepstorff, J Fohlman. Proposal for a common nomenclature for sequence ions in mass spectra of peptides. Biomed Mass Spectrom 11:601, 1984.

27. S Ohyama, N Ishibashi, M Tamura, H Nishizaki, H Okai. Synthesis of bitter peptides composed of aspartic acid and glutamic acid. Agric Biol Chem 52:871–872, 1988.

28. M Tamura, T Nakatsuka, M Tada, Y Kawasaki, H Kikuchiokai. The relationship between taste and primary structure of "delicious peptide" (Lys-Gly-Asp-Glu-Glu-Ser-Leu-Ala) from beef soup. Agric Biol Chem 53:319–325, 1989.

29. S Arai, M Yamashita, M Noguchi, M Fujimaki. Tastes of L-glutamyl oligopeptides in relation to their chromatographic properties. Agric Biol Chem 37:151–156, 1973.

30. EM Frerot, SD Escher, F Naef. The contribution of peptides to savory flavors. Book of Abstracts, 210th ACS National Meeting, Chicago, August 20–24 AGFD-216, 1995.

31. AHA Van den Oord, PD van Wassenaar. Umami peptides: Assessment of their alleged taste properties. Z Lebensm Unters Forsch 205:125–130, 1997.

32. B Gutsche, C Grun, D Scheutzow, M Herderich. Tryptophan glycoconjugates in food and human urine. Biochem J 343:11–19, 1999.

33. VA Yaylayan, A Huyghues-Despointes. Chemistry of Amadori rearrangement products: Analysis, synthesis, kinetics, reactions, and spectroscopic properties. Crit Rev Food Sci Nutr 34:321–369, 1994.

34. H Schlichtherle-Cerny, M Affolter, I Blank, C Cerny, F Robert, E Beksan, T Hofmann, P Schieberle. Flavoring compositions. European patent application EP 1 252 825, 2002.

35. F Roudot-Algaron, L Kerhoas, D Le Bars, J Einhorn, JC Gripon. Isolation of γ-glutamyl peptides from Comte cheese. J Dairy Sci 77:1161–1166, 1994.

36. P Ferranti, E Itolli, F Barone, A Malorni, G Garro, P Laezza, L Chianese, F Migliaccio, V Stingo, F Addeo. Combined high resolution chromatographic techniques (FPLC and HPLC) and mass spectrometry–based identification of peptides and proteins in Grana Padano cheese. Lait 77:683–697, 1997.

26

On-Line Monitoring of the Maillard Reaction

Lalitha R. Sivasundaram, Imad A. Farhat, and Andrew J. Taylor
The University of Nottingham, Loughborough, Leicestershire, England

I. INTRODUCTION

Extensive reports of the Maillard reaction can be found in the scientific literature, most focus on aqueous systems, and few monitor the reaction with time. An interesting system for processing liquid streams of reactants under precise time–temperature conditions was described by Stahl and Parliment [1]. Analysis of the data resulted in a plot of flavor and color generation with time from which the kinetics of the reactions could be easily calculated. Gaining the same sort of data from food matrices that dehydrate on heating is more difficult, and there are few published reports.

The complexity of low-water systems is derived from the interdependence of the time, temperature, and water content parameters, which, in turn, are linked to phenomena such as glass transition and sugar crystallization in some systems. To demonstrate the complexity of these changes, the sequence of events in a typical thermal process (bread baking) is described here. Initially, the dough is an elastic solid, containing the Maillard reactants in aqueous solution. The temperature of the dough rises unevenly, with the outer layer heating faster than the inside. The surface cannot reach temperatures above 100°C until there is sufficient water evaporation, a process that involves both heat and mass transfer and has a strong time dependence. As the surface dehydrates, the matrix in this area becomes hotter and more viscous. The former factor increases the rate of reaction; the latter factor slows the reaction as a result of decreased reactant mobility. With little water present, sugars may crystallize, thereby releasing water in the immediate vicinity of the crystals and, significantly, potentially changing the local water content of the system, given the already low moisture content. Ultimately, at low water content, the matrix may undergo a change from a rubbery state to a glassy state, although these changes usually occur as the product cools post processing.

Gogus and coworkers [2,3] studied these phenomena using a gel-based system to hold Maillard reactants during heating and dehydration of the matrix. These authors showed that dehydration not only caused the changes of temperature, moisture content, and physical state described, but also caused concentration of reactants in specific regions. They postulated that dehydration caused a "front" of water, which carried water-soluble reactants, to travel through the matrix. This effect plus the existence of the different time, temperature, and moisture content conditions in the inner dough region and the outer bread crust region provide an explanation for the distinct color and flavor differences found in these two regions of bread.

Karmas and Karel [4] studied the Maillard reaction as a function of water content and physical state. More recently, Gunning and coworkers [5,6] studied the mobility of encapsulated compounds in glassy matrices. Both groups postulated that the mobility of molecules (and by implication the rate of Maillard reaction) would be dependent on the parameter $T - T_g$ where T is the temperature at which the Maillard reaction is being studied and T_g is the glass transition temperature of that system. This $T - T_g$ parameter is convenient as it allows comparison between systems with different T_g values. However, it is also recognized that T_g may just be a marker of water content rather than the primary factor controlling the reaction. Similar studies have been carried out in the pharmaceutical area [7]. All the literature references cited point to the overriding problem of obtaining suitable measurements so that the effects of time, temperature, water content, and physical state on the mobility and reactivity of compounds can be understood. The work presented in this chapter is part of an ongoing research program to build small-scale reactors that can be fully monitored (and the processing conditions controlled) so we might understand the fundamental factors affecting the course of the Maillard reaction. Using atmospheric pressure ionization mass spectrometry (APIMS) provides real-time information on aroma generation during processing, the use of attentuated total reflectance (ATR) Fourier transform infrared analysis (FTIR) yields data on glass transition of the food matrix, lactose crystallization, and the onset of the Maillard exotherm, which can be correlated with differential scanning calorimetry (DSC) results. With this basic knowledge, it may prove possible to move from our current, rather crude "one-pot" methods of forming flavor in foods (i.e., in which all reactants are placed in a vessel and heated to form flavor under the prevailing conditions) to more sophisticated processing in which the desired reaction pathways can be engineered, leading to the desired color and flavor qualities required for a particular food.

II. MATERIALS AND METHODS

β-Lactose and D-lysine (3:1 mass ratio; Sigma Chemical Co, Poole, UK) were dissolved in water and freeze dried to yield an intimate mixture of the reactants. Skim milk powder (SMP) was purchased from a local supermarket. Moisture contents of the materials were determined by Karl Fischer coulometric titration (Mitsubishi CA-05 moisture meter, Tokyo, Japan) and values were expressed as grams water per 100-g sample. The values for the various systems were as follows: lysine, 1.53%; lactose-lysine, 1.73%; SMP, 3.6%. Details of the packed bed reactor-APIMS system, the ATR FTIR system, and the DSC analyses can be found elsewhere [8,9]. Although DSC is an off-line technique, it is a standard method for determining the behavior of matrices during heating and provided benchmark values against which the later FTIR experiments could be compared. Since the value of T_g depends on the rate of heating, a rate of 3.7°C/min was used throughout this study so the results from the different techniques could be compared directly.

III. RESULTS

A. Differential Scanning Calorimetry Studies

Initially, samples of lactose, lactose-lysine, and skim milk powder (SMP) were analyzed by DSC to determine the glass transition (T_g) of these systems and the points in the thermogram where crystallization and the Maillard reaction took place. Results are shown in Fig. 1. Lactose showed a T_g of 83°C and a strong exotherm at 141°C, which coincided with crystallization. The mixture of lactose-lysine, however, had a T_g of 79°C, and the exotherm event started at a lower temperature and did not exhibit the same behavior as did the lactose crystallization event. The T_g of SMP was 68°C, and the onset temperature of the exotherm was closer to that of lactose than of the lactose-lysine system.

The relationship between these calorimetric events and the Maillard reaction was studied by heating another series of samples on the DSC and studying the color formed. A sample was taken at 30°C (as a control), then other samples were taken just before the onset of the exotherm, at the

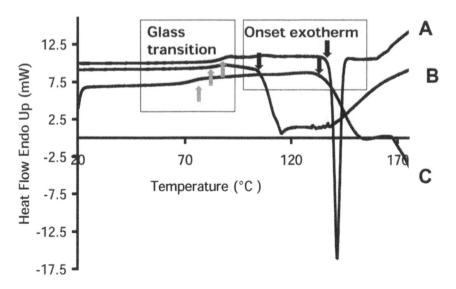

Figure 1 Thermograms from differential scanning calorimetry A, lactose; B, lactose-lysine 3:1 mixture; C, skim milk powder. All thermograms were obtained at 3.7°C/min. The arrows show the glass transition points of the systems as well as the onset exotherms.

maximal value of the exotherm, and at a later point. The temperatures and a qualitative assessment of the colors are shown in Table 1. To determine whether crystallization was associated with color formation, the DSC samples were analyzed by polarized light microscopy, which readily shows the presence of crystals.

Lactose showed no color change during heating from 30°C to 150°C, but the DSC exotherm at 140°C was associated with crystallinity. For the lactose-lysine and SMP samples, there was no significant change in color as they were heated through T_g; nor was there a great increase after the exotherm event. Color formation was significant after the exotherm event with marked changes over a narrow temperature range. This finding supports previous work that indicates that color formation occurs late in the Maillard sequence of events. In the lactose-lysine mixture, crystal formation started at 100°C (the onset temperature of the exotherm) and increased up to 115°C. For SMP, no crystallization was observed at any temperature.

These data indicated no clear relationship between T_g and Maillard browning nor between crystallization and Maillard browning. Instead, the systems seemed to behave differently. The lactose-lysine mixture was chosen as the proportions of the reactants encourage reaction, making observing

Table 1 Relationship of Differential Scanning Calorimetry, Events, Color Formation, and Crystallization in the Samples

Sample/treatment	Lactose	Lactose-lysine	Skim milk powder
Unheated	White	White	White
	30	30	30
	None	None	None
Post T_g	White	White	White
	130	90	100
	None	None	None
Onset exotherm	White	Off-white	White
	136	100	125
	None	Slight	None
Maximal value exotherm	White	Pale brown	Dark brown
	139	115	140
	Increasing	Increasing	None
Post maximal exotherm	White	Dark brown	Dark brown
	142	130	150
	Significant	Significant	None

[a] Color and crystallinity data are qualitative; temperatures of the DSC events are in °C.

the Maillard reaction easier. With a lactose-lysine ratio of 3:1, however, this model system cannot be compared with SMP, for which the ratio of the reactants is 50:3. It is therefore not surprising that the physical and chemical behavior of the two systems is different.

B. Fourier Transform Infrared Studies

The lactose-lysine and SMP samples were heated on the attenuated total reflectance (ATR) system at 3.7°C/min and FTIR spectra recorded at 5°C intervals from 30°C to 140°C. The stacked plots (Fig. 2A) were inspected visually and areas that showed changes around the glass transition temperature identified (see expanded trace in Fig. 2B). The changes in FTIR spectra for T_g were small but more distinct for the events around the exotherm (Fig. 3). The exotherm started at 105°C and continued to a maximal value at 120°C, and there are clear corresponding changes in the FTIR spectra in the 1600–1000/cm region (Fig. 3A). Between 125°C and 140°C, significant browning occurred, and there were again changes in the FTIR spectra in the 1200–1000/cm region (Fig. 3B). The visual inspection of the FTIR data suggested that specific regions of the spectra correlated with the events observed on DSC, so further analysis of the spectra was carried out by using principal component analysis (PCA). The entire FTIR spectra (for

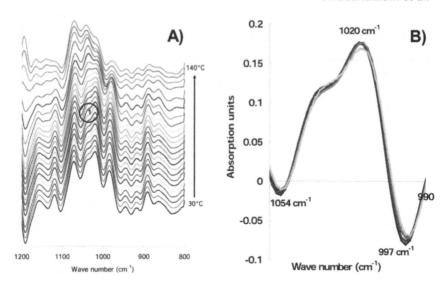

Figure 2 A, Fourier transform infrared (FTIR) traces of lactose-lysine (3:1 mixture; water content, 1.73%) over the temperature range 30°C to 140°C at 5°C intervals. B, Expansion of those areas (region marked 1 in Fig. 2A) showing changes around the glass transition temperature.

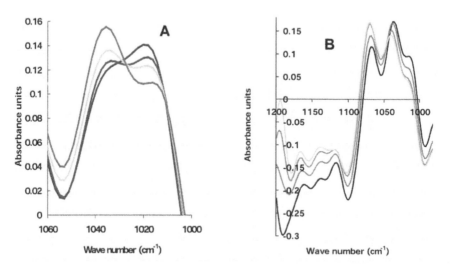

Figure 3 Fourier transform infrared (FTIR) regions correlating with A, the exotherm (105°C–120°C), and, B, the postexotherm (125°C–140°C) events for the lactose-lysine system. Spectra were deconvoluted and normalized. They were obtained at 5°C intervals with the highest temperature at the top of the stack.

all temperatures) were imported into the PCA software, and two principal components emerged. For the lactose-lysine system, PC1 described the exotherm; PC2 seemed to be associated with the T_g event and crystallization. In the SMP system, PC2 showed a maximum at T_g, but both PC1 and PC2 correlated with the exotherm event. The PCA results confirmed that there were correlations between FTIR changes and temperature. Previously [8], we had used visual inspection and ratios of IR spectral data to show a correlation, a process that is subjective. However, the validity of the previous visual approach is confirmed here as PCA shows correlations, and further analysis of the PCA plots confirmed that the spectral regions identified visually (Figs. 2 and 3) were indeed associated with the PC1 and PC2 parameters (Figs. 4 and 5).

C. Combined Studies

The generation of volatile compounds was monitored from the lactose-lysine and SMP systems at 3.7°C/min by using on-line APIMS analysis coupled to a packed bed reactor [8]. The internal temperature of the sample in the reactor was monitored so that data from the different techniques (APIMS, DSC, and FTIR) could be plotted on a common temperature axis so that the relationship between the physical events and aroma generation could be seen (Figs. 6 and 7). For the lactose-lysine mixture, generation of volatile compounds occurred around 130°C, after the onset of the exotherm and after crystallization had taken place. In contrast, the SMP system

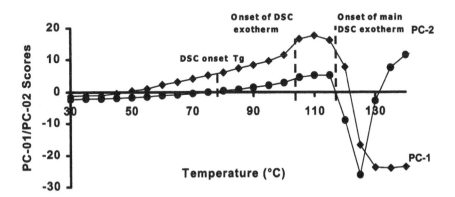

Figure 4 Principal components analysis (PCA) biplot of Fourier transform infrared (FTIR) data from lactose-lysine (3:1 mixture) heated from 30°C to 140°C. Principal components PC1 and PC2 account for 74% and 26% of the variation, respectively.

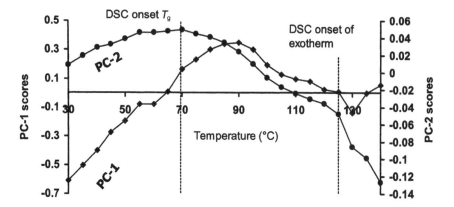

Figure 5 Principal component analysis (PCA) biplot of Fourier transform infrared (FTIR) data from skim milk powder (SMP) heated from 30°C to 140°C. Principal components PC1 and PC2 account for 96% and 4% of the variation, respectively.

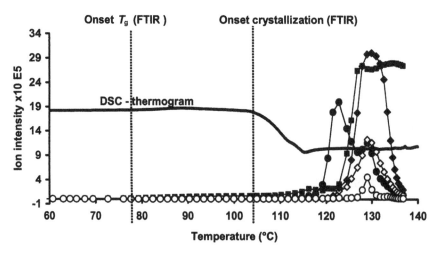

Figure 6 Combined differential scanning calorimetry (DSC) and atmospheric pressure ionization (API) traces showing relationship between volatile compound formation and physical changes in matrix. Dotted lines show Fourier transform infrared (FTIR) measurements of T_g and the onset of crystallization. Volatile compounds were provisionally identified as furfuryl formate/maltol (open circles), acetone (filled circles), furfural (open diamonds), acetylfuran (filled diamonds), and acetic acid (filled squares).

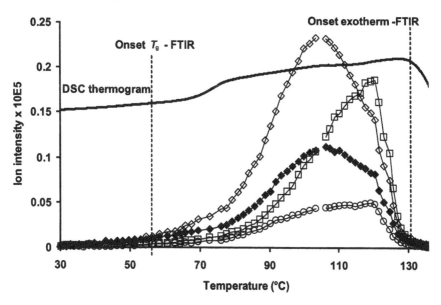

Figure 7 Combined differential scanning calorimetry (DSC) and atmospheric pressure ionization (API) traces showing relationship between volatile compound formation and physical changes in matrix. Dotted lines show Fourier transform infrared (FTIR) measurements of T_g and the onset of crystallization. Volatile compounds were provisionally identified as furfuryl formate/maltol (open circles), furfural (open diamonds), acetylfuran (filled diamonds), and diacetyl (open squares).

generated lower amounts of volatile compounds across a broader temperature range (90°C to 120°C), but volatile compounds were observed well before the onset of the exotherm (130°C). Some caution is required in interpreting the data as all techniques were carried out separately, and, despite attempts to keep heating rates constant and to monitor temperatures carefully, there may be some small changes from system to system. Another issue is the water content of the systems, which differed in this study. Ideally, a study of the different systems over a range of water contents would yield more information from which better interpretations could be drawn. However, these initial results show differences in the patterns of physical change with color and flavor formation in the two systems studied. The lack of a common pattern suggests that each system may have its own "rules," but a systematic study is now needed to confirm this hypothesis.

ACKNOWLEDGMENTS

LRS thanks the Norwegian government and Bjorge Biomarin A/S, Norway, for supporting a studentship. The financial support of Firmenich SA, Switzerland, is also acknowledged, as is the technical assistance of Dr Jenny Turner, Mrs. Val Street, and Ms. Iffat Farhat.

REFERENCES

1. HD Stahl, TH Parliment. Formation of Maillard Products in the Proline-Glucose Model System—High-Temperature Short-Time Kinetics: Thermally Generated Flavors. Washington, D.C.: American Chemical Society, 1994, pp 251–262.
2. F Gogus, BL Wedzicha, J Lamb. Modelling of Maillard reaction during the drying of a model matrix. J Food Eng 35:445–458, 1998.
3. F Gogus, BL Wedzicha, J Lamb. Migration of solutes and its effect on Maillard reaction in an agar microcrystalline cellulose matrix during dehydration. Food Sci Technol 30:562–566, 1997.
4. R Karmas, M Karel. Modeling Maillard browning in dehydrated food systems as a function of temperature, moisture-content and glass-transition temperature. In C-T Ho, CT Tan, CH Tong, eds. Flavor Technology: Physical Chemistry, Modification, and Process. Washington, DC: American Chemical Society, 1995, pp 64–73.
5. YM Gunning, PA Gunning, EK Kemsley, R Parker, SG Ring, RH Wilson, A Blake. Factors affecting the release of flavor encapsulated in carbohydrate matrixes. J Agric Food Chem 47:5198–5205, 1999.
6. YM Gunning, R Parker, SG Ring, NM Rigby, B Wegg, A Blake. Phase behavior and component partitioning in low water content amorphous carbohydrates and their potential impact on encapsulation of flavors. J Agric Food Chem 48:395–399, 2000.
7. SR Byrn, W Xu, AW Newman. Chemical reactivity in solid-state pharmaceuticals: Formulation implications. Adv Drug Deliv Rev 48:115–136, 2001.
8. JA Turner, LR Sivasundaram, M-A Ottenhof, IA Farhat, RST Linforth, AJ Taylor. Monitoring chemical and physical changes during thermal flavor generation. J Agric Food Chem 50:5406–5411, 2002.
9. JA Turner, RST Linforth, AJ Taylor. Real time monitoring of thermal flavour generation. J Agric Food Chem 50:5400–5405, 2002.

27

Optimizing Release of Flavor in Purge and Trap Analysis Using Humidified Purge Gas and Inverse Gas Chromatography

James Castellano and Nicholas H. Snow
Seton Hall University, South Orange, New Jersey, U.S.A.

I. INTRODUCTION

Purge and trap analysis is a widely used technique for qualitative and quantitative flavor analysis. Often it is not known how much flavor or what flavor compounds are contained in the sample. Consequently, it is not known whether all the flavor components that the sample contains were completely recovered. Typically, stepwise changes in conditions are made and the total flavor recovery monitored as well as specific compounds. An assessment is then made as to the extent of flavor recovery that was achieved. There is still the unknown of whether the more aggressive conditions are recovering more of the inherit compounds or whether artifacts are being created by the new conditions. The use of a humidified purge gas can increase the amount of flavor released versus a dry purge gas and becomes another variable in purge and trap analysis. Using the technique of inverse gas chromatography (IGC) provides a means to monitor the amount of flavor released, and as a real-time technique, is used to optimize purge and trap conditions for subsequent flavor analysis [1].

Inverse gas chromatography is typically considered to be a single-solute chromatographic technique. The shape, height, width, and total area of the eluting peak are used to determine the absorption and desorption characteristics of the probe or binding component with the solid absorbent.

In IGC, the sample is the column packing and the injected material is the probe used to study the sample characteristics through measurement of retention time and peak shape [2–4]. The advantage of IGC is that one can replace several purge and trap experiments with one IGC experiment. IGC is a rapid dynamic measurement, which is suited for rapid equilibrium processes. Frontal analysis is a technique whereby a constant vapor concentration of water or an organic is flowed through the sample in conjunction with the carrier gas. This technique can be used to generate isotherms and to study competitive adsorption phenomena, surface areas, acid–base interactions, and porosity. The frontal analysis technique was used here to evaluate the effect of relative humidity (RH) (water vapor pressure) in the purge gas on volatile release. Frontal analysis or finite concentration experiments were used in isotherm generation, competitive adsorption studies, and surface area determinations. An infinite dilution technique was used to determine volatile retention or volatile binding on known flavor components or suitable organic probes. In infinite dilution a small amount of an organic compound, in the vapor phase, is introduced into the column and the retention time of the peak recorded for surface energy experiments as well as peak shape for adsorption and desorption experiments. A small amount of vapor was used, infinitely dilute, to prevent any solvent solvent interactions so that only the solvent solute interactions could be studied. Infinite dilution experiments are typically used to study surface energies, acid–base interactions, and adsorption thermodynamics.

There are several processes involved in the release of flavor compounds due to the introduction of water to the sample vapor via the carrier gas or purge gas. Competition of the water and flavor molecule for binding sites on the surface, morphological changes, and changes in the partition coefficient of the flavor compound into the gas stream due to the addition of water vapor affect flavor release and recovery. An increase in the amount of water introduced into the solid sample as water vapor increases the probability of the water's successfully competing with the flavor molecules in the sample for the binding sites. If the binding of water to sites on the sample, through hydrogen bonding, is energetically favorable, then the water excludes or releases the organics from the surface. An overabundance of water molecules causes the release of flavor by a pure statistical process due to the number of binding sites, relative number of water versus flavor molecules, and typical exchange rate of a substrate on a surface with the surrounding medium. The use of heat overcomes the binding energy of the flavor components to the sample, as well as effecting morphological and phase changes and increasing overall flavor recovery but with the potential downside of creating artifacts and atypical changes to the sample. If at the increased water vapor pressure, the partitioning coefficient increases, it decreases the overall thermodynamic value ΔG for the system. There is a point at which the rate of flavor release is at a maximum for a given water vapor pressure, P_i.

$$\Delta G = -[RT \ \text{In} \ K_{pi}]_{\text{flavor}} - [RT \ \text{In}(P_i/P_o)]_{\text{water}}$$

The partition coefficient, K_{pi}, is at a maximum at that point and can be found by varying P_i/P_o at constant T, where P_i/P_o is the relative humidity of the purge gas and P_o is the partial pressure of water vapor in a water-saturated gas [5–8].

Morphological and phase changes, caused by the increased hydration of the sample, affect the release rate of compounds from glassy matrices or from conformational changes. The introduction of water to the sample lowers the glass transition temperature, increases diffusion rates in glassy matrices, and increases diffusion rates in rubbery matrices. The use of a humidified purge gas has a varied effect on the release of flavor components since the size and functional group of the molecule affect its ability to bind to a surface. The variations in binding energies cause variations in the release rates of flavor compounds [9–11].

In optimizing purge and trap conditions, most investigators focus on temperature, purge gas flow, and time and possibly some specific sample preparation in order to achieve a reproducible analysis. Increasing the temperature and monitoring the volatiles released are common tasks in developing a purge and trap analysis. Determining the optimal temperature may be a subjective assessment since the actual flavor composition of the product is not known so it is not known whether artifacts are formed by using too high a purge temperature. The volume of purge gas used is typically determined by the trapping method used to prevent trap breakthrough or overloading. The IGC procedure can be used to optimize temperature and purge volume with the additional variable of using a humidified purge gas. The problem still exists that the actual flavor composition is still not known. Using this technique, it is possible to optimize analysis conditions quickly; it also offers the advantage of using less harsh conditions by using a humidified carrier gas rather than increasing purge temperature to achieve higher flavor release.

II. MATERIALS AND METHODS

A. Instrumentation

Instrumentation consisted of an Agilent 6890 gas chromatograph interfaced to an Agilent 5973 (Wilmington, DE) mass selective detector

(MSD). A split/splitless injector was used for liquid sample introduction along with either a flame ionization detector or MSD to monitor the effluent. The column consisted of 1/16-in stainless steel tubing approximately 30 cm long connected to a 1/4-in-inner-diameter (i.d.) by 5-cm aluminum tube, which was packed with approximately 300 mg of the solid adsorbent. The aluminum tube was connected to another 30-cm length of 1/16-in stainless steel tubing, which was attached to a 1/16-in stainless steel tee, allowing splitting of the flow of effluent into the flame ionization detector and the mass spectrometer. A 1-m length of 0.25-mm-inner-diameter (i.d.) uncoated capillary tubing was used to limit the flow into the MSD to ~2 mL per minute. The mass spectrometer was used in the multiple ion monitoring (MIM) mode, to be able to measure the absorption/desorption of several probes simultaneously from the sample [1,12]. Ions monitored by MIM were predetermined on the basis of flavor compounds known to exist in the sample. Data were collected by using TurboChrom (Perkin Elmer, Norwalk, CT) chromatography software and ChemStation mass spectrometry/chromatography software (Agilent, Wilmington, DE). Generation and control of the water vapor pressure for the carrier gas stream were accomplished by mixing known volumes of dry and humidified air. A gas stream of dry air was split in two, and a pair of flow controllers were used to ensure consistent and proper flow rates (Model 33116, Cole Parmer, Vernon Hills, IL). One stream was passed through a 1000-mL gas-washing bottle (Atmar, Kennet Square, PA) containing 800 mL water to provide a water vapor–saturated gas. The humidified stream was recombined with the dry stream in various proportions to give the desired total flow and water vapor pressure. The gas stream was plumbed through the injector of the gas chromatograph to allow the injector to be used for the introduction of known compounds onto the sample column. Water vapor pressure (relative humidity) of the gas stream was confirmed by using a digital hydrometer (Model 3313-65, Cole Palmer, Vernon Hills, IL).

For the infinite dilution experiments, a commercially available iGC™ (Surface Measurement Systems, London, UK) was used. A quartz column was packed with the sample and placed into the oven compartment. Instrument was equipped with a 0.25-mL sample loop, and using a series of flow controllers, a measured amount of gas containing a 10% saturated vapor of butanol, hexanol, or octanol (Aldrich, Milwaukee, WI), in a helium carrier, was used to introduce the dilute organic vapor into the carrier gas stream and onto the sample. The net retention times were recorded and plotted as a function of the RH of the purge gas. An increase in the net retention time was indicative of increased binding of the organic compound to the sample.

B. Sample Preparation

Casein and whey were obtained from Sigma (St Louis, MO) and were used as is. Cheese flavored crackers, chocolate cookies, and freeze-dried coffee were commercially obtained products. Samples were ground if necessary and sieved to obtain a particle size range between 50 and 70 mesh (CE Tylor Inc., Mentor, OH). Sample particle size was important since the surface area, amount, and rate of flavor release change with particle size. Samples were protected from moisture changes and stored in an airtight container until used. Samples were used as is and not dried; initial moisture and sample RH measurements were made.

Samples were ground to give a free flowing powder. Particle size should be a small as reasonably possible by hand grinding. For the data collected to be used for determination of release rates or binding capacity, the sample should be sieved to achieve a known particle size range; otherwise sieving is not required. If the sample is not a free flowing powder, then an appropriate amount of an inert material should be added to achieve a free flowing powder. Sample was gravity packed and lightly packed into the column by tapping the column on a firm surface. If the sample was firmly packed into the column, channeling could occur and there would be a loss of signal intensity and reduced sensitivity. Channeling does not harm overall data integrity but does cause a loss of sensitivity. The amount of material packed into the column should be approximately 300 mg or less. If large sample sizes are used, the sample may not reach equilibrium with the RH of the purge gas in the time allotted.

In order to determine and optimize volatile release as a function of RH, a frontal IGC experiment was conducted. This experiment was generally intended for dry products. The experiment consisted of a series of segments in which the RH of the carrier gas was varied during the run. Purge gas was flowed through the sample at a rate of 50 mL per minute and into a flame ionization detector (FID), resulting in a signal proportional to the amount of organics released. Alternatively, an MSD in MIM could be used to record several ions simultaneously, providing several independent data channels capable of monitoring specific compounds or classes of compounds.

III. RESULTS AND DISCUSSION

Data showed an increase in total volatiles released when using a humidified purge gas. Figure 1 shows the relative amount of volatiles released when using a dry purge gas and the subsequent increase in volatiles released as the

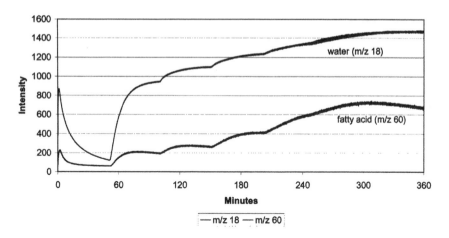

Figure 1 Inverse gas chromatography mass spectrometry (IGCMS) analysis of a cheese flavored cracker and release of free fatty acids as a function of relative humidity (RH) of purge gas as indicated.

RH of the purge gas is increased. This demonstrates the amount of volatiles that may not be released during a typical purge and trap experiment using only a dry purge gas. The amount of volatiles released, using a dry purge, falls off and was fairly constant after 30 min. Once the RH of the purge gas was increased, there was a significant increase in the amount of volatile release. The m/z 18 ion signal of water did not show a linear response with the increase in RH, and this result is attributed to overloading of the ion source and possibly also of the detector. The relative intensity of the m/z 18 ion signal vs. the m/z 60 signal (free fatty acids) was not proportional because of the sampling time of the MSD at each ion, which was 20 times greater for m/z 60 than m/z 18 to achieve sufficient sensitivity. The RH or water vapor pressure of the purge gas had an effect on what was released (i.e., certain compounds were desorbed faster at certain RHs), indicating that there were specific interactions of the water with the surface manifested in the desorption of the volatile compounds from cheese flavored crackers with the analysis monitored by IGC mass spectrometry MIM (IGCMS-MIM) and IGC FID, as shown in Figs. 1 and 2, respectively. The effect of water vapor pressure on the release of specific compounds is listed in Table 1 and shown in Fig. 3. The release of free fatty acids significantly increased over the RH range of 40% to 55% while the saturated and unsaturated aldehydes, represented by m/z 57 (representing saturated aldehydes) and 53 (representing unsaturated aldehydes), respectively, showed an increase in

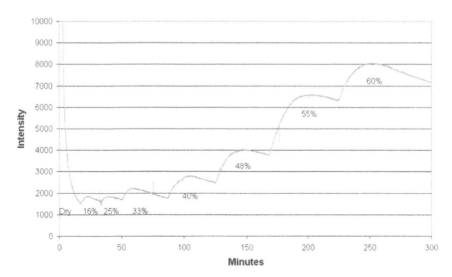

Figure 2 Inverse gas chromatography (IGC) analysis of cheese flavored cracker using a flame ionization detector (FID) showing total organics released as a function of purge gas relative humidity (RH).

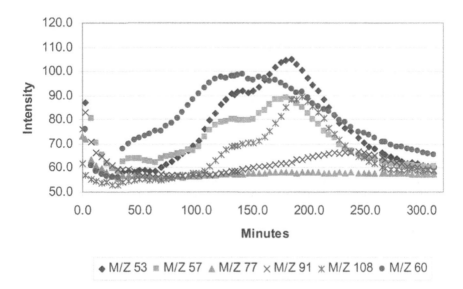

◆ M/Z 53 ■ M/Z 57 ▲ M/Z 77 × M/Z 91 ✳ M/Z 108 ● M/Z 60

Figure 3 Inverse gas chromatography mass spectrometry (IGCMS) data showing the effect of relative humidity (RH) of purge gas on release of specific compound classes. Release rates varied for a range of compounds.

release rates at 55% to 60% RH along with m/z 108 (dimethyl pyrazine). There seems to be a trend in that the more polar compounds are released at lower purge gas RH content than the less polar compounds. This may be due to the free fatty acid retention's being mainly due to hydrogen bonding with the water effectively competing for those binding sites even though the purge gas has a low RH. Less polar compounds are retained by a combination of hydrogen bonding and van der Waals forces such that the competition of the water for the hydrogen bonding sites is not sufficient to affect a release of the organic compound from the surface. The single-ion graphs show how various compounds have differing release rates affected by the RH of the purge gas. Table 1 shows the compounds or class of compounds corresponding to the m/z ratio signals plotted in Fig. 3. The differing release rates point to competitive binding versus morphological effects, since morphological effects would show as noncompound specific increases in release rates. Figure 4 shows a sample of freeze dried coffee in which the volatile release can be attributed to a morphological change in the sample. The sample also had a significant increase in volatiles released using a humidified purge gas. Figure 5 shows a chocolate cookie; the volatile retention mechanism can be attributed to binding since the release of volatiles was dependent upon RH of the purge gas. Both of these samples, freeze dried coffee and chocolate cookie, show significantly more volatile release at a lower purge gas RH than the cheese flavored products. Simply increasing RH to a high level to achieve a better recovery of components may actual result in less volatiles' being recovered.

Conducting an infinite dilution experiment, as shown in Fig. 6, an increase in the alkyl chain length affects the degree of binding, as seen by the change in the relative retention times of the organic probes used. Figure 6 shows a trend, that the higher the RH of the purge gas, the less retention of the compounds occurred, which offers the potential of simplifying

Table 1 m/z Ratios Monitored and Corresponding Compounds or Class of Compounds

m/z Ratio	Compound
53	Unsaturated aldehydes
57	Saturated aldehydes
60	Volatile fatty acids
77	Benzaldehyde
91	Tropyllium ion
108	Dimethyl pyrazine

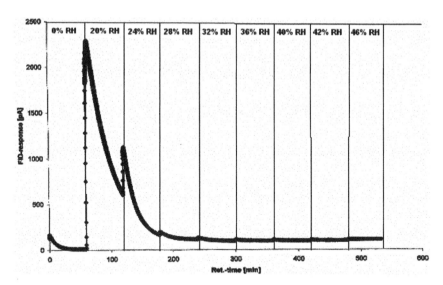

Figure 4 Analysis of freeze-dried coffee and effect of relative humidity (RH) on flavor release due to morphological effects.

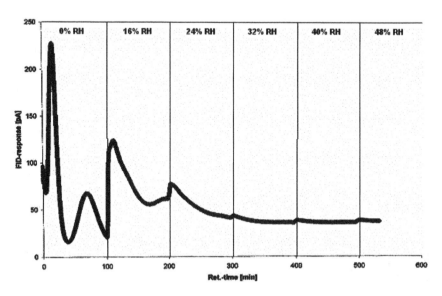

Figure 5 Degree of hydration of a chocolate cookie affects flavor binding/retention ability.

optimization of purge conditions. Figure 7 demonstrates that all solid substrates do not behave similarly, and as a result of the several causes of volatile retention or binding, that variation can make predictions of binding potential difficult. Foods are typically heterogeneous matrices, and this structure only serves to compound the problem, making the optimization of conditions for each food product important. Morphological changes play a role, but it is possible to distinguish between surface effects and bulk effects with additional experiments if desired. The use of a nonpolar probe such as decane along with a polar probe, decanal, could aid in differentiating changes in binding potential from changes in surface area. Conducting an infinite dilution experiment with decane and decanal could show that a surface area change was taking place if the retention times of decane and decanal increase but do not change relative to each other. If the relative retention time of decane and decanal does change, this change would indicate a change in the binding potential on the surface of the sample.

IV. CONCLUSIONS

The IGCMS technique allows for the detection of desorption of one material due to the adsorption of another, water. The interactions among the matrix, water, and volatile component are different for each component and lead to numerous desorption curves, each individually temperature and P_i/P_o-dependent. Use of IGCMS-MIM provides real-time determination of volatile release of several compounds simultaneously. The MIM technique is

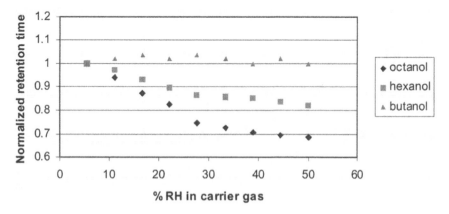

Figure 6 Effect of alkyl chain length on retention of a homologous series of alcohols.

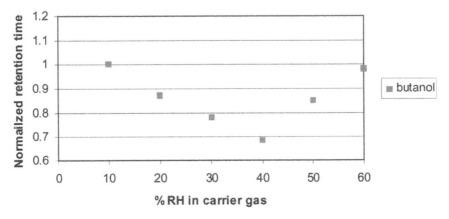

Figure 7 Changes in surface area due to hydration of the sample results in a reversal of binding capacity.

necessary to achieve sufficient sensitivity to monitor the flavor components at the levels found in the products and can also provide data on specific compounds. Relative rate and amount of volatile release can be determined on the basis of the area of the peak and the time for elution. The slope of the signal, signal intensity versus time, can be used to determine the relative rate of loss. Because of the complex and numerous interactions taking place with the sample, predicting the optimal purge and trap conditions is not feasible but rather must be determined experimentally.

REFERENCES

1. J Castellano, NH Snow. Modeling flavor release using inverse gas chromato-graphy/mass spectrometry. J Agric Food Chem 49:4296–4299, 2001.
2. SG Gilbert. Shelf life studies of foods and beverages. In: G Charalambous, ed, Chemical, Biological, Physical and Nutritional Aspects. Vol. 5. Elsevier Science Publishers B.V., 1993, pp 1071–1079.
3. DS Smith, CH Mannheim, SG Gilbert. Water sorption isotherms of sucrose and glucose in inverse gas chromatography. J Food Sci 46:1051–1054, 1981.
4. PG Demertizis, KA Riganakos, PN Giannakokos, MG Kontominas. Study of water sorption behaviour of pectins using a computerized elution gas chromatographic technique. J Sci Food Agric 54:421–428, 1991.

5. RP Danner, F Thiminlioglu, RK Surana, JL Duda. Inverse gas chromato-graphy applications in polymer-solvent systems. Fluid Phase Equilibria 1487: 171–188, 1998.

6. H Grajek, S Neffe, Z Witkiewicz. Chromatographic determination of the physico-chemical parameters of adsorption on activated carbon fibres. J Chromatogr 600:67–77, 1992.

7. KA Katsanos, R Thede, F Roubani-Kalantzopoulou. Diffusion, adsorption and catalytic studies by gas chromatography. J Chromatogr A, 795:133–184, 1998.

8. I Kaya, E Ozdemir. Thermodynamic interactions and characterization of poly(isobornylmethacrylate) by inverse gas chromatography at various temperatures. Polymer 40:2405–2410, 1990

9. I Quiones, G Guiochon. Application of different isotherm models to the description of single-component and competitive adsorption data. J Chromatogr A, 734:83–96, 1996.

10. J Roles, G Guiochon. Experimental determination of adsorption isotherm data for the study of the surface energy distribution of various solid surfaces by inverse gas–solid chromatography. J Chromatogr 591:233–243, 1992.

11. J Roles, G Guiochon. Validity of the model used to relate the energy distribution and the adsorption isotherm. J Chromatogr 591:267–272, 1992.

12. S Panda, Q Bu, B Huang, RR Edwards, Q Liao, KS Yun, JF Farber. Anal Chem 2485–2495, 1997.

28
Novel Mass Spectrometric Techniques for Monitoring Aroma Volatiles

Rob S. T. Linforth and Andrew J. Taylor
The University of Nottingham, Loughborough, Leicestershire, England

I. INTRODUCTION

Several methods for the real-time analysis of volatile compounds by mass spectrometry have been developed. Some of the earliest used atmospheric pressure chemical ionization (APcI) to measure volatiles in the breath during exhalation [1]. Others used electron impact mass spectrometers, interfaced to the gas phase via membranes [2]. More recently, further developments in APcI [3] and the emergence of proton-transfer-reaction mass spectrometry (PTRMS) [4] have further advanced our ability to monitor volatile compounds in real time. The unifying feature of all of these methods is the characteristic that all analytes enter the ionization region simultaneously: there is no chromatographic separation before detection. They all rely on the potential of the mass spectrometer to separate the analytes on the basis of mass to give them selectivity. Consequently, if two or more compounds occur at the same mass, then it may not be possible to distinguish between them by using these techniques. This situation becomes more difficult as the complexity of the system under analysis increases, as in the case of real foods. This can be highlighted by the fact that the majority of the 2500 or more aroma compounds that have been identified have even mass molecular weights between 70 and 190 da.

Clearly there is a need to improve the selectivity of such systems (while maintaining sensitivity and temporal resolution) for the analysis of complex mixtures. It may be possible to use mathematical solutions to deconvolute the data obtained from complex systems. However, it would be far better if the analytical procedure could be improved to generate simple, clear, specific data.

II. DISCUSSION OF TECHNIQUES

The real-time analytical systems are all similar in the basic overall design. Each has an interface, linking the ionization region to the sample (outside world); an ionization region; and a mass separation and detection system (Fig. 1). Selectivity can potentially be obtained at each of the key points in the system.

A. Gas Phase Sampling

For the majority of the systems there is no selectivity in the transfer of volatiles into the ionization region of the mass spectrometer; the gas phase sample is simply drawn into the apparatus intact. Soeting and Heidema [2] used a membrane at the interface between the sample gas stream and the ionization region. This was designed to exclude water and air from the ionization region while allowing volatile compounds in. Such systems have an element of selectivity built into them, since compounds vary in their affinity for the membrane, and hence transmission across it. In this case the designers were not primarily concerned with selectivity; they were more interested in solving problems with temporal resolution, caused by the slow migration of compounds through the membrane.

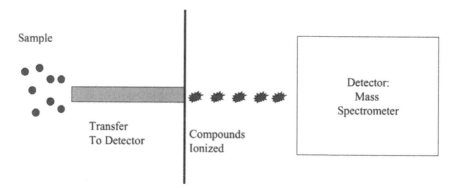

Figure 1   Schematic overview of real-time analysis.

There is clearly, however, the potential to use membrane technology either to introduce compounds into the ionization region selectively or to remove them from the gas stream during introduction. Little has been done to utilize this technology to produce what would effectively be a series of broad-band molecular filters (based on either size or chemical functionality) for mass spectrometry.

B. Ionization Selectivity

The ionization region is the next phase of sample analysis, in which the analyte molecules gain charge. Typically, apparatus has been designed (APcI and PTR) for the ionization of a broad range of compounds: i.e., the goal has been to prevent discrimination rather than develop selectivity. These methods are based on charge transfer reactions utilizing water as an intermediate [Eq. (1)]. Charge transfer occurs when the analyte molecule (M) has a higher gas phase basicity than water [5].

$$H_3O^+(H_2O)_n + M \rightarrow (H_2O)_{n+1} + MH^+ \tag{1}$$

Some selectivity can be achieved by the use of specific reagent ions. For example, ammonia rather than water can be used as the reactant species for charge transfer. In this case, only those molecules with a gas phase basicity greater than that of ammonia become protonated. This allows a broad filter to eliminate the ionization of compounds with low gas phase basicities. Alternatively the ionization mode itself can be changed. Photoionization (PI) uses photons to ionize molecules in the gas phase [Eq. (2)]. The energy of the photon is the key feature affecting ionization selectivity. Compounds with ionization potentials lower than that of the photon can be ionized by that photon, whereas those with higher ionization potentials remain unaffected. The energy of the photons can be varied by the use of different light sources (xenon, 8.4 eV; krypton, 10.0 and 10.6 eV; argon, 11.7 eV) or by the use of lasers.

$$M + h\nu \rightarrow M^+ + e^- \tag{2}$$

It is, however, important to realize that gas phase reactions can utilize many different routes to achieve ionization [6]. Consequently the initial ionization process may have selectivity, but unless the ionization conditions are suitable (pressure, etc.), further analyte–analyte interactions may occur, resulting in further charge transfer. Consequently compounds that were not initially ionized may gain charge, eliminating the initial selectivity.

Hexanal (molecular weight 100 da) can lose water and deprotonate as well as protonate to give a series of ions (m/z 83, 99, and 101, respectively)

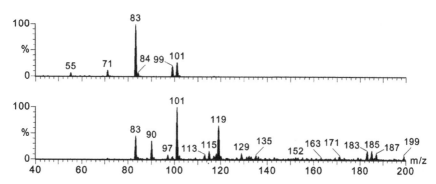

Figure 2 Atmospheric pressure ionization (API) (top) and atmospheric pressure photoionization (PI) (bottom) spectra of hexanal.

during APcI analysis (Fig. 2). A custom-built PI source operating at atmospheric pressure also showed protonated and dehydrated hexanal. In addition, the hexanal water adduct ($M \cdot H_3O^+$) was formed (m/z 119) as a result of poor declustering (typically induced by physical collision with nitrogen) within this source (the other ions observed were artifacts associated with this prototype source). These ions formed in the PI source were generated by analyte–analyte interaction that followed initial ionization. The PI source at atmospheric pressure may therefore offer little selectivity. The development in 2000 of a reduced-pressure PI source for gas phase analyses [7] may reduce these interactions and allow photon-based selectivity to operate. Unfortunately, it was not possible to test the potential for reduced-pressure PI analysis using the apparatus available, as a result of the difficulties of adding a low-pressure region to existing hardware.

C. Detector Selectivity

After ionization, the volatile compounds enter the high-vacuum region of the mass spectrometer. Here the compounds are focused and separated on the basis of their mass-to-charge ratio. In addition to simply detecting the ions at their nominal masses, further options exist for increasing selectivity.

D. Accurate Mass Analysis

Accurate mass analysis can be used to determine the elemental composition of ions. The masses of the ions detected (measured to four decimal places) are adjusted relative to a "lock ion," a compound (typically a compound added to the sample) of known molecular weight present in the spectrum.

The adjustment involves the assignment of the molecular weight(s) of ions at a nominal mass, followed by a correction for the difference between the true mass of the lock ion and its mass in the spectrum. Once the spectrum has been adjusted, the most likely elemental composition can be determined.

Because coffee has a complex volatile profile containing a large number of compounds [8], the certainty that a given ion corresponds to a specific compound or even one of its isomers is substantially reduced. Part of the spectrum of coffee headspace (Fig. 3) showed ions at virtually every mass, as expected for a complex mixture. The compound at nominal mass m/z 134 was cymene, which was added to the sample (no significant amounts of ions were detected at this mass before cymene addition); it has an accurate mass of 134.1096. The spectrum was adjusted to the mass of this compound to allow the elemental analysis of the other ions present.

Elemental composition analysis mathematically compares the mass observed with those calculated for elemental compositions that would generate similar nominal masses. The quality of the match is determined by the difference between these values. Analysis of the ion at m/z 123.0935 (Fig. 3) showed that the most likely molecular formula was $C_7H_{11}N_2$ (Table 1), which would have a molecular weight 0.0013 da lower than that observed. Other near matches also suggested that the compound contained nitrogen. The possibility that this compound contained no nitrogen atoms,

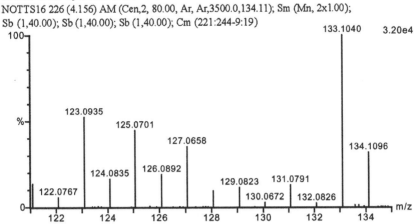

coffee with cymene lock mass

NOTTS16 226 (4.156) AM (Cen,2, 80.00, Ar, Ar,3500.0,134.11); Sm (Mn, 2x1.00);
Sb (1,40.00); Sb (1,40.00); Sb (1,40.00); Cm (221:244-9:19)

Figure 3 Mass spectrum (m/z 121–135) from the accurate mass analysis of coffee headspace. The masses of the ions were adjusted to the correct weight of the lock ion cymene (m/z 134.1096).

Table 1 Predicted Elemental Composition for the Ion of Molecular Weight 123.0935

Observed mass	Calculated mass	Millidalton[a]	Formula
123.0935	123.0922	1.3	$C_7H_{11}N_2$
	123.0882	5.3	$C_2H_{11}N_4O_2$
	123.0994	− 5.9	$CH_{11}N_6O$
	123.1048	− 11.3	$C_8H_{13}N$
	123.0810	12.5	$C_8H_{11}O$
	123.0796	13.9	$C_6H_9N_3$
	123.0770	16.5	$C_3H_{11}N_2O_3$
	123.0756	17.9	$CH_9N_5O_2$
	123.1174	− 23.9	C9 H15

[a] The millidalton difference between the observed mass and that calculated on the basis of the molecular formula.

or was purely made of carbon and hydrogen, was highly unlikely on the basis of its mass.

Elemental composition analysis (and some idea of the volatile composition of coffee) strongly suggested that this ion corresponded to a pyrazine. It did not, however, allow us to distinguish between trimethyl pyrazine and ethyl methyl pyrazine, isomers that consequently have the same molecular weight. Accurate mass may therefore allow us to discriminate between some molecules or molecular classes but will be of little value in the analysis of mixtures of isomers, such as systems containing a wide range of terpenes.

E. MS/MS

Mass spectrometry mass spectrometry (MS/MS) allows additional information to be gained about the nature of analytes. From the spectral profile an ion can be isolated and then further fragmented to give a series of ions, wich depends on the structure and composition of the original molecule. Libraries of MS/MS data can be produced [9] and may even allow spectral screening for identification. Indeed, techniques such as APcI and PTR, in which the compound remains predominantly intact during ionization (producing an MH^+ ion), should be ideal for linking to MS/MS. There may even be additional benefits with increased sensitivity.

However, closer examination of the results of Baumann and co-workers [9] may suggest limitations. The compounds with masses >400 da showed good MS/MS spectra with a considerable number of ions (>10) for spectral comparison. Compounds with a mass of ca. 300 produced only three or four ions, and with masses below that only two main ions were formed. With molecules of mass <200 (typical aroma compounds) there may not be much potential for fragment formation because of the limited size of the structure. Indeed, MS/MS studies of a range of environmental volatile organic compounds (VOCs) showed limited fragment formation and no improvement in sensitivity, despite the high efficiency of parent-to-daughter ion conversion [10].

Two cases were investigated: first, the esters ethyl pentanoate and isoamyl acetate (both with molecular weight 130); second, the terpenes limonene and α-pinene (molecular weight 136). Both of the esters ionized to produce a dominant MH$^+$ ion at m/z 131; these were isolated and fragmented further. In both cases there was only one main fragment, formed by the cleavage of the bond between the acid and the alcohol moiety. This resulted in daughter ions at m/z 71 and 103, for isoamyl acetate and ethyl pentanoate, respectively. Therefore, it should be possible using MS/MS analysis to discriminate between these two compounds.

Limonene and α-pinene are also very different structurally since limonene is monocyclic and α-pinene bicyclic. Both compounds ionize to form MH$^+$ and M·H$_3$O$^+$, m/z 137 and 155, respectively (Fig. 4). However, on fragmentation the MS/MS spectra are virtually identical for the two

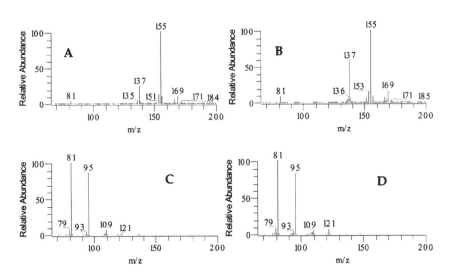

Figure 4 Full-scan mass spectra of limonene (A) and α-pinene (B) and the MS/MS spectra of limonene (C) and α-pinene (D) formed from ion 137 (MH$^+$).

compounds, such that if the two compounds were present in a mixture it would not be possible to analyze them separately.

Consequently in some instances MS/MS may allow the measurement of two compounds simultaneously, even though their MH^+ ions have the same mass, whereas in others the broad similarity of the structures results in similar losses and hence fragment formation, making it difficult to differentiate them.

III. CONCLUSIONS

Options exist for increasing the selectivity of real-time volatile analysis. Modifying ionization conditions may allow selective ionization of groups of compounds, depending on the potential for further analyte–analyte interactions and charge transfer. More precise selectivity may be obtained at the detector (mass spectrometer) level itself, via the use of accurate mass analysis and MS/MS.

Potentially these techniques can be combined to enhance the selectivity and specificity of analyses without the need for chromatographic separation.

ACKNOWLEDGMENT

The authors would like to thank Firmenich for funding this research.

REFERENCES

1. FM Benoit, WR Davidson, AM Lovett, S Nacson, A Ngo. Breath analysis by atmospheric pressure ionization mass spectrometry. Anal Chem 55:805–807, 1983.
2. WJ Soeting, J Heidema. A mass spectrometric method for measuring flavour concentration/time profiles in human breath. Chem Senses 13:607–617, 1988.
3. RST Linforth, AJ Taylor. Apparatus and methods for the analysis of trace constituents of gases. European Patent, EP 0819 937 A2, 1998.
4. W Lindinger, A Hansel, A Jordan. Proton-transfer-reaction mass spectrometry (PTR-MS): On-line monitoring of volatile organic compounds at pptv levels. Chem Soc Rev 27:347–354, 1998.
5. J Sunner, G Nicol, P Kebarle. Factors determining relative sensitivity of analytes in positive mode atmospheric pressure ionisation mass spectrometry. Anal Chem 60:1300–1307, 1988.
6. RK Mitchum, WA Korfmacher. Atmospheric pressure ionization mass spectrometry. Anal Chem 55:1485–1499, 1983.

7. JA Syage, MD Evans, KA Hanold. Photoionization mass spectrometry. Am Lab, 32:24, 2000.
8. I Flament. Coffee, cocoa, and tea. Food Rev Int 5:317–414, 1989.
9. C Baumann, MA Cintora, M Eichler, E Lifante, M Cooke, A Przyborowska, JM Halket. A library of atmospheric pressure ionization daughter ion mass spectra based on wideband excitation in an ion trap mass spectrometer. Rapid Commun Mass Spectrom 14:349–356, 2000.
10. SM Gordon, PJ Callahan, DV Kenny, JD Pleil. Direct sampling and analysis of volatile organic compounds in air by membrane introduction and glow discharge ion trap mass spectrometry with filtered noise fields. Rapid Commun Mass Spectrom 10:1038–1046, 1996.

29

Identification of Volatile Compounds Using Combined Gas Chromatography Electron Impact Atmospheric Pressure Ionization Mass Spectrometry

Andrew J. Taylor, Lalitha R. Sivasundaram, Rob S. T. Linforth, and S. Surawang
The University of Nottingham, Loughborough, Leicestershire, England

I. INTRODUCTION

Atmospheric pressure ionization mass spectrometry (APIMS) was developed for the direct analysis of volatile compounds in the gas phase [1], and the technology was adapted in our laboratory for the measurement of flavor release in vivo on a breath-by-breath basis [2]. Proton-transfer-reaction mass spectrometry (PTRMS) uses the same basic principles as API and has also found applications in flavor analysis [3,4]. Both techniques sample volatile compounds directly from the gas phase into the ionization source with no prior chromatographic separation. Ionization conditions are set so that formation of the protonated molecular ions is favored, as extensive fragmentation makes interpretation of the resulting ion trace extremely difficult. Maintaining consistent ionization in the source so that all compounds are ionized is a potential problem with all charge transfer ionization processes. Providing the analyte concentration is within the linear range, reliable quantification can be achieved [2], but there are certain conditions that can lead to selective suppression of ionization, leading to inconsistent and nonquantitative results. One example is the analysis of beer and wine samples in which the high ethanol content (about 4% and 12% v/v, respectively) interferes with the ionization of aroma compounds present at trace levels (typically milligrams per kilogram or micrograms per kilogram).

Another challenge affecting quantification in direct MS is that several compounds can produce molecular ions with the same ion mass, and it is not possible to determine the relative contribution of each compound to that ion mass. Table 1 gives some examples relevant to flavor analysis. Some resolution can be achieved in APIMS by controlling the cone voltage parameter. For instance, hexenol and hexanal both have a molecular weight (MW) of 100 and potentially give a protonated ion $[M + H]^+$ at m/z 101. To differentiate the compounds, hexenol can be dehydrated to produce an ion at m/z 83, thus allowing resolution of the two compounds [2]. A complementary study describing the fragmentation of volatile compounds under proton-transfer reaction MS (PTRMS) conditions showed how, in some cases, fragments could be used to identify isobaric and isomeric compounds [5]. The present chapter describes an approach that can assign compounds in a sample to the ion masses observed on APIMS. It is then possible to determine the certainty with which quantitative data can be interpreted.

II. MATERIALS AND METHODS

A. Headspace Sampling

Skim milk powder (SMP) was heated for 6 hr at 100°C at a moisture content of 8.1 g/100 g dry solids in a glass bottle with the lid loosely sealed. After

Table 1 Aroma Compounds with the Same Mass That May Produce the Same m/z Value on Ionization

Mass	Compounds
59	Acetone
	Propanal
86	Pentanal
	2-Methylbutanal
	3-Methylbutanal
	Diacetyl
96	Furfural
	Dimethyl furan
127	Furfuryl formate
	Maltol

cooling to 40°C, headspace was collected onto a Tenax™ trap (SGE, Milton Keynes, UK) by flushing with nitrogen at 30 mL/min for 15 min [6]. Whole tomatoes (about 50 g) or cucumber pieces (about 50 g) were placed in a modified blender at room temperature, macerated for 10–20 sec, then headspace collected onto a Tenax trap, using a stream of nitrogen (30 mL/min) for 5 min [7,8].

B. The Gas Chromatography Electron Impact Atmospheric Pressure Ionization Mass Spectrometry Procedure

The Tenax traps were thermally desorbed onto a BP-1 column (SGE; 30 m × 0.25 mm; film thickness 1 μm) and volatile compounds were cryotrapped by placing a section of the column in liquid nitrogen. Helium was the carrier gas, with a head pressure of 20 psi. The temperature program was as follows: 30°C for 2 min, 5°C/min to 100°C, 10°C/min to 200°C, hold for 2 min. The exit of the column was split by using a Y piece (SGE) and equal lengths of deactivated fused silica (SGE) led to the EI-MS and the APIMS sources. The EIMS was operated in full-scan mode, monitoring ions over a mass range of 25–200. Gas phase APIMS was operated in positive ion full-scan mode for the mass range 30–200 (scan time 0.4 sec; interscan delay 0.02 sec). Compounds were identified by using retention indices when authentic standards were available and/or mass spectral matching (NIST library).

C. Results

1. Maillard Reaction in Skim Milk Powder

The SMP system was chosen as it is a real food ingredient that is involved in the Maillard reaction but is relatively simple (compared to that of coffee, for instance). Thus it represents the lower end of the scale of complexity of food Maillard reactions. The headspace from the heated (and cooled) SMP sample was analyzed by direct API monitoring (Fig. 1). Visual inspection of Fig. 1 reveals about 20 compounds, but the only information on their identity is provided by the ion mass. Since this trace was obtained in scan mode using a general set of ionization parameters, there may also be some fragment ions present. For known compounds, molecular ion formation is favored by using specific conditions for each compound [2] when the APIMS is operated in selected ion mode.

To determine the relationship between ion mass in Fig. 1 and the actual compounds present in the SMP headspace, GC-EI-APIMS was carried out, and the two traces obtained are shown in Fig. 2. Generally the

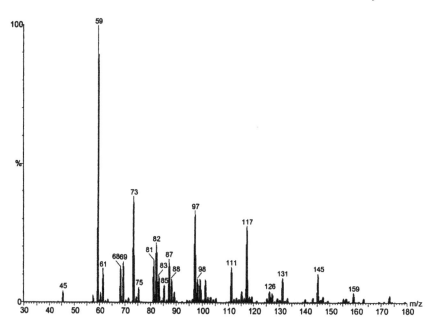

Figure 1 Atmospheric pressure ionization mass spectrum obtained by direct sampling from headspace above a Maillard system (skim milk powder heated for 1 hr at 100°C, then cooled to 40°C for sampling).

traces were very similar in terms of both the compounds detected and the relative intensities of the compounds. The EI trace was analyzed to identify the compounds present by retention parameters and spectral match, leading to a list of 18 compounds (Table 2). The API trace was then analyzed by extracting ion profiles, and a typical result is shown in Fig. 3, in which ion 59 is found at only one location on the GC trace, and this coincides with acetone (identified by GC-EIMS). For ion mass 97 on the API trace, there are two occurrences related to furfural and furfuryl alcohol on the EIMS trace. For ion mass 87, three compounds were identified with a fourth unidentified compound.

The results from the combined analysis of the SMP headspace demonstrate that some ion masses can be attributed to a single compound and therefore quantified reliably. Other ion masses, however, correspond to several compounds and quantification is at best an estimate. This experiment shows that the strength of APIMS lies in its ability to monitor volatile compounds on-line in real time, but its weakness in food flavor systems is its inability to resolve the compounds with the same ion mass.

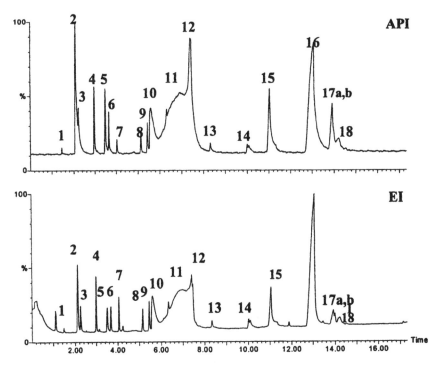

Figure 2 Traces obtained from simultaneous gas chromatography analysis of Maillard headspace, using both API and EIMS detectors. API, atmospheric pressure ionization; EI, electron impact; MS, mass spectrometry.

Despite this limitation, much useful information can be obtained from simple Maillard systems [6,9]. Some further resolution of the compounds might be achieved through the use of MSMS, but with low-molecular-weight compounds, there are limits on what can be achieved.

2. Monitoring Volatile Compounds Released from Macerated Tomato

The headspace from a macerated tomato was collected onto Tenax and analyzed by the combined GC-EI-APIMS technique to yield the traces in Fig. 4. Again there were similarities between the two traces, and 13 compounds were identified by conventional analysis of the EI trace (Table 3). There were two pairs of compounds that nominally had the same molecular weight and could have interfered with quantification. However, in both cases it was possible to obtain a unique ion (Table 3) when

Table 2 Compounds Identified in Headspace of Heated Skim Milk Powder by Gas Chromatography Electron Impact Atmospheric Pressure Ionization Mass Spectrometry

Peak no.	Atmospheric pressure ionization	Molecular weight	Identification	Confirmation
1	45	44	Acetaldehyde	**
2	59	58	Acetone	**
3	68	68	Furan	***
4	73	72	2-Methyl propanal	***
5	87	86	Diacetyl	***
6	73 (102)	72	2-Butanone	***
7	83	82	2-Methyl furan	***
8	87 (85, 69)	86	3-Methyl butanal	***
9	87 (85)	86	2-Methyl butanal	***
10	47	60	Oxybis methane	*
11	61	60	Acetic acid	***
12	75	74	1-Hydroxy-2-propanone	***
13	89	88	Acetoin	***
14	101	100	2-Methyl dihydrofuranone	*
15	97	96	Furfural	***
16	98 (81)	98	Furfuryl alcohol	***
17a	87 (85, 114)	?	Unknown	—
17b	126	126	Furfuryl formate (overlapping peaks)	***
18	111	110	2-Acetyl furan	***

***, confirmation of identity by retention index and mass spectrum of authentic compound; **, confirmation by mass spectrum of authentic standard; *, library mass spectral match only. Peak number refers to Fig. 2.

the API was operated in selected ion mode by choosing an appropriate cone voltage. For tomato, therefore, it is possible to monitor all 13 compounds of interest and quantify the amounts in the headspace by calibration with authentic standards. The technique has been used to study the formation of aroma in tomatoes subject to different pre- and postharvest treatments [7,8].

Although the data in Table 3 suggest that quantification can be accomplished, translating the information about headspace concentrations to actual contents expressed as milligram volatile compound per kilogram fresh tissue requires some extra calculation steps and some literature values with which to compare the values obtained. Tomato volatile data are available, but the values are very variable (see, for example, Refs. 10 and

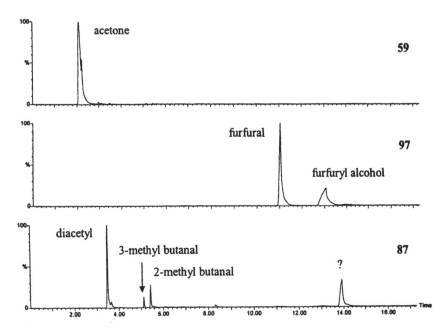

Figure 3 Extracted ion profiles from the atmospheric pressure ionization mass spectrometry detector with identities established by EI analysis of the same peaks.

11), and tomato is therefore not a good product with which to test the quantification process.

Cucumber fruits were studied by using the same analytical approach as they contain only five key volatiles and their concentration in cucumber tissue seems to be less variable than in tomato (see, for example, Ref. 12). Table 4 shows the EI and API correlations for the five compounds; each one could be attributed to a single ion mass on the API, and calibration with authentic standards allowed conversion of the ion signal into concentration units (parts per billion by volume). The amount of the C_9 volatiles present in the macerated tissue was estimated by microwaving a sample of cucumber to inactivate the enzyme system that produces C_6 and C_9 volatile compounds. Inactivation was confirmed by APIMS analysis of the headspace above the treated samples. Microwaved samples were macerated after spiking them with known amounts of the C_9 volatiles, then measuring volatile compound release in the blender apparatus. The values obtained from the spiked standards were then compared with the release traces from cucumber samples and the amounts of nonenal and nona-2,4-dienal estimated as 5 and 8 mg/kg fresh tissue, respectively. These values compare well with the

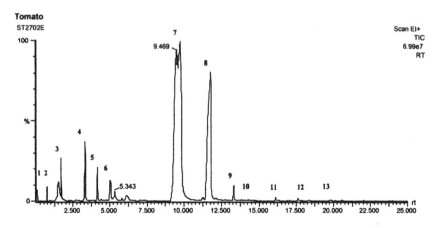

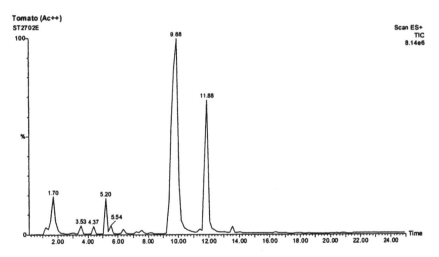

Figure 4 Gas chromatography traces obtained with the EI detector (top trace) and the atmospheric pressure ionization detector (bottom trace) for headspace analysis of macerated tomato.

published values of 8 mg/kg for nonenal and 4 to 13 mg/kg for nonadienal [12].

III. CONCLUSIONS

By using combined GC-EI-APIMS those compounds that are associated with the ion masses observed in APIMS could be identified through GC

Table 3 Key Volatiles Identified in the Headspace from Macerated Tomato.

Peak no.	Compound (LRI and EI match)	Molecular weight	Atmospheric pressure chemical ionization, m/z value	Gas chromatography mass spectrometry major ion(s)
1	Acetaldehyde	44	45	44
2	Acetone	58	59	
3	Pentane	72	73	57
4	Ethyl acetate	88	89	70
5	3-Methylbutanal	86	87	71 (69)
6	1-Penten-3-one	84	85	55
7	Hexanal	100	101	100 (82)
8	Z-3-Hexenal	98	99	83 (97)
9	Hexenol	100	83	69
10	3-Methylbutanol	88	71	70
11	Z-2-Heptenal	112	113	83
12	6-Methyl-5-hepten-2-one	126	127	108
13	2-Isobuthylthiazole	141	142	99

[a]Peak numbers refer to Fig. 4.
LRI, linear retention index; EI, electron impact.

retention parameters and by EIMS spectral matching. For the tomato and cucumber samples, each major constituent was represented by a single ion mass and quantification could be achieved. However, in heated SMP (a simple Maillard system) the chemical similarity of some products was such that several compounds contributed to an ion mass, and reliable

Table 4 Key Volatiles Identified in the Headspace from Macerated Cucumber

Compound identity (LRI and EI match)	Molecular weight	Atmospheric pressure chemical ionization mass spectrometry ion	Gas chromatography mass spectrometry major ion(s)
Hexanal	100	101	100
2-Hexenal	98	99	83
Nonanal	142	143	114 (124)
2,6-Nonadienal	138	139	136
2-Nonenal	140	141	138 (122)

LRI, linear retention index; EI, electron impact.

quantification of all compounds was not possible. In some cases, a common ion was indicative of a related group of compounds, e.g., ion mass 97 for sugar degradation products, and could be used to monitor this particular pathway. In other cases, e.g., ion mass 87, products were formed from several pathway, e.g., Strecker degradation (the methyl butanals) as well as from sugar degradation (diacetyl).

ACKNOWLEDGMENTS

LRS thanks the Norwegian government and Bjorge Biomarin A/S, Norway, for supporting a studentship. SS acknowledges financial support from the Thai government. The financial support of Firmenich SA, Switzerland, is also acknowledged, as is the technical expertise provided by Dr. F. Boukobza and Dr. J. Turner.

REFERENCES

1. FM Benoit, WR Davidson, AM Lovett, S Nacson, A Ngo. Breath analysis by atmospheric pressure ionisation mass spectrometry. Anal Chem 55:805–807, 1983.
2. AJ Taylor, RST Linforth, BA Harvey, A Blake. Atmospheric pressure chemical ionisation for monitoring of volatile flavour release in vivo. Food Chem 71:327–338, 2000.
3. F Gasperi, G Gallerani, A Boschetti, F Biasioli, A Monetti, E Boscaini, A Jordan, W Lindinger, S Iannotta. The mozzarella cheese flavour profile: A comparison between judge panel analysis and proton transfer reaction mass spectrometry. J Sci Food Agric 81:357–363, 2001.
4. RST Linforth. Developments in instrumental techniques for food flavour evaluation: Future prospects. J Sci Food Agric 80:2044–2048, 2000.
5. K Buhr, SM van Ruth, CM Delahunty. Analysis of volatile flavour compounds by proton transfer reaction–mass spectrometry: Fragmentation patterns and discrimination between isobaric and isomeric compounds. Int J Mass Spectrom 221:1–7, 2002.
6. JA Turner, RST Linforth, AJ Taylor. Real time monitoring of thermal flavour generation. J Agric Food Chem 50:5400–5405, 2002.
7. F Boukobza, P Dunphy, AJ Taylor. Measurement of lipid oxidation–derived volatiles in fresh tomatoes. Postharvest Biol Technol 23:117–131, 2001.
8. F Boukobza, AJ Taylor. Effect of post harvest treatment on flavour volatiles of tomatoes. Postharvest Biol Technol 25:321–331, 2002.
9. JA Turner, LR Sivasundaram, M-A Ottenhof, IA Farhat, RST Linforth, AJ Taylor. Monitoring chemical and physical changes during thermal flavor generation. J Agric Food Chem 50:5406–5411, 2002.

10. MS Brauss, RST Linforth, AJ Taylor. Effect of variety, time of eating, and fruit-to-fruit variation on volatile release during eating of tomato fruits (*Lycopersicon esculentum*). J Agric Food Chem 46:2287–2292, 1998.
11. DJ Stern, RG Buttery, R Teranishi, L Ling, K Scott, M Cantwell. Effect of storage and ripening on fresh tomato quality. 1. Food Chem 49:225–231, 1994.
12. RH Buescher, RW Buescher. Production and stability of (E,Z)-2,6-nonadienal, the major flavor volatile of cucumbers. J Food Sci 66:357–361, 2001.

30

Variables Affecting Solid-Phase Microextraction Headspace Analysis of Orange Juice Volatiles

Robert J. Braddock, Renée M. Goodrich, and Charles R. Bryan
University of Florida, Lake Alfred, Florida, U.S.A.

I. INTRODUCTION

A. Liquid/Vapor Properties of *d*-Limonene

Important to any measurement of citrus juice volatile flavor components is the presence of *d*-limonene, since this compound is naturally present as the most concentrated component in all of the natural citrus oils. Also, the solubility of *d*-limonene in aqueous media must be considered, since after liquid phase saturation, the headspace concentration remains constant. It has long been established for *d*-limonene and similar nonpolar flavor compounds over water that meaningful headspace measurement techniques [e.g., solid-phase microextraction (SPME)] require equilibrium of the vapor and liquid phase concentrations. Equilibrium may take a number of hours for static (unstirred) experiments and less than 1 hr for stirred systems. These conditions have been discussed elsewhere, and solubility and activity coefficients of *d*-limonene in water and sucrose solutions have been determined [1,2]. More recently, the chemical and physical properties as well as citrus industry applications of *d*-limonene and other citrus essential oils have been compiled [3]. Although not specific to *d*-limonene, important relationships affecting behavior of flavor release and partitioning between the headspace and the liquid phase of a number of food systems have also been discussed [4].

B. Solid-Phase Microextraction Headspace Criteria

The theory of SPME for headspace applications has been described in many publications. Some studies described conditions for measurement of citrus volatiles, particularly referencing *d*-limonene. Headspace extraction efficiencies of different fibers for some flavors were compared with the correct recognition that mixtures in water were within the solubility limits [5]. Another study concluded that SPME effectively measured headspace volatiles (e.g., *d*-limonene, myrcene, pinene, and terpinene) of centrifuged orange juice; however, these authors did not report the actual concentrations of compounds in the whole juice [6]. Not knowing a compound's juice concentration prevents quantitative measurement of the partitioning among the complex system matrices of fiber, headspace, juice, and pulp.

Method developments for food headspace SPME analyses have considered some, but not all, important issues related to equilibria and capacities of the various fibers. Dilutions for gas chromatography olfactometry (GCO) were monitored as a result of compound mass distribution between the headspace and liquid; however, there was no indication of how to deal with liquid phase saturation [7]. The requirement for calibration of fibers by use of standard solutions has also been addressed, considering the various parameters needed to calculate supposed headspace concentrations for dilution analysis [8]. Comparing static headspace vapor sampling with SPME for citrus volatile compounds in standard solutions below the solubility limits [parts per billion (ppb)] of most of the compounds resulted in linear calibration curves; however, curves were nonlinear at [parts per million (ppm)] concentrations above solubility limits [9]. This study [9] determined that there were analytical differences between actual juices, but these differences were nonquantitative.

Mass transfer after equilibrium of components into the fiber upsets the headspace equilibrium (sample, fiber, headspace); thus, long sampling times (15–30 min) can affect the gas chromatographic (GC) profile. Because of this, short time exposure (1 min) of SPME fibers to the headspace components is useful to provide a better representation of the true concentration, especially for compounds with high concentrations or high fiber affinities [10]. The extracted mass of the compounds into the fiber affects the mass transfer into the headspace from the liquid. This process is further complicated by stirring versus static systems and by the relative volatility, diffusion, and concentrations of the various components, requiring much care for high reproducibility and precision [11]. Other complicating factors affect headspace concentration of volatiles in the actual flavor-containing matrix. These factors include beverage base ingredient interactions with volatiles, as well as varying physical conditions related to headspace component equilibrium [12].

II. EXPERIMENT

A. Orange Juice Volatiles Concentration

Variability of quantitative data for compounds measured in orange juice headspace analyses prompted a study to determine the effect of juice pulp in binding certain components. Significant results of this research indicated that nonpolar volatiles (limonene, pinene, myrcene, and valencene) were prominently distributed in the pulp fraction; more polar volatiles (ethyl butyrate, octanal, linalool) remained in the pulp-free serum phase [13]. Terpenes and nonpolar oxygenates from mango juice also partitioned more into the pulp fraction than the serum [14]. Pulp adsorption of nonpolar volatiles lowers the juice serum and headspace concentrations of these compounds. The liquid and vapor phase concentrations of water-soluble compounds such as esters are dependent on the presence of other water-soluble components and may not show significant equilibrium differences, unless they interact with other juice or beverage components [15]. It should be noted that any standard solutions for quantitative purposes must use the exact same liquid matrix (e.g., sugars, pulp, ingredients, and type of process) as the beverage being studied.

Studies of orange juice headspace composition are many and have been based on a need to compare processes, products, packages, storage conditions, flavor packages, etc. No differences in processes were detected in nonpolar components such as limonene and pinene [16]. This study used statistical analysis, coupled with a sensitive GC detector, to allow comparisons of fresh versus pasteurized orange juice based on determinations of short-chain alcohols and aldehydes. Another quantitative study used radiolabeled standards and a trapping technique to concentrate components for GC analysis [17]. Their estimates of headspace concentrations of ethyl butyrate, acetaldehyde, and hexanal were dependent on the method used for processing orange juice, although they reported no concentration differences for limonene, myrcene, and pinene. By comparison, it is possible to show differences in the composition of nonpolar compounds (and other flavor components) in various juices, if the compounds are solvent-extracted from the juice matrix and not analyzed by a headspace technique [18]. The effectiveness of extraction methods is based on recovery of the compounds adsorbed to the pulp as well as the other juice matrix compounds.

A study using principal component analyses of headspace volatiles attempted to classify three categories of orange juice (unpasteurized, pasteurized, and pasteurized-from-concentrate) based on purge and trap GC and standard compounds added to unflavored juice [19]. It was possible to distinguish unpasteurized juice from the other two categories; however, considerable overlap between the two pasteurized juices probably resulted from the difficulty in selecting the proper components for analysis and errors in estimating juice concentrations from headspace measurements. Linear regression lines for a number of orange juice flavors were determined by SPME headspace analysis of standard solutions prepared in a deodorized juice matrix [20]. This study reported that limonene equilibrium was different in water and in the juice matrix; however, quantitative equilibrium data and liquid saturation were not reported. Headspace SPME also was used in a nonquantitative study attempting to measure odor intensities by GCO of heated and unheated orange juice [21].

B. Analytical Conditions

Gas chromatographic analyses for standard curves and all SPME fiber studies were generally as follows: Hewlett Packard 5890 GC with flame ionization detector (FID), 260°C; column, DB5, 0.53 mm × 30 m, 1.5-μm film; carrier gas, H_2 at 5.25 mL/min; oven, 60°C for 2 min, to 195°C, 5°C/min; injector, splitless, 250°C, purge on at 2 min, SPME desorbed for 2 min.

The SPME fibers were inserted in static samples in glass bottles with Teflon® septa. Most common analyses used 1-min sampling times according to the need not to destabilize the equilibrium [10]. Long equilibria for d-limonene and other compounds necessitated overnight holding of samples. Attempts to shorten equilibrium time and reduce sample size by heating and stirring samples in small vials increased error and gave poor results for water-insoluble compounds such as d-limonene, which tended to adsorb to the glass above the liquid phase, giving anomalous headspace concentrations. Reproducibility studies of small, stirred samples involved preparing standard water solutions (200 mL) of limonene at 4, 10, 15, 20, and 30 ppm. Samples (3 mL) were pipetted to 6-mL stirred vials with Teflon septa and stirbars in a round holder. The headspace was sampled at 25°C after 2 hr with 100 μm polydimethyl siloxane (PDMS) for 1 min. With three replications at each concentration, the standard error was 30%. This error was mostly due to nonhomogeneous distribution of insoluble limonene adhering to the glass and inaccurate transfer from the standard solutions, which was due to biphasic limonene. We also found poor quantitative reproducibility between fibers for the divinylbenzene (DVB)/Carboxen/PDMS fibers, and they were less durable than the PDMS fibers.

By use of standard curves of peak areas versus quantity with solutions of known concentration, it was decided that data reported as microgram component on the SPME fiber gave reliable results and were readily derived from the equilibrium headspace data. Some other reports of SPME using certain model citrus components have included myrcene as one of the components [20]. We opted not to include it, since purchasing pure myrcene standards is not possible. Myrcene polymerizes readily (unstable) and may be difficult to separate from octanal on very nonpolar GC columns. Although it is very aromatic, it is not a strong citrus aroma impact compound.

III. RESULTS AND DISCUSSION

A. Fiber Comparison

Pure d-limonene (99%), 4.5 mL in 9-mL vials at 26°C, equilibrated overnight, was used to determine the capacities of 100-μm PDMS and 50/30-μm DVB/Carboxen/PDMS fibers. Values in Table 1 indicate that the PDMS fiber reached saturation of 77 μg d-limonene in 30 min; the DVB/carboxen/PDMS fiber was saturated (35 μg d-limonene) in about 5 min. The higher capacity of the PDMS fiber for citrus juice components allowed more sensitivity for GC analysis, especially for compounds in lower headspace concentration. Thus, the PDMS fiber was preferred for work with citrus volatiles.

Table 1 Headspace Equilibrium of d-Limonene with 100 μm Polydimethyl Siloxane and 50/30-μm Divinylbenzene/Carboxen/Polydimethyl Siloxane Fibers[a]

Sampling time (min)	d-Limonene, μ, g on PDMS	d-Limonene, μ, g DVB/Carboxen/PDMS
1	11.8 ± 0.2	17.7 ± 0.7
5	33.5 ± 1.6	31.3 ± 0.6
10	48.5 ± 0.5	34.0 ± 0.3
20	60.9 ± 1.6	35.3 ± 0.9
30	77.3 ± 0.6	35.0 ± 0.9
40	78.1 ± 1.4	—
50	77.9 ± 1.5	—

[a] Samples were headspace over 4.5 mL pure d-limonene in 9-mL vials equilibrated at 26°C overnight. Values are mean ± s.d., $n = 3$. PDMS, polydimethyl siloxane; DVB, divinylbenzene.

B. Headspace Equilibration of Volatiles

We conducted studies to determine the equilibration time of limonene in the headspace over water and deodorized orange juice diluted to 11.8 °Brix from 65 °Brix evaporator concentrate. After many trials, the most consistent procedure involved use of a static (unstirred) system at room temperature (24°C–25°C), spiked with the proper amount of d-limonene. Heating samples was considered to shorten equilibrium time, but because of the wide variation in component vapor pressures, we did not see any advantage for the static system. Figure 1 illustrates that headspace equilibrium time for d-limonene (below solution saturation) at room temperature was approximately 13 hr for either water or orange juice. The higher values earlier in the timed experiments result from limonene's contacting the glass and evaporating directly to the headspace before equilibrium. Once this time (13 hr) was established, results to determine equilibrium time (approximately 20 min) of PDMS SPME fibers could be determined; they are presented in Fig. 2. Since long fiber-sample headspace contact times affect the headspace equilibrium, SPME fiber contact times for most of our experimental work were 1–2 min. Compounds have different fiber affinities and are present in the liquid and headspace at different concentrations and may not be adsorbed linearly with time. Thus, a short adsorption time was

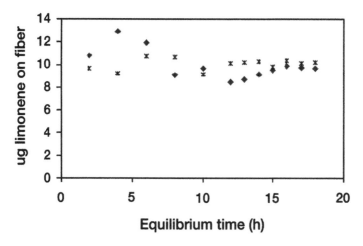

Figure 1 Headspace equilibrium time for d-limonene in water (12 ppm) and 11.8 °Brix, 10% suspended pulp orange juice (200 ppm). Sample (125 mL) in 250-mL capped (Teflon septa) bottle held up to 18 hr at 26°C. Fiber (100 μm PDMS) contact with sample headspace for 1 min. PDMS, polydimethyl siloxane. *, water; ♦, juice.

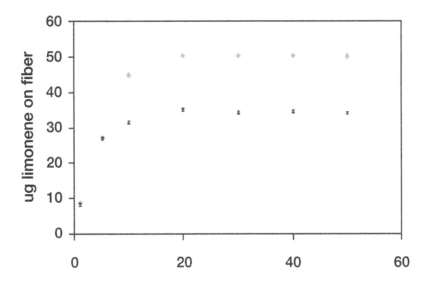

Figure 2 PDMS SPME fiber equilibrium times for *d*-limonene in water (12 ppm) and 11.8 °Brix, 10% suspended pulp orange juice (200 ppm). Sample (125 mL) in 250 mL capped (Teflon septa) bottle held overnight at 26°C. PDMS, polydimethyl siloxane; SPME, solid-phase microextraction. *, water; ♦, juice.

chosen to be close to linearity and have minimal effect on equilibrium conditions [10].

Headspace saturation of *d*-limonene in water followed an established solubility limit of approximately 15 ppm limonene in water (Fig. 3). Suspended pulp content dramatically affects the apparent solubility of *d*-limonene in juice. For juice with 10% v/v suspended pulp [22], headspace saturation occurred when the juice contained approximately 400–450 ppm *d*-limonene (Fig. 4). The result of partitioning *d*-limonene into the pulp was to keep the juice headspace concentration below the solubility of limonene in water. Headspace saturation did not occur until the pulp became saturated.

Because measurement of suspended pulp as volume/volume (v/v) in citrus juice is not precise, a mass relation was established by ultracentrifugation (250,000 × g in 2-mL minitubes) of juice to recover pulp (90% moisture) and remove all of the cloud. This pulp was added to pulp-free juice as percentage by weight of juice and the saturation point of *d*-limonene in 11.8 °Brix juice at different pulp levels determined. Using a standard

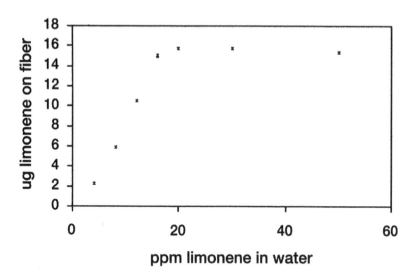

Figure 3 Headspace saturation of *d*-limonene in water. Sample (125 mL) in 250 mL capped (Teflon septa) bottle held overnight at 26°C. Fiber (100 μm PDMS) contact with sample headspace for 1 min. PDMS, polydimethyl siloxane.

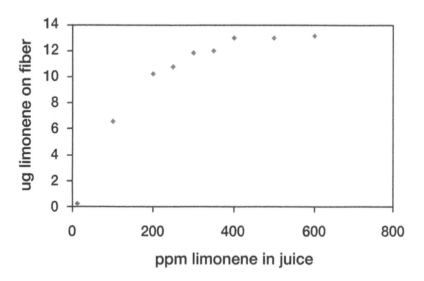

Figure 4 Headspace saturation of *d*-limonene in 11.8° Brix, 10% suspended pulp orange juice. Sample (125 mL) in 250 mL capped (Teflon septa) bottle held overnight at 26°C. Fiber (100 μm PDMS) contact with sample headspace for 1 min. PDMs, polydimethyl siloxane.

curve based on the amount of d-limonene added, the mass distribution between pulp and serum at saturation was determined; the amount of d-limonene in the serum was constant at 2.5 mg/125 mL (Table 2). About 4% pulp by weight was equivalent to 10% v/v suspended pulp.

C. Headspace Saturation of Volatiles

1. Water

On the basis of amounts in orange juice [17], samples of water (125 mL) in 250-mL bottles with Teflon septa were fortified with increasing amounts of a mixed standard containing ethyl butyrate (99% pure, 133 µg/µL), α-pinene (90% pure, 91 µg/µL), octanal (95% pure, 147 µg/µL), linalool (90% pure, 129 µg/µL), decanal (95% pure, 148 µg/µL), and valencene (78% pure, 135 µg/µL). The water samples were spiked with increasing amounts of the mixed standard at 4, 8, 16, 32, 64, and 128 µL with a constant amount of d-limonene (99% pure) at 20 µg/mL. Concentrations were based on calculated 100% purity. The concentrations of the mixed standard were selected to allow the headspace concentration to be high enough to detect headspace saturation. The concentrations are approximately 100 times greater than one might expect in an actual juice. This study determined the headspace saturation (as micrograms on the fiber) of different components in water with d-limonene constant as well as the effect on d-limonene of increasing amounts of the other components. The results in Table 3 indicate

Table 2 Headspace Saturation of d-Limonene at Different Pulp Contents in 11.8 °Brix Orange Juice Diluted from Aroma-Free 65 Brix Evaporator Pump-out, Spiked with d-Limonene[a,b]

Suspended pulp, %, v/v	Pulp, % w/w[b]	Limonene saturation, ppm	Serum limonene, mg/mL	Pulp limonene, mg/g dry pulp
0	0	20	0.02	0
1	0.7 ± 0.1	100	0.02	110 ± 4
2	1.4 ± 0.1	200	0.02	116 ± 5
4	1.5 ± 0.1	300	0.02	142 ± 3
7	1.9 ± 0.4	400	0.02	108 ± 7
10	3.4 ± 0.6	500	0.02	107 ± 4
15	5.4 ± 0.1	600	0.02	104 ± 5

[a] Pulp-adjusted juice (125 mL) in 250-mL bottle, overnight equilibrium at 26°C, PDMS fiber for 1 min. PDMS, polydimethyl siloxane.
[b] Pulp at 90% moisture. Values are mean ± s.d., $n = 3$.

Table 3 Headspace Saturation of Compounds in Water with d-Limonene Constant (20 ppm)[a]

Compound	Parts per million in water with spike of					
	4 µL	8 µL	16 µL	32 µL	64 µL	128 µL
Ethyl butyrate, 133 µg/µL	4.3	8.6	17.2	34.4	68.8	137.6
α-Pinene, 91 µg/µL	2.9	5.8	11.6	23.2	46.4	92.8
Octanal, 147 µg/µL	4.7	9.4	18.8	37.6	75.2	150.4
d-Limonene		20	20	20	20	20
Linalool, 129 µg/µL	4.1	8.2	16.4	32.8	65.6	131.2
Decanal, 148 µg/µL	4.7	9.4	18.8	37.6	75.2	150.4
Valencene, 135 µg/µL	4.3	8.6	17.2	34.4	68.8	137.6
	Micrograms on fiber at each concentration					
Ethyl butyrate	0.75	0.78	0.83	0.93	1.11	1.51
α-Pinene	3.84	5.91	8.66	10.74	12.09	12.70
Octanal	0.83	1.07	1.51	2.18	3.33	4.78
d-Limonene	25.93	19.62	13.63	7.56	4.02	2.26
Linalool	0.57	0.59	0.63	0.69	0.82	1.05
Decanal	1.21	1.72	2.18	2.47	2.64	2.58
Valencene	0.76	0.81	0.82	0.82	0.82	0.81

[a] Overnight equilibration at 26°C. Effects of increasing concentration (ppm) of a standard mixture spiked into water on microgram on PDMS fiber, sampled for 5 min. Values are mean, $n = 2$.

that at the concentrations studied, ethyl butyrate, octanal, and linalool did not reach saturation; α-pinene saturated at 12 µg; decanal, at 2.5 µg; and valencene, at 0.8 µg. Competition with other compounds resulted in decreased adsorption of d-limonene on fiber sites as other compounds increased.

2. Juice

A similar experiment involved spiking 11.8 °Brix, 10% suspended pulp juice with increasing amounts of mixed standards. Because of the adsorptive effect of the pulp, to obtain measurable headspace amounts on the fiber it was necessary to increase the amounts spiked from the standard orange juice solution, which contained ethyl butyrate (118 µg/µL), α-pinene (80 µg/µL), octanal (130 µg/µL), d-limonene (86 µg/µL), linalool (115 µg/µL), decanal (131 µg/µL), and valencene (127 µg/µL). As for Table 3, the concentrations were selected in order for compounds to be concentrated enough for

Table 4 Headspace Saturation of Compounds in 10% Suspended Pulp 11.8 Brix Juice (125 mL in 250-mL Bottle)[a]

Compound, concentration	Parts per million in juice with spike of						
	60 µL	125 µL	250 µL	500 µL	700 µL	850 µL	1000 µL
Ethyl butyrate, 118 µg/µL	57	117	236	471	660	801	943
α-Pinene, 80 µg/µL	39	80	161	322	450	547	644
Octanal, 130 µg/µL	62	130	260	520	728	884	1040
d-Limonene, 86 µg/µL	41	86	171	343	480	582	685
Linalool, 115 µg/µL	55	115	229	458	642	779	917
Decanal, 131 µg/µL	63	132	263	526	736	894	1052
Valencene, 127 µg/µL	61	127	254	509	712	865	1017
	Micrograms on fiber at each concentration						
Ethyl butyrate	0.85	1.41	2.49	4.09	5.02	5.45	5.86
α-Pinene	7.90	9.62	10.12	9.90	9.76	9.46	9.23
Octanal	1.63	2.85	4.15	5.02	5.31	5.29	5.32
d-Limonene	5.63	7.24	7.75	7.55	7.52	7.31	7.26
Linalool	0.44	0.68	1.06	1.37	1.49	1.53	1.61
Decanal	0.92	1.31	1.58	1.55	1.56	1.53	1.58
Valencene	0.49	0.52	0.50	0.45	0.45	0.44	0.47

[a] Overnight equilibration at 26°C. Effects of increasing concentration (ppm) of a standard mixture spiked into juice on microgram on polydimethyl siloxane (PDMS) fiber, sampled for 5 min. Values are mean, $n = 2$.

detection of headspace saturation by the fiber. Results in the juice (Table 4) differed from those in water, in that ethyl butyrate was the only compound at these higher concentrations that did not reach headspace saturation. The situation for d-limonene indicated that because of adsorption by the pulp, a constant low level was present in the headspace. Although values vary widely, the concentrations of the stated compounds in processed orange juice have been reported [23]: ethyl butyrate (0.5 ppm), α-pinene (0.4 ppm), octanal (0.9 ppm), d-limonene (135–180 ppm), linalool (0.6 ppm), and decanal (0.8 ppm), and valencene (1 ppm).

Another factor complicating use of SPME headspace measurement to quantify the amount of components in juice is the variability of fiber equilibration times for the different compounds. Samples of the standard mixture in water (discussed previously) were allowed to equilibrate 16 hr (26°C) and then sampled for fiber contact times up to 120 min. The amounts of the various compounds and their equilibration times with the PDMS fiber

Table 5 Fiber Equilibrium of Water Solutions of Compounds in Headspace with 100 μm Polydimethyl Siloxane[a]

Sampling time, min	Compound on fiber, μg						
	Ethyl butyrate	Pinene	Octanal	Limonene	Linalool	Decanal	Valencene
1	0.83	5.09	1.09	4.44	0.57	0.97	0.53
5	0.84	8.89	1.60	9.11	0.64	2.26	0.88
10	0.84	9.22	1.69	10.20	0.67	3.38	1.29
20	0.85	9.84	1.77	11.03	0.71	4.93	2.09
30	0.85	10.10	1.81	11.50	0.71	5.89	2.87
40	0.85	10.39	1.87	12.00	0.74	6.55	3.64
50	0.86	10.98	1.94	12.57	0.75	7.11	4.36
60	0.86	11.01	1.95	12.59	0.75	7.33	4.89
70	0.86	11.04	1.97	12.57	0.76	7.51	5.46
80	0.87	10.98	2.01	12.64	0.76	7.75	6.61
90	0.86	10.91	2.03	12.57	0.76	7.99	6.85
120	0.86	10.95	1.96	12.58	0.76	8.06	7.97

[a] Samples contained 21 μL mixed standard (same as in Table 4) in 125 mL water equilibrated at 26°C overnight. Values are mean, $n = 3$.

are listed in Table 5. Significant results were the short equilibration times of ethyl butyrate (5 min), the long equilibration of decanal (90 min), and the finding that valencene did not reach equilibrium with the fiber in 120 min.

IV. CONCLUSIONS

In summary, we believe these results indicate that use of SPME headspace analysis of citrus juices is limited to nonquantitative data acquisition, specifically to determination of the qualitative absence or presence of certain components. The SPME headspace measurements used to compare juice processes (e.g., thermal vs. pulsed electric field pasteurization), storage stability, etc., may have little value if the absolute amount of the components in the juice is not easily determined or is unknown. Also, many studies neglect to consider various equilibrium conditions, the effect of juice pulp, and interactions of components on headspace concentrations. For many juice product/process comparisons, thorough sensory analyses would provide more useful information than SPME analysis.

ACKNOWLEDGMENT

This research was supported by the Florida Agricultural Experiment Station and approved for publication as Journal Series No. R-09117.

REFERENCES

1. HA Massaldi, CJ King. Simple technique to determine solubilities of sparingly soluble organics: Solubility and activity coefficients of *d*-limonene, *n*-butylbenzene and *n*-hexylacetate in water and sucrose solutions. J Chem Eng Data 18(4):393–397, 1973.
2. CJ King. Physical and chemical properties governing volatilization of flavor and aroma components. In: M Peleg, EB Bagley, eds. Physical Properties of Foods. Westport, CT: Avi, 1983, pp 399–421.
3. RJ Braddock. Essential oils and essences: *d*-Limonene. In: Handbook of Citrus By-Products and Processing Technology. New York: John Wiley & Sons, 1999, pp 149–190.
4. AJ Taylor. Physical chemistry of flavor. Int J Food Sci Technol 33:53–62, 1998.
5. AD Harmon. Solid-phase microextraction for the analysis of flavors. In: R Marsili, ed. Techniques for Analyzing Food Aroma. New York: Marcel Dekker, 1997, pp 81–112.
6. A Steffen, J Pawliszyn. Analysis of flavor volatiles using headspace solid-phase microextraction. J Agric Food Chem 44:2187–2193, 1996.
7. KD Deibler, TE Acree, EH Lavin. Solid phase microextraction application in gas chromatography/olfactometry dilution analysis. J Agric Food Chem 47:1616–1618, 1999.
8. KD Deibler, EH Lavin, TE Acree. Solid phase microextraction application in GC/olfactometry dilution analysis. In: JF Jackson, HF Linskens, eds. Analysis of Taste and Aroma: Molecular Methods of Plant Analysis, Vol. 21. Berlin: Springer, 2002, pp 239–248.
9. MM Miller, JD Stuart. Comparison of gas-sampled and SPME-sampled static headspace for the determination of volatile flavor components. Anal Chem 71:23–27, 1999.
10. DD Roberts, P Pollien, C Milo. Solid-phase microextraction method development for headspace analysis of volatile flavor compounds. J Agric Food Chem 48:2430–2437, 2000.
11. K Sukola, J Koziel, F Augusto, J Pawliszyn. Diffusion-based calibration for SPME analysis of aqueous samples. Anal Chem 73:13–18, 2001.
12. KD Deibler, T Acree. Effect of beverage base conditions on flavor release. ACS Symp Ser 763 (Flavor Release), 2000, pp 333–341.
13. T Radford, K Kawashima, PK Friedel, LE Pope, MA Gianturco. Distribution of volatile compounds between the pulp and serum of some fruit juices. J Agric Food Chem 22:1066–1070, 1974.

14. SE El-Nemr, A Askar. Distribution of volatile aroma compounds between pulp and serum of mango juice. Deutsch Lebensm-Rundschau 82(12):383–386, 1986.
15. JM Conner, L Birkmyre, A Paterson, JR Piggott. Headspace concentrations of ethyl esters at different alcoholic strengths. J Sci Food Agric 77:121–126, 1998.
16. RS Carpenter, DR Burgard, DR Patton, SS Zwerdling. Application of multivariate analysis to capillary GC profiles: Comparison of the volatile fraction in processed orange juices. In: G Charalambous, G Inglett, eds. Instrumental Analysis of Foods, Vol. 2. New York: Academic Press, 1983, pp 173–186.
17. PA Rodriguez, CR Culbertson. Quantitative headspace analysis of selected compounds in equilibrium with orange juice. In: G Charalambous, G Inglett, eds. Instrumental Analysis of Foods, Vol. 2. New York: Academic Press, 1983, pp 187–195.
18. R Marsili, G Kilmer, N Miller. Quantitative analysis of orange oil components in orange juice by a simple solvent extraction–gas chromatographic procedure. LC-GC 7(9):778, 780–783, 1989.
19. MG Moshonas, PE Shaw. Dynamic headspace gas chromatography combined with multivariate analysis to classify fresh and processed orange juices. J Essent Oil Res 9:133–139, 1997.
20. M Jia, H Zhang, DB Min. Optimization of solid-phase microextraction analysis for headspace compounds in orange juice. J Agric Food Chem 46:2744–2747, 1998.
21. R Bazemore, K Goodner, R Rouseff. Volatiles from unpasteurized and excessively heated orange juice analyzed with solid phase microextraction and GC-olfactometry. J Food Sci 64:800–803, 1999.
22. JB Redd, CM Hendrix Jr, DL Hendrix. Quality control manual for citrus processing plant, Vol. 1. Method 8. Suspended Pulp. Safety Harbor, FL: Intercit, Inc., 1986, p 37.
23. PE Shaw. Fruits II. In: H Maarse, ed. Volatile Compounds in Foods and Beverages. New York: Marcel Dekker, 1991, pp 305–327.

31

Analysis of Microbial Volatile Metabolites Responsible for Musty–Earthy Odors by Headspace Solid-Phase Microextraction Gas Chromatography/Mass Spectrometry

Henryk H. Jeleń, Małgorzata Majcher, and Erwin Wąsowicz
Agricultural University of Poznań, Poznań, Poland

I. INTRODUCTION

Musty taints and off-flavors in foodstuffs have been a source of concern in the food industry for many years. Although a list of potential compounds causing this defect is long [1], the main ones responsible for it are 2-methoxy-3-isopropylpyrazine (IPMP), 2-methoxy-3-isobutylpyrazine (IBMP), 2,4,6-trichloroanisole (TCA), 2-methylisoborneol (MIB), and geosmin (GEO) (Fig. 1).

2-Methoxy-3-alkylpyrazines are often associated with musty–earthy and potato-like odors. 2-methoxy-3-isopropylpyrazine is responsible for the musty, potato-bin odor and contributes to off-flavor notes in wines; 2-methoxy-3-isobutylpyrazine is a main constituent of green bell pepper and jalapenios flavor. It is known that alkylpyrazines are formed in thermal reactions but can be also produced by microorganisms, mainly bacteria, such as *Pseudomonas* species [2].

Chloroanisoles are associated with musty taints and were detected as responsible compounds in outbreaks of this taint in chickens [3], in "Rio off-

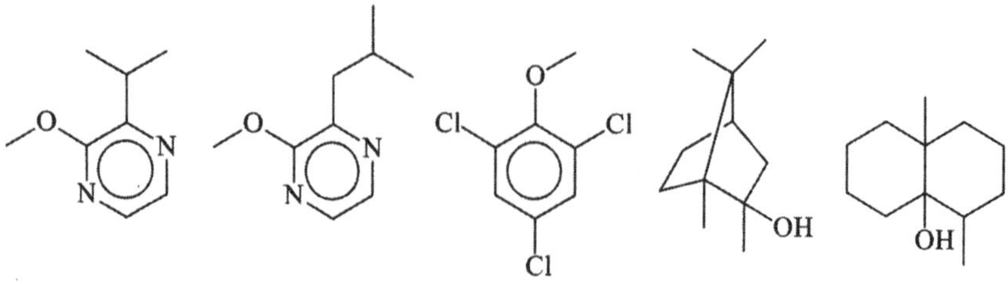

Figure 1 Structures of investigated compounds produced by microorganisms responsible for occurrence of musty flavors in foods. 1, 2-Methoxy-3-isopropylpyrazine; 2, 2-methoxy-3-isobutylpyrazine; 3, 2,4,6-trichloroanisol; 4, 2-methylisoborneol; 5, geosmin.

flavor" of Brazilian green coffee [4], and in various products such as dried fruit packed in cardboard boxes or wooden containers. Moldy–musty off-odor known as a *cork taint* is one of the most serious problems affecting the wine industry. The estimates of cork taint incidence range from 2% to 6% in wines that are rejected. The major cause of this taint is 2,4,6-trichloroanisole [5]. This compound is usually formed as a result of microbial degradation and methylation of pentachlorophenol—a component used in some wood preservatives or insectides—or from 2,4,6-trichlorophenol during hypochloride treatment of the cork. A lot of microorganisms, such as *Aspergillus, Penicillium, Trichoderma,* or *Streptomyces* species, can catalyze these reactions. Geosmin and 2-methylisoborneol are probably the most well known and explored compounds of musty–earthy character because of the role they play in development of taints in water. Geosmin was identified as an actinomycetes metabolite responsible for an earthy odor in water supplies. It is produced by *Streptomyces* species, and by strains of *Nocardia, Micromonospora, Microbispora, Oscillatoria,* and *Phormidium* species [6]. It has been also identified as a metabolite of *Penicillium* species [7] and many genera of algae in drinking water. Methylisoborneol was identified in musty smelling natural waters, where it occurs as a metabolite of actinomycetes, blue algae, *Oscillatoria tenuis,* and *Phormidium* species [6]. Musty and earthy off-odors are also a concern for the grain industry. Improperly stored grain may develop substantial levels of storage microflora, mainly *Aspergillus* and *Penicillium* species, and as a result of this, musty-off flavors can occur [8].

Progress in analytical chemistry, especially in mass spectrometry, capillary gas chromatography, and other hyphenated techniques, has allowed the identification of many compounds responsible for off-odors. For malodor analysis, apart from the separation and detection tools and techniques required, the sample preparation process plays the most crucial role. Because of the very low concentrations in which these compounds occur in food matrices, and their extremally low odor treshholds (Table 1), enrichment of analytes is required in the process of their isolation. Determination of these compounds employs various extraction techniques, such as liquid/liquid extraction followed by Kuderna Danish concentration, closed loop stripping analysis, simultaneous distillation extraction, and purge and trap. Solid-phase microextraction (SPME) offers a solventless and rapid alternative to laborious methods of extraction based on dynamic headspace, vacuum distillation, or simultaneous distillation extraction. Developed at the beginning of the 1990s, SPME found applications in analysis of food volatile compounds present in trace concentrations [19]. It has been used to analyze alkylpyrazines from must and wine [20,21] and 2,4,6-trichloroanisole from wine [22], and for determination of geosmin and 2-methylisoborneol in water [23,24]. However, data on extraction of these compounds from solid matrices are very limited [25–27].

Our goal was to elaborate a method of analysis of musty–earthy compounds in a solid matrix based on solid-phase microextraction. Wheat grain was chosen for method development since musty–earthy off-odors are

Table 1 Odor Descriptions and Odor, and Taste Thresholds of Investigated Compounds[a]

Compound	Odor description	OT, µg/L	TT, µg/L	Reference
2-Methoxy-3-	Potato bin, earthy,	0.002		9
isopropylpyrazine	musty	0.002		10
(IPMP)		0.0002	0.02	11
2-Methoxy-3-	Green bell pepper,	0.002		12
isobutylpyrazine	earthy–musty	16		13
(IBMP)		0.001	0.003	14
2-Methylisoborneol	Earthy–musty,	0.029		14
(MIB)	camphorous	0.015	0.018	11
Geosmin (GEO)	Earthy–musty	0.01		15
			0.05	16
		0.0038	0.0016	11
2,4,6-Trichloroanisole	Earthy–musty	0.000003		17
(TCA)		0.000003		3
		0.0003		18
		0.0009	0.050	11

[a] OT, odor threshold; TT, taste threshold.

often a concern in grain quality assesment. Grain samples are typically evaluated for the presence of off-odors by inspectors who use sensory methods. Instrumental analysis of compounds responsible for specific off-odor compounds is too sophisticated to be done routinely at the grain elevators and storage facilities. Crucial steps in SPME method development are discussed in this chapter.

II. MATERIALS AND METHODS

Begra variety sound wheat grain, of 13% water content, supplied by the Agricultural University of Poznan experimental station Zlotniki, was used in the model experiments. Malodorous grain samples were received from farmers' granaries. Wheat (50 g) was placed into 100-mL bottles (Supelco, Bellefonte, PA). The matrix was spiked with a mixture of standards in methanol to obtain a concentration of analytes equal to 1 ppb; for fiber selection experiments the concentration of standards was 6 ppb. All standards were purchased from Supelco and were of 99% gas chromatography (GC) purity. After spiking with standards, bottles were capped with Teflon lined caps and stored at room temperature (20°C) until the next day, when they were analyzed. For all experiments, except the fiber choice study, 50/30-μm Carboxen divinylbenzene polydimethyl siloxane (C/DVB/PDMS) fiber was used. For the fiber choice experiment the following fibers were examined: 100 μm polydimethyl siloxane (PDMS), 75 μm Carboxen/poly-dimethyl siloxane (C/PDMS), and 65 μm divinylbenzene/polydimethyl siloxane (DVB/PDMS). Fibers were supplied by Supelco and preconditioned before analyses according to the manufacturer's recommendations. All SPME extractions and injections were performed manually. Fiber dwell time in the injection port was 5 min. For GCMS analyses a Trace 2000 gas chromatograph coupled to a Finnigan Polaris Q ion trap mass spectrometer (ThermoFinnigan, Austin, TX) was used. The mass spectrometer had a programmed temperature vaporization (PTV) injection port working at a constant temperature (260°C), splitless mode (purged at 1 min), and an MS^n option. Compounds were resolved on a HP-5 column (30 m $\times$ 0.25 mm $\times$ 0.25 μm, Agilent Technologies, Palo Alto, CA). A programmed temperature was used for compound resolution: 40°C for 3 min, then an increase of 8°C/min to 280°C. Helium was used as the carrier gas at 0.8-mL/min flow. Identification of compounds of interest in the matrix was based on the comparison of retention time and mass spectra with those of authentic standards. Mass spectrometer was run in full-scan, multiple ion monitoring (MIM) and MS/MS modes. In full-scan mode a mass range of 40 to 340 amu was applied. For MIM three ions (one target and two qualifiers) were selected for each of compounds: $m/z = 124$, 137, 152 for 2-isopropyl-3-methoxypyrazine; $m/z = 94$, 124, 151 for 2-isobutyl-3-methoxypyrazine; $m/z = 195$, 197, 210 for 2,4,6-trichloroanisole; $m/z = 95$, 107, 168 for 2-methylisoborneol; $m/z = 112$, 125, 182 for geosmin. For MS/MS experiments the following parent ions were selected for subsequent CID collisions: $m/z = 137$ for 2-isopropyl-3-methoxypyrazine, $m/z = 124$ for 2-isobutyl-3-methoxypyrazine, $m/z = 195$ for 2,4,6-trichloroanisole, $m/z = 95$ for 2-methylisoborneol, and $m/z = 112$ for geosmin.

III. RESULTS AND DISCUSSION

A. Influence of Matrix on the Extraction Process

Spiking the matrix with a known amount of analyte provides a consistent sample for method elaboration; however, one must remember that this procedure is a simplified approximation, as release of naturally bound microbial volatiles in matrix can involve different mechanisms. In the preliminary experiment the matrix was spiked with a mixture of standards at 1 ppb each. Extraction with C/DVB/PDMS fiber for 30 min was performed from randomly chosen bottles immediately after spiking (day 0), and after 1, 2, and 3–6 days. The highest responses for all analyzed compounds were noted on day 0. When extraction was performed 1 day after spiking, recoveries of compounds were approximatelly 30%–40%, compared to those of day 0 extraction. The amount of extracted analytes in days 2–4 was similar to that of 1 day and decreased another 15%–20% on days 5 and 6. Relative intensities of extracted compounds did not vary from day to day. To minimize variations caused by adsorption on the matrix in all procedures, samples were spiked 1 day before analyses.

As grain water content influences its physical and therefore sorptive parameters, four spiked matrices that differed in water content (w.c.) were prepared. Wheat grain (13% w.c.) was moisturized with deionized water to the required water content, left overnight to absorb it, and then spiked. Figure 2 shows extraction efficiency of analytes, which depended on water content in wheat grain matrix. Grain subjected to storage usually has less than 13% water content. Either harvested in conditions that provide this moisture content or dried after harvest, grain must have the lowest possible water content to prevent development of microorganisms. The recoveries of analytes were highest for sound grain of 13% water content. A higher level of water, especially over 20%, resulted in a dramatic decrease of compounds extracted from it, though theorethically higher water content in the matrix should favor partitioning of these compounds into the gas phase, since they

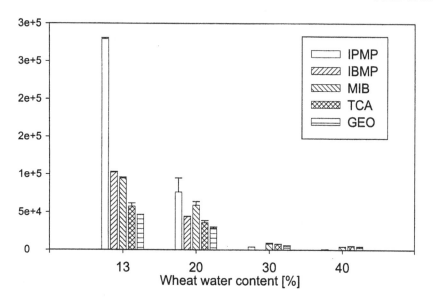

Figure 2   Influence of water content in the wheat grain matrix on the amounts of compounds isolated by solid-phase microextraction (SPME). IPMP, 2-methoxy-3-isopropylpyrazine; IBMP, 2-methoxy-3-isobutylpyrazine; MIB, 2-methylisoborneol; TCA, 2,4,6-trichloroanisole; GEO, geosmin.

have a low water solubility. In samples where water content in grain was 40%, the amount of extracted compounds was from 12 (GEO, TCA), 23 (MIB), to 310 times (IPMP) lower than from sound 13% w.c. grain samples. Moreover, it was impossible to detect 1 ppb of IBMP from wheat containing 30% or 40% water. However, determination of compounds responsible for off-odors in grain with water content exceeding 20% may be of little practical use, as such moist grain rapidly deteriorates and cannot be stored. These data indicate that matrix composition influences its sorptive characteristics to a great extent, and in practice the method should be elaborated for the specific matrix and validated each time the composition of the matrix investigated changes.

B.   Fiber Selection

Selectivity in SPME analysis is dependent on the fiber coating selected. Initially only two types of fiber coatings were manufactured—PDMS and polyacrylate (PA). Now various fiber coatings are available, among them porous particles dispersed in PDMS polymer (Carboxen/PDMS, DVB/ Carboxen/PDMS, and DVB/PDMS). These fibers have been designed for the analysis of low-molecular-weight compounds present in trace concentrations.

For the analysis of compounds investigated, four fibers were tested: PDMS (P), Carboxen/PDMS (C/P), DVB/Carboxen/PDMS (D/C/P), and DVB/PDMS (D/P); Fig. 3 shows the detector response for all compounds extracted at 50°C for 30 min. Comparison of these data reveals that porous particle-based fibers extract the selected compounds much more efficiently than PDMS. The most efficient mechanism for binding these volatiles into fiber coating is adsorption, a characteristic of all Carboxen and DVB fibers, not a partition characteristic of PDMS fiber. By comparing these two coatings (in fibers D/P, C/P) it can be concluded that divinylbenzene is more appropriate for analysis of the selected microbial off-odorants than Carboxen. This suggests that divinylbenzene mesopores trap volatiles

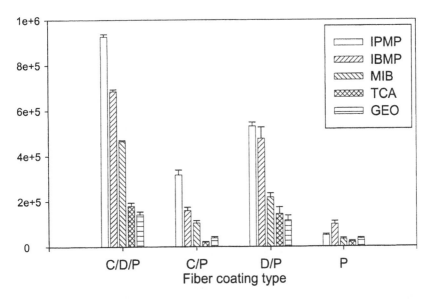

Figure 3 Influence of fiber coating type on the amount of extracted compounds. C/ D/P, Carboxen/divinylbenzene/polydimethyl siloxane fiber; C/P, Carboxen/PDMS; D/P, divinylbenzene/PDMS; P, polydimethyl siloxane (PDMS) fiber. IPMP, 2-methoxy-3-isopropylpyrazine; IBMP, 2-methoxy-3-isobutylpyrazine; MIB, 2-methylisoborneol; TCA, 2,4,6-trichloroanisole; GEO, geosmin.

more efficiently than Carboxen micropores. The best overall responses, however, were noted for Carboxen/DVB/PDMS (C/D/P) fiber, probably because the coating is twice as long as other investigated fibers. Watson and associates [14], comparing different fibers for geosmin and MIB extraction from water, noted the highest recoveries for both when PDMS/DVB fiber was used. Two centimeters of PDMS/C/DVB yielded slightly lower recoveries for both compounds. The PA and PDMS fibers had higher recovery for geosmin than for MIB. However, recoveries obtained by using the latter two fibers were lower than the recoveries using porous polymer-based fibers.

C. Extraction Curves

To check the adsorption profiles of analytes on fiber surface, extractions from spiked wheat were performed from 1 min up to 120 min (Fig. 4). No achievement of equilibrium state was observed for any of analyzed compounds in 120-min extraction. In fibers where capillary condensation occurs, as in Carboxen coating, a state of equilibrium often cannot be reached [28]. The extraction curve slope indicates that detection limits can be substantially improved when longer extraction times are applied. For most of the compounds, prolonging extraction times from 30 min to 60 min resulted in twice the peak areas. However, to decrease total analysis time, extractions are often performed for shorter periods, before the equilibrium state is reached. Therefore, a 30-min extraction time was chosen in the remaining experiments, providing sufficient sensitivity and relatively short extraction time, comparable to GC run time. The extraction process of analytes from wheat grain was also compared to extraction from water. The comparison was done under the same conditions within a 120-min period. Similarly to the extraction from grain, no equilibrium was observed for any of the analyzed compounds. As no adsorbing matrix was present, lack of equilibrium state is due mainly to the porous adsorptive nature of polymer coating on the fiber surface. The IPMP and IBMP behaved similarly, and higher recoveries from grain compared to those from water were observed for these compounds. Contrary to these results, TCA and GEO were more easily extracted from water than from grain, providing higher recoveries from this medium. Extraction curves had a similar appearance for MIB. Two examples of this behavior (IPMP and TCA) are presented in Fig. 5.

In another experiment, extraction over very long periods was evaluated to check whether equilibrium can be reached after several hours and whether displacement of one analyte by another took place. Peak areas

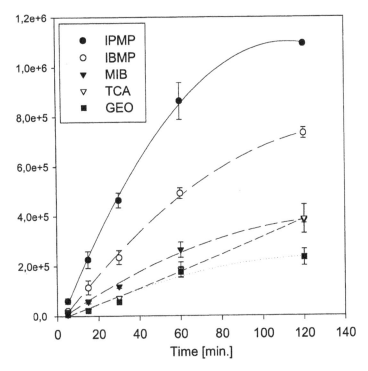

Figure 4 Extraction curves for analyzed compounds isolated from grain. Extraction was performed at 50°C for times ranging from 1 min. to 120 min. IPMP, 2-methoxy-3-isopropylpyrazine; IBMP, 2-methoxy-3-isobutylpyrazine; MIB, 2-methylisoborneol; TCA, 2,4,6-trichloroanisole; GEO, geosmin.

of compounds extracted for 30 min, 6 hr, and 12 hr were compared (Fig. 6). Relative abundance of peaks differed, especially when 30-min extraction was compared to 6-hr or 12-hr extraction. No increase of IPMP and MIB peak areas was observed for 6 and 12 hr. For the remaining compounds, peak areas were increasing throughout the extraction period, indicating that even after such a long extraction time no equilibrium was reached in a system. Moreover, displacement of one compound by molecules of another could potentially take place with adsorption-type fibers, such as Carboxen or divinylbenzene, but no dramatic decrease in one compound concentration in favor of another was observed. Long extraction times have the advantage that more analytes are adsorbed on the fiber surface, thereby improving limits of detection. Watson and colleagues [24] observed reaching

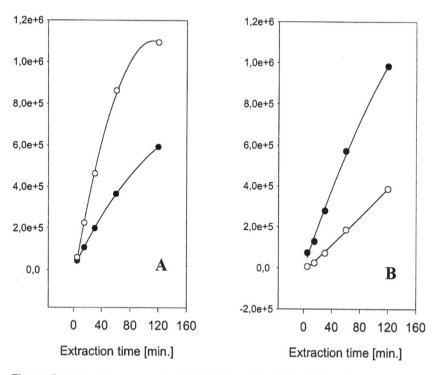

Figure 5 Extraction curves for A, IPMP, and B, TCA, isolated from grain and water (white and black dots on the graph, respectively). Extraction was performed at 50°C four times ranging from 1 min to 120 min. IPMP, 2-methoxy-3-isopropyl-pyrazine; TCA, 2,4,6-trichloroanisole.

of equilibrium for MIB after 1 hr and for geosmin after 2 hr of extraction using PDMS/DVB fiber. Lloyd and Grimm [25] observed that a 20-min extraction was sufficient when PDMS fiber was used. After a 4-min extraction, peak areas were an order of magnitude higher than after 1 min, whereas peak areas after 20-min extraction were only twice as high as after 4 min.

D. Extraction Temperature

Extraction temperature influences distribution of volatile compounds in solid and headspace phases. As temperature increases, air–matrix partition coefficient increases, resulting in more compounds in headspace. However, increasing sampling temperature can decrease the fiber–air partition

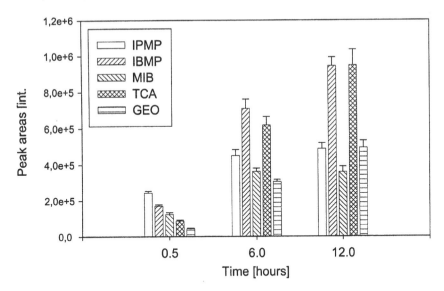

Figure 6 Influence of extraction time on the amount of volatiles extracted from spiked wheat grain. IPMP, 2-methoxy-3-isopropylpyrazine; IBMP, 2-methoxy-3-isobutylpyrazine; MIB, 2-methylisoborneol; TCA, 2,4,6-trichloroanisole; GEO, geosmin.

coefficient, so migration of analytes from the fiber coating takes place. Another factor that must be considered in developing the SPME method is the thermal lability of analyzed, compounds, as is the lability of the matrix. Figure 7 shows detector responses for analyzed compounds with increased extraction temperatures. For GEO, TCA, and IBMP, increasing extraction temperature resulted in peak area increase, whereas for MIB and IPMP, rapid decrease in adsorbed compound amount was observed when extraction temperature was changed from 75°C to 100°C. For IPMP, the amount of adsorbed analyte at 100°C was even lower than that adsorbed at 50°C. This result is probably caused by desorption of analytes from the fiber surface. Another explanation may be the depletion of IPMP and MIB by other compounds present in the matrix. When extraction is performed at elevated temperatures, those temperatures obviously have an impact on the amounts of compounds isolated, and on their profile. Thermal lability of a matrix is a limiting factor, and heating food material to temperatures exceeding 100°C is not practical because of matrix decomposition or creation of artifacts. Therefore, to minimize the influence of coextracted compounds from the matrix (replacement of compounds on fiber, signal-to-

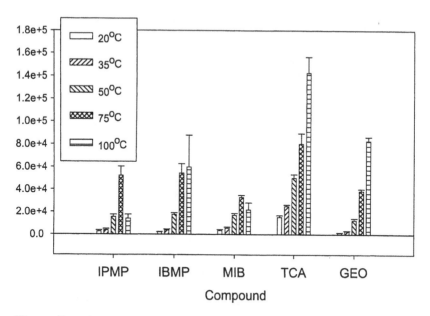

Figure 7 Influence of extraction temperature on the amount of extracted compounds from spiked wheat grain. IPMP, 2-methoxy-3-isopropylpyrazine; IBMP, 2-methoxy-3-isobutylpyrazine; MIB, 2-methylisoborneol; TCA, 2,4,6-trichloroanisole; GEO, geosmin.

noise influence), the lowest extraction temperature guaranteeing sufficient limits of detection is advisable. Watson and coworkers noted [24] that when extraction temperatures increase from 20°C to 80°C, MIB response gradually decreases. For geosmin, the highest recovery was observed at 40°C. At 20°C and 60°C the recoveries were almost equal, whereas at 80°C recovery was lowest.

E. Limits of Detection, Reproducibility, and Linearity

In the process of method development, detection, quantitation limits, reproducibility, and the linear range are determined to evaluate the method. In the analysis of compounds responsible for musty off-odors the limits of detection are of special importance, as the odor thresholds for these compounds are usually very low. The SPME procedure allows sample enrichment on the fiber surface and has been shown in many instances to be a method well suited to trace analysis. In our investigations the optimal sampling parameters included preheating sample at 50°C for 15 min, followed by a sampling time of 30 min. Conditions were chosen as a compromise between high sample throughput and relatively high sensitivity. Utilization of the ion trap mass spectrometer as a detector in our experiments resulted in high selectivity and low detection limits. All modes of operation on the mass spectrometer were evaluated. Detection limits were lowest in multiple ion monitoring (MIM) mode and ranged from 0.001 to 0.003 ppb, depending on the compound. In scan mode limits of detection were 0.005 to 0.01 ppb, and in MS/MS the range was 0.01 to 0.3 ppb. Limits of detection were calculated on the basis of an extracted ion in MIM and scan mode; the TIC of the resulting fragmentation of parent ion in MS/MS mode was used to achieve signal-to-noise ratio of 3:1. Calculated limits of detection were based on signal-to-noise ratio, which may be influenced by several factors—baseline noise, tuning, detector voltage of mass spectrometer, proper mass detector maintenance, and quality of fiber used. The advantage of the ion trap detector is the relatively high sensitivity in scan mode, which allows collection of full spectra of compounds present in a matrix in parts per trillion (ppt) concentrations. The MS/MS options adds confidence in the identification by using secondary fragmentation of the trapped parent ion. In our experiments the most abundant ions were used for collision-induced dissociation (CID). The utilization of MS/MS improves chromatography as the baseline is not disturbed by other ions. Precision estimated at 1 ppb was characterized by RSD values of 4.39% to 8.56%, and linearity for compounds in a concentration range of 0.01–10 ppb was satisfactory ($R^2 > 0.997$ for all compounds except geosmin, for which $R^2 = 0.992$). Evans and associates [22] observed RSD < 13% when using SPME for quantification of TCA in wines. Ming-Liang Bao and coworkers [29] reported low detection limits for MIB and geosmin (1 ng/L) in reagent (ultrapure) water when using ions 95 and 112, respectively, for quantification. For the recovery of analytes they extracted a spiked 1-L water sample with hexane and used an ion trap detector. Using PDMS/DVB for extraction of MIB and geosmin from water, Watson and associates [24] quantified MIB at 1.2 ng/L and geosmin at 3.3 ng/L. Juanola and coworkers [30] used shake flask and Soxhlet extractions to isolate TCA from cork with subsequent purification on a silica gel column, and the LODs reported were 1.19 ppb. Palmentier and colleagues [31] achieved detection limits of 2.0 ng/L for geosmin and 2-methylisoborneol, using Ambersorb extraction and high-resolution mass spectrometry.

The developed method was applied to the analysis of grain with sensorially detectable musty–earthy off-odors, which was obtained from farmers' granaries and bins. Quantification of analyzed musty–earthy compounds was based on external standard calibration elaborated in the process of method development. Figure 8 shows chromatograms of a wheat

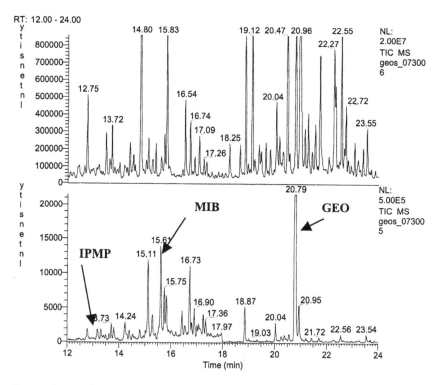

Figure 8 Total ion current chromatograms of volatile compounds isolated from a stored wheat grain sample characterized by a musty–earthy off-odor. The upper chromatogram was acquired in scan mode, whereas the lower one was acquired in multiple ion monitoring mode (see Sec. II). IPMP, 2-methoxy-3-isopropylpyrazine; IBMP, 2-methoxy-3-isobutylpyrazine; MIB, 2-methylisoborneol; TCA, 2,4,6-tri-chloroanisole; GEO, geosmin.

sample that had been stored for 11 m. The upper chromatogram was acquired in a scan mode, whereas the lower one represents total ion current of selected ions in a MIM run. Three of the volatile compounds responsible for off-odors were identified as geosmin (7.57 ppb), 2-methylisoborneol (0.14 ppb), and 2-methoxy-3-isopropylpyrazine (0.01 ppb).

IV. CONCLUSIONS

Solid-phase microextraction proved to be an efficient and robust method for the isolation of microbial volatile compounds in wheat grain. Matrix specificity is a crucial factor in method development. Choosing porous polymer-based fibers (Carboxen or divinylbenzene) results in low detection limits, good reproducibility, and linearity. The elaborated method allows rapid and reliable quantification of volatiles in wheat grain that are responsible for musty, earthy, and moldy off-odors and are usually present in trace concentrations.

ACKNOWLEDGMENT

This work was supported by the State Committee for Scientific Research, grant 5 P06G 034 18.

REFERENCES

1. D Kilkast. Sensory evaluation of taints and off-flavours. In: MJ Saxby, ed. Food Taints and Off-Flavours. Glasgow: Chapman & Hall, 1993, pp. 1–32.
2. RC McIver, GA Reineccius. Synthesis of 2-methoxy-3-alkylpyrazines by *Pseudomonas perolens*. In: C Parliament, R Croteau, ed. Bioformation of Flavours, ACS Symposium Series 1989, pp 266–274.
3. RF Curtis, DG Land, NM Griffiths, MG Gee, D Robinson, JL Peel, JM Gee. 2,3,4,6-Tetrachloroanisole association with musty taint in chickens and microbial formation. Nature 235:223–224, 1972.
4. JC Spadone, G Takeoka, R Liardon. Analytical investigation of Rio off-flavour in green coffee. J Agric Food Chem 38:226–229, 1990.
5. HR Buser, C Zanier, H Tanner. Identification of 2,4,6-trichloroanisol as a potent compound causing cork taint in wine. J Agric Food Chem 30:359–362, 1982.
6. IH Suffet, J Ho, J Mallevialle. Off-flavours in raw and potable water. In: MJ Saxby, ed. Food Taints and Off-Flavours. Glasgow: Chapman & Hall, 1993, pp 89–116.
7. JP Mattheis, RG Roberts. Identification of geosmin as a volatile of *Penicillium expansum*. Appl Environ Microbiol 58:3170–3172, 1992.
8. E Wasowicz, E Kaminski, H Kolmannsberger, S Nitz, RG Berger, F Drawert. Volatile components of sound and musty wheat grains. Chem Microbiol Technol Lebensm 11:161–168, 1988.
9. RM Seifert, RG Buttery, DG Guadagni, DR Black, JG Harris. Synthesis of some 2-methoxy-3-alkylpyrazines with strong bell pepper–like odors. J Agric Food Chem 18:246–249, 1970.
10. A Miller III, RA Scanlan, JS Lee, LM Libbey, ME Morgan. Volatile compounds produced in sterile fish muscle (*Sebastes melanops*) by *Pseudomonas perolens*. Appl Microbiol 25:257–261, 1973

11. WF Young, H Horth, R Crane, T Ogden, M Arnott. Taste and odor threshold concentrations of potential potable water contaminants. Water Res 30:331–340, 1996.

12. RG Buttery, RM Seifert, RE Lundin, DG Guadagni, L Ling. Characterization of an important aroma component of bell peppers. Chem Ind (London) 15:490–491, 1969.

13. HJ Takken, LM Van der Linde, M Boelens, JM Van Dort. Olfactive properties of a number of polysubstituted pyrazines. J Agric Food Chem 23:638–642, 1975.

14. PE Persson. The source of muddy odor in bream (*Abramis brama*) from the Porvoo Sea area (Gulf of Finland). J Fish Res Board Can 36:883, 1979.

15. BCJ Zoeteman, GJ Piet. Cause and identification of taste and odour compounds in water. Sci Total Environ 3:103–115, 1974

16. LL Medsker, D Jenkins, JF Thomas. Odorous compounds in natural waters: An earthy-smelling compound associated with blue–green algae and actino-mycetes. Environ Sci Technol 2:461, 1968.

17. H Maarse, LM Nijsen, J Jetten. Chloroanisoles: A continuing story. In: RG Berger, S Nitz, P Schreier, eds. Topic in Flavour Research. H Eichhorn D-Marzling-Hangenham 25, 1985, pp 241–253.

18. MJ Saxby. A survey of chemicals causing taints and off-flavours in foods. In: MJ Saxby, ed. Food Taints and Off Flavours. Glasgow: Chapman & Hall, 1993, pp 35–62.

19. J Pawliszyn. Solid phase microextraction: Theory and practice. Wiley-VCH, New York, 1997.

20. C Sala, M Mestres, MP Marti, O Busto, J Guasch. Headspace solid phase microextraction method for determining 3-alkyl-2-methoxypyrazines in musts by means of polydimethylsiloxane-divinylbenzene fibers. J Chromatogr A 880:93–99, 2000.

21. C Sala, M Mestres, MP Marti, O Busto, J Guasch. Headspace solid phase microextraction analysis of 3-alkyl-2-methoxypyrazines in wines. J Chromatogr A 953:1–6, 2002.

22. TJ Evans, CE Butzke, SE Ebeler. Analysis of 2,4,6-trichloroanisole in wines using solid phase microextraction coupled to gas chromatography–mass spectrometry. J Chromatogr A 786:293–298, 1997.

23. SW Lloyd, JM Lea, PV Zimba, CC Grimm. Rapid analysis of geosmin and 2-methylisoborneol in water using solid phase microextraction procedures. Water Res 32:2140–2146, 1998.

24. SB Watson, B Brownlee, T Satchwill, EE Hargesheimer. Quantitative analysis of trace levels of geosmin and MIB in source and drinking water using headspace SPME. Water Res 34:2818–2828, 2000.

25. SW Lloyd, CC Grimm. Analysis of 2-methylisoborneol and geosmin in catfish by microwave distillation–solid-phase microextraction. J Agric Food Chem 47:164–169, 1999.

26. H Jelen, E Wasowicz. Application of solid phase microextraction (SPME) for the determination of fungal volatile metabolites. In: S Bielecki, J Tramper, J

Polak, eds. Progress in Biotechnology. 17. Food Biotechnology. Amsterdam: Elsevier, 2000, pp 369–379.

27. T Nilsson, TO Larsen, L Montarella, JO Madsen. Application of headspace solid-phase microextraction for the analysis of volatile metabolites emitted by *Penicillium* species. J Microbiol Methods 25:245–255, 1996.

28. T Górecki. Solid versus liquid coatings. In: J Pawliszyn, ed. Applications of Solid Phase Microextraction. Cambridge: RSC Chromatography Monographs, 1999, pp 92–111.

29. ML Bao, K Barbieri, D Burrini, O Griffini, F Pantani. Determination of trace levels of taste and odor compounds in water by microextraction and gas chromatography–ion trap detection–mass spectrometry. Water Res 31:1719–1727, 1997.

30. R Juanola, D Subira, V Salvado, JA Garcia Reguiero, E Antico. Evaluation of an extraction method in the determination of the 2,4,6-trichloroanisole content of tainted cork. J Chromatogr A 953:207–214, 2002.

31. JPFP Palmentier, VY Taguchi, SWD Jenkins, DT Wang, KP Ngo, D Robinson. The determination of geosmin and 2-methylisoborneol in water using isotope dilution high resolution mass spectrometry. Water Res 32:287–294, 1998.

32

Analysis of Off-Aromas in Wines Caused by Aging

Katherine M. Kittel, Edward H. Lavin, Kathryn D. Deibler, and Terry E. Acree
Cornell University, Geneva, New York, U.S.A.

Audrey Maillard
Université de Bourgogne, Dijon, France

I. INTRODUCTION

Unlike effects due to slow oxidative changes in typical wine aroma compounds over several years, the "atypical aging" effect can show up in wine just a few months after fermentation. This defect was named *untypischen alterungsnote* (UTA) after its initial description in white *Vitis vinerifera* wines of Germany. These wines are initially acceptable after bottling and can fetch a reasonable price on the market. However, if UTA develops, these wines are difficult to drink and hard to sell. This chapter discusses the nature of UTA and a possible similar phenomenon in the American wine industry.

Untypischen alterungsnote has been found throughout Europe in most types and styles of wine [1]. Historically dubbed "vintage tone," "naphthalene note," or "hybrid note," the odor character of this phenomenon has been described as naphthalene, floor polish, wet wool, mothball, and acacia blossom [1–3]. As well as the presence of the defect, UTA wines are characterized by a noticeably decreased perception of varietal character.

Several chemical compounds have been linked with UTA as either indicator or causative compounds. *ortho*-Aminoacetophenone (OAP) is of primary interest, as it has been named the causative compound of the UTA aroma, though skatole and indole may also be involved [2,4,5]. The odor character of OAP has been described as foxy (along with methyl anthranilate, this compound characterizes the *Vitis labrusca* grape aroma), whereas skatole and indole have aroma characteristics atypical of grape aromas [6]. Skatole is described as having a mothball or fecallike aroma, whereas indole is characterized as burned mothball, fecal, or jasmine [5,7]. These compounds are amino acid metabolites; several studies have investigated their possible routes of formation as well as methods for their detection [3,8–12]. Indole acetic acid, glutathione, tryptophan, and kynurenine have been proposed as precursors of OAP [1,8,13–15]. Anecdotal reports from winemakers noting a correlation between the formation of UTA in wine and the environmental conditions of the affected vintages have been supported by statistical evaluation [16–18]. There has been a rise in UTA-affected wines since the widespread fertilization of vineyards was stopped in Europe [18,19]. Potential sources of UTA and/or higher levels of OAP in wine include the viticultural stress of early harvest, high yields, low nitrogen content (of the soil and/or the vine), water stress, wind stress, or high levels of solar radiation [1,16,20–22]. Microbiological links have also been investigated. Spontaneous or wild fermentations with non-Saccharomyces yeast, such as *Kloeckera* or *Metschnikowia* species, have produced higher levels of OAP in wines [23]. Some sulfur compounds have also been associated with UTA and are thought to accompany this defect, rather than cause the UTA aroma [19,24,25].

In 1999, during a tasting by a group of European wine researchers, several American Finger Lakes Rieslings and Chardonnays had perceived UTA-like aromas. According to Thomas Henick-Kling's notes from the tasting, "The group identified [an UTA-like] defect in some wines in each set of wines ... also (just as important) each set contained several older wines which had aged beautifully!" (T. Henick-Kling, personal communication). This was confirmed in 2002, when an informal tasting of New York wines by nine European wine researchers characterized several of these wines as having a strong "UTA-like" aroma. This led to an exploration into an American version of the UTA phenomenon called *atypical aging* (ATA); it is not yet apparent whether these phenomena are related. The ATA wines have been described as having a dishrag, wet towel, or musty aroma, although descriptors of UTA have also been applied. Environmental conditions producing the UTA problem, such as drought, water stress, and nitrogen status, have been linked to the occurrence of ATA [17,18]. It was unclear, however, whether the chemical compounds identified as indicator or causative compounds of UTA were involved in ATA. These observations led to the following questions: Is there a relationship between OAP and the intensity of ATA character in affected wine? In regard to the sensorial loss of varietal components in these wines, is this a masking phenomenon by a strong defect, or a true loss of varietal aroma compounds during the chemical process of ATA formation?

II. MATERIALS AND METHODS

To characterize possible similarities between ATA and UTA, several preliminary tests were run on four American wines from the New York State Finger Lakes region. The wines were a 1999 dry Riesling, 1999 semidry Riesling, and two 2000 dry Rieslings. All wines were from different wineries, grapes, vineyards, and winemakers. These wines were ranked and numbered 1 through 4 (low to high intensity), as judged by an informal panel of American wine researchers for the intensity of the ATA character of the wine. To investigate the varietal character of the most affected ATA wine (1999 semidry Riesling), another semidry Riesling was used from the same vines, winery, and winemaker, but from 2001.

Compounds identified as indicator or causative compounds of UTA, as well as varietal aroma compounds, were extracted from American wines using Twister^{TM} (Gerstel, Inc., Baltimore, MD, U.S.A.), a form of stir-bar sorption extraction (SBSE) (reviewed by Baltussen and associates [26]). The Twister, a glass-encapsulated magnetic stir bar coated with a cylinder of polydimethyl siloxane (5-mm film thickness, 10-mm length), was spun at room temperature in a 25-mL wine sample in a 50-mL glass beaker for 2 h, at a spin rate of 180 rpm, then thermally desorbed at 200°C for 5 min onto a portion of a DB-5 column (width 250 μm, length 30 m) cooled to −40°C (cryo-cooled by liquid N_2) in a gas chromatograph mass spectrometer (GCMS) (HP 6890 Series GC system/5973 Mass Selective Detector). The GC had an initial temperature of 35°C, which was held for 3 min before ramping up to 225°C at a rate of 4°C per minute, where it was held for a further 3 min. The GC was run at constant pressure. The MS was set to multiple ion monitoring for the detection of OAP (120 m/z ion), skatole (130 m/z ion), and indole (117 m/z ion). Varietal aroma compounds were detected in scan mode. An internal standard of 1 ppm (1 mg/L) 2,6-dichloroaniline was added to each wine sample to monitor extraction efficiency of the Twister. The retention time for each volatile was converted to Kovats retention indices using 7 to 20 carbon normal paraffins [27].

III. RESULTS AND DISCUSSION

A comparison of ATA sensory level and OAP level (the causative compound of UTA) in the investigated wines is shown in Table 1. A base wine free of the compounds of interest (i.e., below 0.1 µg/L for each chemical), spiked with 1 µg/L of OAP, skatole, and indole, gave a signal-to-noise ratio of 25 to 1, indicating an extraction limit near 0.1 µg/L for Twister extraction. There were no measurable (>0.1 µg/L) amounts of skatole or indole extracted by Twister in any of the American ATA wines, well below the published thresholds of 3 and 90 µg/L for skatole and indole, respectively, in aqueous solutions [28]. An extraction limit of 0.1 µg/L for OAP extracted by Twister was determined by using an external calibration curve in base wine. The OAP was extracted from the four ATA affected wines, at levels above the detection limit, yet below the published threshold of 0.5–1.5 µg/L in wine [29,30]. A Spearman rank correlation comparing the rank ATA intensity to OAP level had an r_s coefficient of −0.39, indicating a poor rank correlation.

These tests, characterizing the possible similarities of ATA and UTA, indicate that there is no simple correlation between OAP, skatole, or indole level with the intensity of ATA character in the affected wine. Although OAP may be an indicator compound for the chemical changes that occur in ATA and perhaps UTA wines, it does not seem to be the primary impact or causative chemical for the ATA defect. It is possible that a similar underlying chemical phenomenon exists for both ATA and UTA.

Table 1 Level of Perceived Atypical Aging Character Compared to o-Aminoacetophenone Level in New York Wines from the Finger Lakes[a]

Wine	ATA level[b]	OAP, level µ/L[c]
1. 2000 Dry Riesling	Low	0.4
2. 2000 Dry Riesling	⇓	0.3
3. 1999 Dry Riesling		0.3
4. 1999 Semidry Riesling	High	0.3
5. 2001 Semidry Riesling	None	No measurable amount

[a] Wines 1–4 are from different wineries, vines, and winemakers; wines 4 and 5 are from the same winery, vines, and winemaker, but from different years. There were no extractable amounts of skatole or indole in any of the wines using Twister. ATA, atypical aging; OAP, o-aminoacetophenone.
[b] Intensity of ATA in wine, as judged by an informal panel of American wine researchers.
[c] Measured by GCMS, extracted by stir-bar sorption extraction.

Another, related situation involved the appearance of ATA with increasing severity over three consecutive vintages (2001–1999) in semidry Riesling wines from the same grapes, vines, and winemakers. In the 1999 vintage, ATA aroma was obvious, and there was hardly any varietal character perceptible in the wine. The ATA was also detectable in the 2000 wine, yet this vintage still expressed some varietal character. The 2001 wine had no trace of any off-aromas and contained many varietal aromas. The 1999 wine contained a measurable amount of OAP, but OAP was not detected in the 2001 wine (wine 5, Table 1). A comparison of the volatile components from the 1999 and 2001 vintages using GCMS showed a decrease of 4 to 12 times the level of several terpene-containing aroma components, including limonene, α-terpinolene, γ-terpinene, and β-ocimene (Table 2). Some terpenes, particularly linalool and α-terpinolene, have been characterized by gas chromatography olfactometry (GCO) as important contributors to the varietal character in Riesling wines [31]. From the data obtained it is still unknown whether this is a masking defect or a true loss of volatile chemicals.

IV. CONCLUSIONS

Future work will include experiments using the 2001 Finger Lakes semidry Riesling and GCO to monitor changes in the odor components of this wine over the next two years. It appears that UTA or a similar character may occur in American wines, but this is not due to detectable levels of OAP, skatole, or indole. Although controlled experiments have yet to be

Table 2 Gas Chromatography Mass Spectrometry Peak Area Counts of Varietal Aroma Compounds from Twister Extraction of Wine Samples of Wines from the Same Winery, Vines, and Winemaker

	Peak area count, x10^{-6}				
	1999	2001	Factor of increase	Ion	Retention index
Limonene	0.094	0.75	8	136	13.50
α-Terpinolene	6.7	82	12	TIC	15.86
γ-Terpinene	21	84	4	TIC	14.7
β-Ocimene	2.3	20	8.7	TIC	13.97

[a] Internal standard of 2,6-dichloroaniline varied by 5% between the two samples. Ion values indicate which ions were used to identify the compound. TIC, total ion chromatogram. Hydrocarbon retention index values are for a DB-5 column.

completed, progress so far indicates that something like UTA is a threat to the American wine industry. The use of the stir-bar extraction technology (Twister^{TM}, Gerstel Inc.) combined with GCMS and GCO has the sensitivity necessary to determine the role of OAP, skatole, and indole in ATA wines.

REFERENCES

1. A Rapp, G Versini. Occurrence, origin and possibilities for a decrease of atypical ageing (ATA) in wine—a survey. In: International Association of Enology, Management and Wine Marketing; 13th International Enology Symposium. J Gafner, ed. Montpellier, France: TS Verlag, 2002, pp 285–310.
2. A Rapp, G Versini, H Ullemeyer. 2-Aminoacetopheno. Verursachende Komponente der "untypischen Alterungsnote" ("Naphthalinton," "Hybridton") bei Wein. Vitis 32:61–62, 1993.
3. K Hoenicke, TJ Simat, H Steinhart, N Christoph, M Gessner, H-J Köhler. "Untypical aging off-flavor" in wine: Formation of 2-aminoacetophenone and evaluation of its influencing factors. Anal Chim Acta 458:29–37, 2002.
4. T Hühn, WR Sponholz, A Friedmann, G Hess, H Muno, W Fromm. The influence of high-energy short-wave radiation and other environmental factors on the genesis of compounds affecting the wine quality in *Vitis vinifera L., c.v. Müller-Thurgau*. Wein-Wissenschaft 54:101–104, 1999.
5. WR Sponholz. The atypical ageing—a Survey. In: 31st Annual New York Wine Industry Workshop. T Henick-Kling, ed. Geneva, NY: New York State Agricultural Experiment Station, 2002, pp 75–78.
6. TE Acree, EH Lavin, N Ritsuo, S Watanabe. *O*-Aminoacetophenone, the "foxy" smelling component of Labruscana grapes. In: 6th Weurman Symposium. Y Bessiere, AF Thomas, eds. Wädenswil, Switzerland: Wiley, 1990, pp 49–52.
7. H Arn, TE Acree. Flavornet: A database of aroma compounds based on odor potency in natural products. In: Food Flavors: Formation, Analysis and Packaging Influences. ET Contis, C-T Ho, CJ Mussinan, TH Parliment, F Shahidi, AM Spanier, eds. Amsterdam: Elsevier Science B.V., 1998, pp 27–28.
8. N Christoph, M Bauer-Christoph, M Gessner, H-J Köhler, TJ Simat, K Hoenicke. Formation of 2-aminoacetophenone and formylaminoacetophenone in wine by reaction of sulfurous acid with indole-3-acetic acid. Wein-Wissenschaft 53: 79–86, 1998.
9. F Mattivi, U Vrhovsek, G Versini. Determination of indole-3-acetic acid, tryptophan and other indoles in must and wine by high-performance liquid chromatography with fluorescence detection. J Chromatogr A 855: 227–235, 1999.
10. N Christoph, M Bauer-Christoph, M Gessner, K Hoenicke, H-J Köhler, TJ Simat. Untypical aging off-flavour in wine: Formation of 2-aminoacetophe-

none in white wine by radical cooxidation of indole-3-acetic acid and sulfite. In: Frontiers of Flavour Science; Proceedings of the Weurman Flavour Research Symposium. P Schieberle, K-H Engel, eds. Freising, Germany: Deutsche Forschungsanstalt fuer Lebensmittelchemie, 2000, pp 561–568.

11. K Hoenicke, TJ Simat, H Steinhart, H-J Köhler, A Schwab. Determination of free and conjugated indole-3-acetic acid, tryptophan, and tryptophan metabolites in grape must and wine. J Agric Food Chem 49:5494–5501, 2001.

12. K Hoenicke, O Borchert, K Grüning, TJ Simat. "Untypical aging off-flavor" in Wine: Synthesis of potential degradation compounds of indole-3-acetic acid and kynurenine and their evaluation as precursors of 2-aminoacetophenone. J Agric Food Chem 50:4303–4309, 2002.

13. A Rapp, G Versini, L Engel. Determination of 2-aminoacetophenone in fermented model wine solutions. Vitis 34:193–194, 1995.

14. B Dollmann, A Schmitt, H Köhler, P Schreier. Formation of the "untypical aging off-flavor" in wine: Generation of 2-aminoacetophenone in model studies with *Saccharomyces cerevisiae*. Wein-Wissenschaft 51:122–125, 1996.

15. D Dubourdieu, V Lavigne-Cruege. Role of glutathione on development of aroma defects in dry white wines. In: J Gafner, ed. International Association of Enology, Management and Wine Marketing; 13th International Enology Symposium. Montpellier, France: TS Verlag, 2002, pp 331–347.

16. AL Schwab, N Christoph, H-J Köhler, M Gessner, TJ Simat. Influence of viticultural treatments on the formation of the untypical aging off-flavour in white wines. Part 1. Influence of the harvest time. Wein-Wissenschaft 54:114–120, 1999.

17. AN Lakso, RM Pool, L Cheng, T Martinson, K-T Li. Drought and water stress in New York vineyards and the potential for atypical aging of New York wines. In: 31st Annual New York Wine Industry Workshop. T Henick-Kling, ed. Geneva: New York State Agricultural Experiment Station, 2002, pp 94–98.

18. L Cheng, T Martinson, T Henick-Kling, A Lakso. Nitrogen management to improve vine N status and reduce atypical aging of wine in New York. In: 31st Annual New York Wine Industry Workshop. T Henick-Kling, ed. Geneva: New York State Agricultural Experiment Station, 2002, pp 99–102.

19. D Rauhut, H Kürbel, K Schneider, M Grossmann. Influence of nitrogen supply in the grape must on the fermentation capacity and the quality of wine. Acta Hortic 512:93–100, 2000.

20. O Löhnertz, HR Schultz, B Hünnecke, A Linsenmeier. Precautionary measures against ATA in viticulture. In: International Association of Enology, Management and Wine Marketing; 13th International Enology Symposium. J Gafner, ed. Montpellier, France: TS Verlag, 2002, pp 215–228.

21. T Hühn, S Cuperus, M Pfliehinger, WR Sponholz, K Bernath, W Patzwahl, M Grossmann, R Amadó, J Galli, A Friedmann. Influence of environment and the effects of substrates on the value of wine components. In: International Association of Enology, Management and Wine Marketing; 13th International Enology Symposium. J Gafner, ed. Montpellier, France: TS Verlag, 2002, pp 313–328.

22. HR Schultz, O Löhnertz, B Hünnecke, A Linsenmeier. Viticulture and atypical aging. In: 31st Annual New York Wine Industry Workshop. T Henick-Kling, ed. Geneva: New York State Agricultural Experiment Station, 2002, pp 83–85.
23. WR Sponholz, T Hühn. Aging of wine: 1,1,6 Trimethyl-1,2-dihydronaphthalene (TDN) and 2-aminoacetophenone. In: Proceedings of the Fourth International Symposium on Cool Climate Viticulture and Enology. T Henick-Kling, TE Wolf, EM Harkness, eds. Rochester: New York State Agricultural Experiment Station, 1996, pp VI-37–57.
24. D Rauhut, H Kürbel. Formation of reduced sulphur aroma and/or atypical aging defect: A possible differentiation. In: International Association of Enology, Management and Wine Marketing; 13th International Enology Symposium. J Gafner, ed. Montpellier, France: TS Verlag, 2002, pp 371–383.
25. D Rauhut. Volatile sulfur compounds—impact on "reduced sulfur" flavor defects and "atypical ageing" in wine. In: 31st Annual New York Wine Industry Workshop. T Henick-Kling, ed. Geneva: New York State Agricultural Experiment Station, 2002, pp 103–109.
26. E Baltussen, P Sandra, F David, C Cramers. Stir bar sorptive extraction (SBSE), a novel extraction technique for aqueous samples: Theory and principles. J Microcolumn Separations 11:737–747, 1999.
27. E Kovats. Gas chromatographic characterization of organic substances in the retention index system. Adv Chromatogr 1:229–247, 1965.
28. B Moss, S Hawe, N Walker. Sensory thresholds for skatole and indole. In: Production and utilization of meat from entire male pigs. M Bonneau, ed. Roskilde, Denmark: Institut national de la recherche agronomique, Paris, 1992, pp 63–68.
29. N Christoph, C Bauer-Christoph, M Gessner, HJ Köhler. Die "Untypische Alterungsnote" im Wein Teil 1. Rebe Wein 48:350–356, 1995.
30. M Gessner, HJ Köhler, N Christoph, C Bauer-Christoph, R Miltenberger, A Schmitt. Die "Untypische Alterungsnote im Wein" Tiel 2. Rebe Wein 48:388–394, 1995.
31. MG Chisholm, LA Guiher, TM Vonah, JL Beaumont. Comparison of some French–American hybrid wines with white Riesling using gas chromatography-olfactometry. Am J Enol Viticult 45:201–212, 1994.

33

Analysis of Important Flavor Precursors in Meat

**Donald S. Mottram, Georgios Koutsidis,
Maria-Jose Oruna-Concha, Maria Ntova,
and J. Stephen Elmore**
The University of Reading, Reading, England

I. INTRODUCTION

A. Meat Flavor Formation

Meat flavor is thermally derived, as uncooked meat has little or no aroma and only a bloodlike taste. During cooking, a complex series of thermally induced reactions between nonvolatile components of lean and fatty tissues occur, resulting in a large number of reaction products. The volatile compounds formed in these reactions are largely responsible for the characteristic flavors associated with cooked meat. An examination of the literature relating to the volatile compounds found in meat indicates that more than 1000 volatile compounds have been identified. A much larger number have been identified in beef than in other meats, but this is reflected in the much larger number of publications for beef compared with pork, sheep meat, or poultry [1,2]. The thermally induced reactions occurring during heating that provide meat flavor are principally the Maillard reaction, between amino acids and reducing sugars, and the degradation of lipid. Both types of reaction involve complex reaction pathways, leading to a wide range of products, which account for the large number of volatile compounds found in cooked meat. Heterocyclic compounds, especially those containing sulfur, are important flavor compounds produced in the Maillard reaction; they provide savory, meaty, roast, and boiled flavors. Lipid degradation produces compounds that give fatty aromas to cooked meat and compounds that determine some of the aroma differences between meats from different species.

B. Water-Soluble Flavor Precursors

The main water-soluble precursors in raw meat, which interact on heating to produce aroma volatiles, are amino acids, peptides, thiamine, and carbohydrates. Sulfur-containing volatile compounds are considered to make a particularly important contribution to the characteristic aromas of cooked meat [2]. These include thiol-substituted furans, mercaptoketones, and their disulfides, which have meaty aroma characters and exceptionally low odor threshold values. An important route to these compounds is the Maillard reaction between reducing sugars and sulfur-containing amino acids, such as cysteine or methionine. Meat contains significant quantities of ribose, which is a pentose sugar, and its reaction with cysteine, in model systems, has been shown to give meatlike aromas [3,4]. The reaction is widely used in the preparation of reaction-product flavorings with meatlike characteristics.

The main sources of ribose in meat are inosine 5'-monophosphate (IMP) and smaller quantities of ribose 5-phosphate and free ribose. The IMP is formed in muscle post slaughter from the enzymatic dephosphorylation and deamination of adenosine triphosphate (ATP), the ribonucleotide that is essential to muscle function in the live animal [5]. Further enzymatic breakdown of IMP may lead to hypoxanthine, ribose, and ribose 5-phosphate (Fig. 1), although most of the ribose in meat remains bound within IMP.

5'-Monophosphate is well recognized as a flavor potentiator and is associated with the taste sensation known as *umami* [6,7]. However, IMP may also provide a source of ribose for Maillard reactions that occur during the cooking of meat. The *N*-glycoside link between ribose and the base, hypoxanthine, involves the reducing group of the sugar and, therefore, Maillard-type reactions do not occur until this link is hydrolyzed. Although IMP appears to have relatively high thermal stability, some hydrolysis does occur on heating in aqueous solution, and this is enhanced under acidic conditions [8]. In a comparison of heated model systems containing cysteine and different sources of ribose it was shown that IMP was relatively unreactive compared with ribose, and ribose 5-phosphate was the most reactive [9]. It was suggested that the enhanced reactivity of ribose 5-phosphate with cysteine occurred because it was readily dephosphorylated to give reactive intermediates. It is, therefore, likely that meat flavor is related to the concentrations of these sugars and their derivatives present in the raw meat.

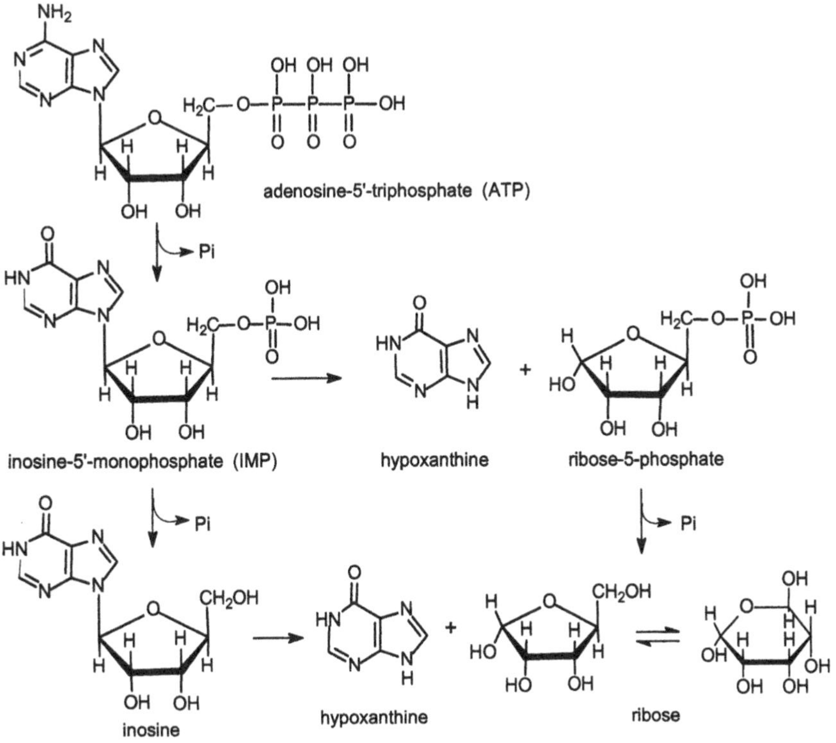

Figure 1 Breakdown of adenosine triphosphate in postmortem muscle.

A major challenge to meat flavor research is to understand how factors in meat production and processing can be used to optimize the desirable flavor of cooked meat. An understanding of changes in water-soluble flavor precursors brought about by production and processing factors could provide key information about the control of flavor in meat. This chapter reports on the analysis of sugars and ribonucleotides in a range of beef samples.

C.   Determination of Water-Soluble Meat Flavor Precursors

Some of the early work on meat flavor chemistry examined potential water-soluble precursors in raw meat and changes that occurred during cooking. The earliest report of the analyses of sugar in meat was that of Tarr and associates in 1954 [10], who used paper chromatography to identify glucose and ribose in beef muscle and estimated the concentrations to be 77 and 20 mg/100 g for glucose and ribose, respectively. Other early work by Macy and colleagues [11], again using paper chromatography, reported similar levels of glucose, but lower ribose levels. Using gas chromatography, after derivatization to silylated sugars, Jaboe and Mabrouk [12] reported small amounts of fructose and glucose phosphates as well as glucose, fructose, and ribose.

Sugars are present only at low levels in meat, and high-performance liquid chromagraphy (HPLC) using a refractive index detector is not sufficiently sensitive, and the lack of adsorption in the ultraviolet (UV) range makes UV detection impractical. In 2002, Aliani and Farmer [13] used HPLC involving postcolumn reaction with tetrazolium blue for the analysis of reducing sugars in chicken meat. Phosphorylated sugars were also analyzed by dephosphorylation using alkaline phosphatase followed by similar postcolumn derivatization. As well as glucose, fructose, and ribose, the corresponding phosphorylated sugars were identified in chicken and beef.

This chapter reports on the use of gas chromatography mass spectrometry (GCMS) and capillary electrophoresis (CE) for analysis of sugars and ribonucleotides in a range of beef samples.

II. EXPERIMENTAL PROCEDURES

A. Materials

Beef samples were supplied by the Division of Meat Animal Science at the University of Bristol. They were sourced from an experiment designed to show the effect of feeding on quality. The Aberdeen Angus and Hereford × Friesian steers had been fed on either grass silage or concentrate from weaning to slaughter at an age of 24 m. After slaughter the meat was conditioned on the carcass for 14 days. Samples of m. longissimus lumborum were taken, frozen in a blast freezer, and stored at − 18°C until analyzed.

Sugars, sugar phosphates, and IMP were purchased from Sigma (Dorset, UK). Silylation reagents, hexamethyldisilazane and trimethylsilyl-chlorosilane, were purchased from Fluka (Dorset, UK), cyclohexane (anhydrous) and dimethyl sulfoxide (anhydrous) were purchased from Aldrich (Dorset, UK).

B. Extraction Procedure

Water-soluble components of meat were extracted by using water and subsequent ultrafiltration. Fifty grams of lean meat was weighed and homogenized in a commercial mixer. Aliquots (3 g) were weighed accurately and extracted with 10 mL of cold water, using rhamnose as the internal standard. The extract was centrifuged at 10,000 g for 20 min at 4°C, and the residue was reextracted with 5 mL of cold water. The two extracts were combined and filtered through a 0.2 μm disk filter. Ten milliliters of the extract was transferred into an ultrafiltration tube (3000 MWCO Vivaspin, Vivascience Ltd., Lincoln, UK) and centrifuged at 3000 g for 4 hr at 4°C. The extracts were stored at −80°C until further analysis.

C. Gas Chromatography and Capillary Electrophoresis

Portions of the extract (1 mL) were freeze-dried and the residue silylated according to the method described by Leblanc and Ball [14]. Gas chromatographic analyses were performed using a HP 5890 series II GC coupled to a HP 5972 mass spectrometer. The capillary column used was a DB-17 fused silica column (30 m × 0.25 mm; J & W Scientific) with 1.5 m of deactivated fused silica as a retention gap [0.25 mm (i.d.)] and helium as the carrier gas at 1.0 mL/min. Direct splitless injections were performed at 250°C. The initial oven temperature (60°C) was held for 2 min and then increased to 170°C at 30°C/min, held for 5 min, then increased at 2°C/min to 220°C, and finally to 300°C at 4°C/min.

Capillary electrophoresis (CE) was carried out on the aqueous extract using a Hewlett Packard HP[3D] CE with diode array detection and a HP[3D] Chemstation for instrument control. The method for CE was based on that described by Uhrova and coworkers [15] using phosphate-borate buffer (0.02 M) adjusted to pH 9.2 and detection at 254 nm. Purine was used as the internal standard.

III. RESULTS AND DISCUSSION

Gas chromatography mass spectrometry analysis of the silylated meat extracts showed the presence of ribose, fructose, glucose, galactose, myoinositol, and the sugar phosphates fructose 6-phosphate and glucose 6-phosphate. No ribose 5-phosphate was present in any of the chromatograms; however, a peak in the GCMS chromatogram was tentatively identified as ribulose 5-phosphate, which is an isomer of ribose 5-phosphate. The most abundant sugar in the meat extract was glucose. The degradation of glycogen is the most significant process that leads to the accumulation of reducing sugars, namely, glucose, in the meat tissue.

In the capillary electrophoresis of aqueous extract of muscle the main components were inosine 5′-monophosphate, its nucleoside (inosine), and free base (hypoxanthine). Very small quantities of other ribonucleotides and their derivatives were also isolated.

Sugar and ribonucleotide-related compounds were examined in beef from a study in which steers were fed grass silage or concentrate diets for 24 m. The animals were all slaughtered under similar conditions and the meat was conditioned for 14 days. The quantities of sugars in the meat showed similar mean values for the two feeding groups and analysis of variance showed no significant differences at $p = 0.05$ (Fig. 2). The overall mean values for ribose and fructose, 13 and 30 mg/100 g, respectively, are of a similar magnitude to values reported in other work [13,16]. However, the mean glucose concentration of 218 mg/100 g was higher than previously published values. The concentrations of glucose and fructose phosphates were considerably higher than previously reported values, and the meat from grass-fed animals showed significantly higher values for glucose phosphate than that from the concentrate-fed animals ($p < 0.05$). The high levels of glucose may reflect high residual glycogen in the animals at slaughter.

Inosine 5′-monophosphate and the corresponding nucleoside, inosine, and the free base, hypoxanthine, were the major ribonucleotide-derived compounds found in the aqueous extracts of the muscle. No differences in mean concentration were found in any of these compounds in the meat

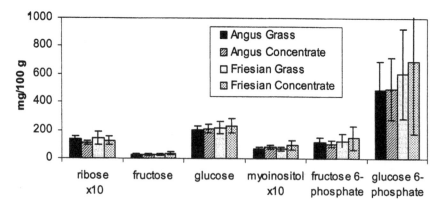

Figure 2 Comparison of sugars and sugar phosphates in meat from two beef breeds fed on grass silage or concentrate ($n = 6$). Values for ribose and myoinositol have been multiplied by 10.

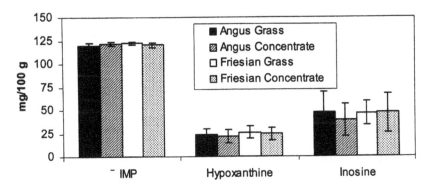

Figure 3 Comparison of inosine 5′-monophosphate (IMP) and related compounds in meat from two beef breeds fed on grass.

samples, which all had similar postslaughter conditioning (Fig. 3). The level of IMP is known to decrease during the conditioning of meat, and in 2002 we showed significant decreases in IMP in meat conditioned for 7, 14, and 21 days with corresponding increases in ribose concentrations [16]. Degradation of IMP to hypoxanthine during postmortem conditioning is well known [17], and release of ribose may be contributed to enhanced flavor in conditioned meat.

In the data, the range of values for each mean was particularly high (Table 1). Each value was the mean of two replicate analyses. These

Table 1 Concentrations of Some Water-Soluble Flavor Precursors in Beef from Animals Raised on Grass and Concentrate Diets, Showing the Range of Values ($n = 24$)

Compound	Milligrams/100 g meat		
	Mean	Maximum	Minimum
Glucose	218	133	306
Fructose 6-phosphate	121	45	310
Glucose 6-phosphate	571	187	1540
Ribose	13	6	28
Fructose	30	15	58
Hypoxanthine	24	15	34
Inosine	45	26	61
Inosine 5′-monophosphate	121	72	179

replicates showed very good agreement; thus the variation was not due to experimental error. All the animals were handled and slaughtered under very similar conditions, so such wide variations are rather surprising. They reflect wide variation between individual animals and necessitate further investigation. However, this may be a factor that contributes to inherent variability in meat flavor quality.

IV. CONCLUSIONS

Whereas there is much information on the volatile composition of cooked meat and the relative importance of different compounds in meat aroma, very little is known about the relationship between these compounds and their flavor precursors. Trace amounts of sugars, sugar phosphate, and ribonucleotides are important for the formation of meaty aroma. These precursors appear to be variable in concentration in meat, and this could lead to variation in the sensory quality of meat. A major challenge for future research in meat flavor is to relate such changes in precursors to sensory quality and the formation of aroma volatiles.

REFERENCES

1. H Maarse, CA Visscher. Volatile Compounds in Food: Qualitative and Quantitative Data. Zeist, The Netherlands: TNO-CIVO Food Analysis Institute, 1996.
2. DS Mottram. Meat. In: H Maarse, ed. Volatile Compounds in Foods and Beverages. New York: Marcel Dekker, 1991, pp 107–177.
3. ID Morton, P Akroyd, CG May. Flavoring substances and their preparation. GB Patent 836694, 1960.
4. LJ Farmer, DS Mottram, FB Whitfield. Volatile compounds produced in Maillard reactions involving cysteine, ribose and phospholipid. J Sci Food Agric 49:347–368, 1989.
5. RA Lawrie. Meat Science. Oxford: Pergamon, 1992.
6. JA Maga. Flavor potentiators. Crit Rev Food Sci Nutr 18:231–312, 1983.
7. YH Sugita. Recent developments in umami research. In: GG Birch, MG Lindley, eds. Developments in Food Flavours. London: Elsevier Applied Science, 1986, pp 63–79.
8. T Matoba, M Kuchiba, M Kimura, K Hasegawa. Thermal degradation of flavor enhancers, inosine-5′-monophosphate and guanosine-5′-monophosphate in aqueous solution. J Food Sci 53:1156–1159, 1988.

9. DS Mottram, ICC Nobrega. Formation of sulfur aroma compounds in reaction mixtures containing cysteine and three different forms of ribose. J Agric Food Chem 50:4080-4086, 2002.
10. HLA Tarr. The Maillard reaction in flesh foods. Food Technol 8:15–19, 1954.
11. RL Macy, HD Naumann, ME Bailey. Water-soluble flavor and odor precursors of meat. I. Qualitative study of certain amino acids, carbohydrates, non–amino acid nitrogen compounds, and phosphoric acid esters of beef, pork, and lamb. J Food Sci 29:136–141, 1964.
12. JK Jarboe, AF Mabrouk. Free amino acids, sugars and organic acids in aqueous beef extracts. J Agric Food Chem 8:494–498, 1974.
13. M Aliani, LJ Farmer. Postcolumn derivatization method for determination of reducing and phosphorylated sugars in chicken by high performance liquid chromatography. J Agric Food Chem 50:2760–2766, 2002.
14. DJ Leblanc, AJS Ball. A fast one-step method for the silylation of sugars and sugar phosphates. Anal Biochem 84:574–578, 1978.
15. M Uhrova, Z Deyl, M Suchanek. Separation of common nucleotides, mono-, di- and triphosphates by capillary electrophoresis. J Chromatogr 681:99–105, 1996.
16. G Koutsidis, DS Mottram, JS Elmore, MJ Oruna-Concha. Sugars and related compounds as flavour precursors in meat. Proceedings of 10th Weurman Flavour Research Symposium, Beaune, France, 2002.
17. DN Rhodes. Nucleotide degradation during the extended storage of lamb and beef. J Sci Food Agric 16:447–451, 1965.

34

The Role of 3-Methyl-2-Butene-1-Thiol in Beer Flavor

David Komarek
Nestle Purina PetCare, PTC, St. Louis, Missouri, U.S.A.

Klaus Hartmann and Peter Schieberle
Technical University of Munich, Garching, Germany

I. INTRODUCTION

Flavor staling in beer has been and still is one of the key concerns in the brewing industry. Although the crucial role of oxygen has been confirmed by numerous studies focused on natural aging of beer, exposure to light, in particular, has been recognized as the reason for formation of the so-called sunstruck flavor. This type of flavor, which had already been mentioned by Lintner in 1875 (cf. review in Ref. 1) is, however, reduced, when higher levels of oxygen are present in beer [2,3]. By 1961, 3-methyl-2-butene-1-thiol (MBT), exhibiting an intense sulfury aroma, was suggested as the main aroma compound responsible for the sunstruck flavor [3], and a very low odor thresholds of 4 to 35 ng/L has been reported [4].

Sensory experiments have shown that the odor of this compound was similar after appropriate dilution, but not identical, when evaluated in comparison to the sunstruck flavor in beer by a trained sensory panel [5]. It was, therefore, assumed that, besides MBT, further odorants contributed to the flavor difference observed in beer during exposure to light [6]. However, this assumption has not yet been confirmed by sensory studies and/or quantitative data. Since its detection by means of headspace gas chromatography was not successful [5], numerous methods have been proposed for MBT enrichment and quantation [7–10], and concentrations between 0.1 and 1 µg/L have been reported [5,6,8] in beers eliciting sunstruck flavor. Because MBT is labile and is present in beer in only trace amounts, the methods necessary for its quantitation must be selective and sensitive, and a stable isotope dilution assay would be the method of choice [11].

Dilution to odor threshold techniques, such as aroma extract dilution analysis (AEDA) [11], has previously been applied to characterize the most odor-active aroma compounds in fresh beer [12]. However, such methods have not yet been used to study odorants formed during light exposure of beers. The aim of the present investigation was, therefore, to study the differences in aroma compounds of fresh beer and beer exposed to daylight. Special emphasis was given to the evaluation of the role of 3-methyl-2-butene-thiol in the generation of sunstruck flavor.

II. MATERIALS AND METHODS

A commercial Pilsner-type beer bottled in green glass was purchased from a local shop in Germany, and a pale lager beer bottled in brown glass was from a Dutch brewery. Pilsner-type beers are characterized by a more intense hoppy aroma, because higher amounts of hops are used in the brewing process when compared to those in normal pale lagers. For light treatment, the beer sample was transferred into a white glass bottle and flushed with pure nitrogen to exclude any influence of oxygen. The samples were subsequently exposed to sunlight at $12°C \pm 2°C$ for the times given in the tables. For the isolation of the volatiles, the beer sample (500 mL) was extracted with diethyl ether and the volatile compounds were separated from the nonvolatile material by solvent-assisted flavor evaporation (SAFE) distillation [13]. The labeled $[^2H_8]$-3-methyl-2-butene-1-thiol was synthesized as reported previously [14]. Further details on the experimental procedures applied are published elsewhere [15].

III. RESULTS

A. Characterization of Differences in Aroma Compounds Caused by Light Exposure

In a first experiment, the Pilsner-type sample (green bottle) was transferred to a white glass bottle and exposed to daylight on 6 cloudy days in spring for a total of 60 hr. The illuminated sample showed a very pronounced sulfury, meatlike odor quality when compared to the nonilluminated sample.

To reveal which aroma-active compounds may cause this flavor difference, the volatile fractions from both beer samples were isolated by careful high-vacuum distillation and the odor-active constituents were detected by applying a comparative aroma extract dilution analysis [11]. The results revealed a total of 39 odor-active volatiles in the flavor dilution (FD) factor range of 16 to 4096 in the sample exposed to daylight. The identification experiments performed by means of reference compounds [15] resulted in 11 compounds exhibiting the highest FD factors, $\geqslant 1024$. Their structures are displayed in Fig. 1. Besides nine compounds previously established by us as key beer odorants, such as 3-methylbutanol, 2-phenyl-1-ethanol, or 4-hydroxy-2,5-dimethyl-3(2H)-furanone [12], 3-methyl-2-butene-1-thiol was present as an additional odorant with a high FD factor of 4096. However, MBT was not present among the odorants of the fresh pale lager beer previously analyzed [12]. The results of the entire identification experiments are given elsewhere [15].

In the nonilluminated sample, the same odorants were detected with similar FD factors, except for MBT [15]. The sulfury smelling thiol showed a much lower FD factor in the nonilluminated beer when compared to the illuminated sample (Table 1).

Furthermore, the results revealed that, in addition to MBT level, 3-(methylthio)propanal and phenylacetaldehyde levels were significantly increased in the illuminated sample (Table 1). The formation of such Strecker aldehydes induced by light has also been observed earlier in model systems containing catechin, riboflavin, and the respective precursor amino acids [16] and has been reported for light-treated skim milk [17]. However, the mechanism of Strecker aldehyde formation induced by light is still unclear.

To confirm the data obtained for the Pilsner-type beer, another beer sample (pale lager beer), originally bottled in brown glass, was also illuminated under the same conditions in white glass. The results revealed that, in particular, levels of MBT and the two Strecker aldehydes were also much increased after exposure to light [15]. But interestingly the FD factor of MBT in the nonilluminated pale lager beer was much lower than that in the nonilluminated sample of the Pilsner type, which was traded in green bottles. The data suggest that probably trading and storage in green glass lead to the formation of MBT.

B. Quantitative Studies

Sulfur compounds are known to be easily oxidized [18] and, furthermore, very often occur in foods in only trace amounts. In quantitative studies it is, therefore, important to use careful methods for enrichment and internal standards that have similar physicochemical properties. As previously shown by our group, stable isotope dilution assays are very useful tools in

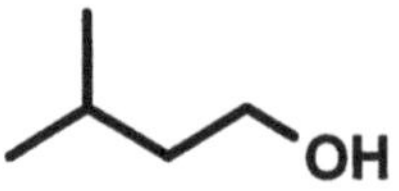

(3-methylbutanol;
malty)

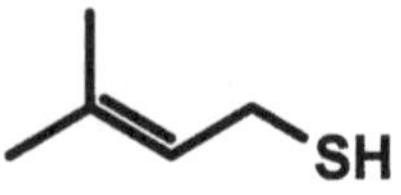

(3-methyl-2-buten-1-thiol;
sulfury)

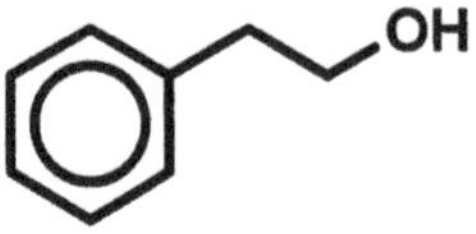

(2-phenyl-1-ethanol;
flowery)

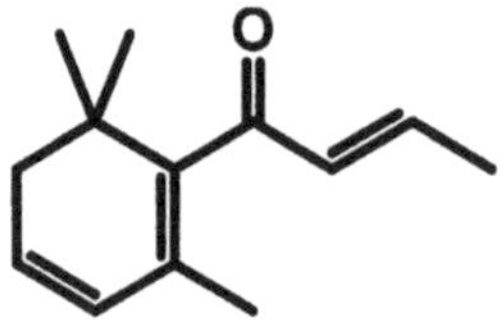

((E)-ß-damascenone;
cooked apple)

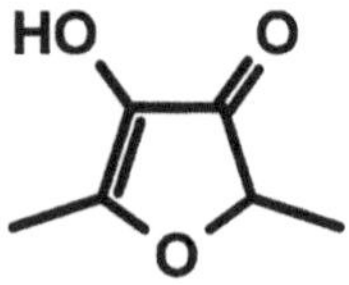

(4-hydroxy-2,5-dimethyl-3(2H)-
furanone; caramel-like)

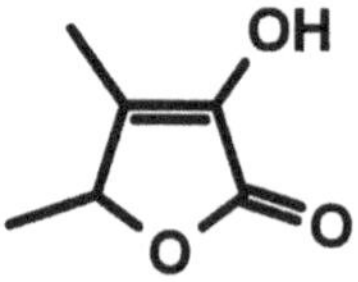

(4,5-dimethyl-3-hydroxy-
2(5H)-furanone; seasoning-like)

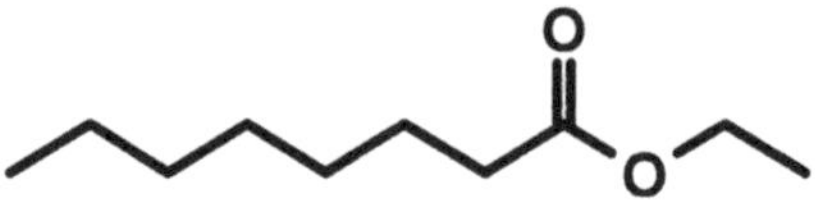

(ethyl octanoate; fruity)

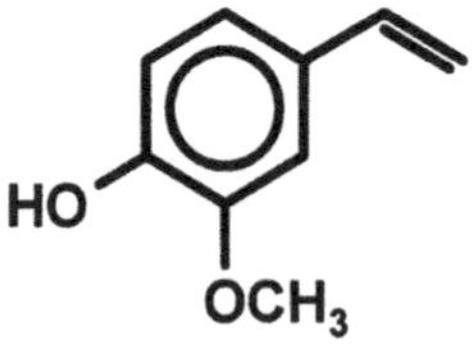

(4-vinyl-2-methoxyphenol; clove-like)

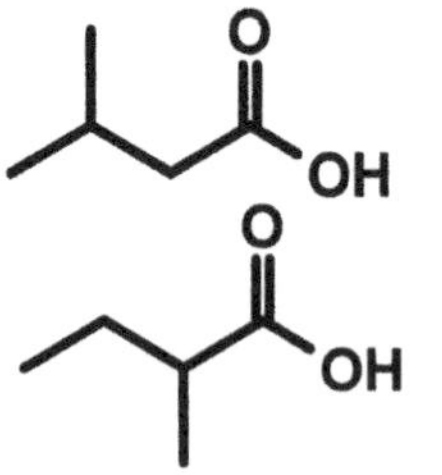

(3- and 2-methylbutanoic acid; sweaty)

Figure 1   Key odorants (flavor dilution [FD] factors ≥ 1024) identified in a Pilsner-type beer exposed to daylight (60 hr at about 12°C). Aroma attributes perceived at the sniffing port are given in parentheses.

Table 1 Potent Odorants FD ≥ 128 in at Least One Sample in the Pilsner-Type Beer Showing Significant Differences in Their Factors Before and After Exposure to Sunlight[a]

Odorant	FD factor in an extract from	
	A	B
3-Methyl-2-butene-1-thiol	256	4096
3-(Methylthio)propanal	16	256
Phenylacetaldehyde	16	256

[a] A, before exposure to sunlight; B, after exposure to sunlight; FD, flavor dilution.

the quantitation of such aroma compounds [11]. The principle of the method applied is shown in Fig. 2: A solvent extract was prepared from beer that had been spiked with the labeled internal standard [^{2}H$_8$]-3-methyl-2-butene-1-thiol. To isolate the thiol and the corresponding internal standard selectively, p-hydroxy-mercury benzoic acid is used [19]. The thiols trapped by the reagent are transferred into the aqueous phase and can thus be separated from the bulk of volatiles, which are present in the solvent phase. For mass spectrometric (MS) analysis, the thiols are finally released from the mercury salt by using an excess of cysteine [15].

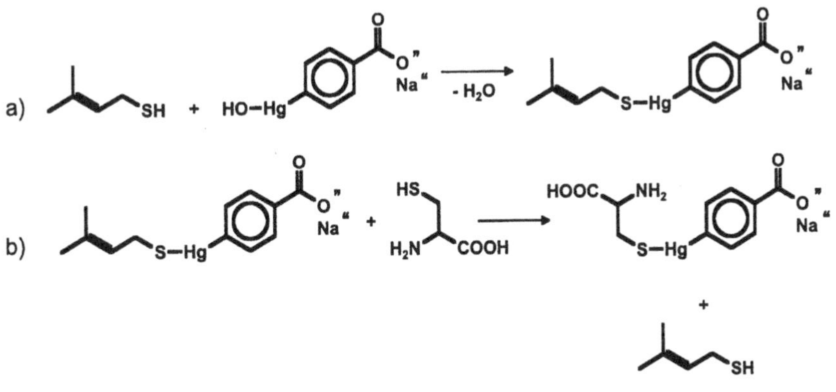

Figure 2 Principle of the method used to enrich 3-methyl-2-butene-1-thiol from distillates containing beer volatiles. a, Binding and transfer into the aqueous phase; b, release and transfer to solvent.

In a sample of fresh, unhopped beer as well as in the nonilluminated pale lager beer traded in brown glass bottles, the concentrations of MBT were extremely low and were clearly below the odor threshold (experiments 1 and 2, Table 2). However, light exposure of the pale lager beer for only 30 min (8-W light tubes) resulted in significant MBT formation when filled into a green glass bottle (experiment 3, Table 2).

The Pilsner-type beer, which was commercially sold in green glass, already showed a quite high concentration of MBT before illumination. However, interestingly, neither the illuminated pale lager beer (experiment 3, Table 2) nor the Pilsner-type sample was judged by the sensory panel to have a "sunstruck" off-flavor. The flavors of both beers were ranked as typical. These data suggest that obviously MBT positively contributed to the samples analyzed.

C. The Role of Riboflavin in MBT Formation

It has been reported in the literature that riboflavin is a crucial factor in MBT formation, and a mechanism involving isohumulone and cysteine as precursors for isoprenyl radicals and an SH radical has been proposed (Fig. 3) [5]. To indicate the influence of the riboflavin content present in beer, in a model study the natural concentrations were enhanced by a factor of 3 using pure riboflavin. The amounts of MBT formed after 9 hr at daylight were then compared to those formed in the unspiked, but also illuminated sample. In both illuminated samples significant amounts of the flavor compound were formed after light exposure (Table 3) and a clear sulfury, "sunstruck" off-flavor was detectable. Although the amounts were always higher in the samples spiked with riboflavin (experiments 2 and 3,

Table 2 Amounts of 3-Methyl-2-Butene-1-Thiol[a] in Different Beer Samples

Experiment	Beer sample	Concentration, µg/L
1	Unhopped beer[b]	<0.03
2	Pale lager beer (brown glass bottle; not illuminated)	<0.03
3	Pale lager beer (illuminated in green glass)	0.18
4	Pilsner type (green glass bottle; not illuminated)	0.26

[a] The odor threshold of 3-methyl-2-butene-1-thiol when added to fresh pale lager beer was determined to be 0.03 µg/L.
[b] The beer was supplied by Dr. D. Kaltner, Institute of Brewing Technology, Technical University of Munich.

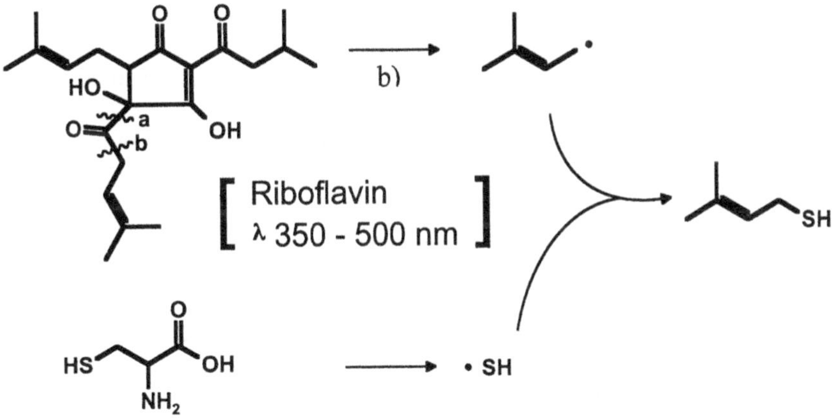

Figure 3 Proposed formation pathway for 3-methyl-2-butene-1-thiol from isohumulone and cysteine when exposed to light in the presence of riboflavin. (Data from Ref. 5.)

Table 3), the increase in riboflavin did not have a significant effect on MBT formation, as reported earlier for a model study [7].

To point out the role of riboflavin in MBT generation clearly, the following model experiment was performed: an aqueous solution (4% ethanol) of isohumulone (isolated from hops; 90% purity) and cysteine was illuminated for 60 hr at daylight either in the presence or in the absence of

Table 3 Influence of Riboflavin on the Formation of 3-Methyl-2-Butene-1-Thiol in Beer and a Model Solution[a]

		Concentration, µg/L	
Expt.	Exposure to daylight, hr	Without	With added riboflavin[b]
1	0	<0.03	<0.03
2	3	0.30	0.50
3	9	0.60	0.80
4[a]	60	1.0	2.0

[a] Isohumulone (25 mg) and cysteine (5 mg) were dissolved in water/ethanol (960 mL + 40 mL), the pH was adjusted to 4.3, the solution was flushed with nitrogen for 20 min and then exposed to daylight.
[b] The beer (500 mL) or the model solution (1 L) was spiked with riboflavin (300 µg).

riboflavin. Although double the amount of the flavor compound was formed in the model solution containing riboflavin, the results suggested (Table 3) that riboflavin is not a key catalyst in MBT formation, as assumed in the literature.

In summary, the results indicate that 3-methyl-2-butene-1-thiol may be a key odorant in certain types of beer that contributes to their typical aroma. The data suggest that the "tolerable" amount for a positive aroma contribution is about 0.2 µg/L, but this has to be proved by further sensory experiments.

Because isohumulone and cysteine, but not riboflavin, could be established as key precursors/catalysts in MBT formation, further studies are necessary to clarify the formation pathway of MBT in order to influence its concentration in beer by technological steps or the wort recipe.

REFERENCES

1. J Templar, K Arrigan, WJ Simpson. Formation, measurement and significance of lightstruck flavor in beer: A review. Brewers' Digest 18–25, 1995.
2. J de Clerck: pH And its applications in brewing. J Inst Brewing 40:407–419, 1934.
3. Y Kuroiwa, N Hashimoto. Composition of sunstruck flavor substance and mechanism of its evolution. Proc Am Soc Brew Chem 28–26, 1961.
4. AJ Irwin, L Bordeleau, RL Barker. Model studies and flavor threshold determination of 3-methyl-2-butene-1-thiol in beer. J Am Soc Brew Chem 51:1–3, 1993.
5. F Gunst, M Verzele. On the sunstruck flavor of beer. J Inst Brewing 84:291–292, 1978.
6. U Kattein. H Miedaner, L Narziss. The problem of lightstruck flavor in beer. Monatsschrift fuer Brauwissenschaft 41:205–208, 1988.
7. S Sakuma, Y Rikimaru, K Kobayashi, M Kowaka. Sunstruck flavor formation in beer. J Am Soc Brew Chem 49:162–165, 1991.
8. TL Peppard. Development of routine investigational tools for the study of sulfury flavors in beer. J Inst Brewing 91:364–369, 1985.
9. H Goldstein, S Rader, AA Murakami. Determination of 3-methyl-2-butene-1-thiol in beer. J Am Soc Brew Chem 51:70–74, 1993.
10. S Masuda, K Kikuchi, K Harayama, K Sakai, M Ikeda. Determination of light-struck character in beer by gas chromatography–mass spectroscopy. J Am Soc Brew Chem 58:152–154, 2000.
11. P Schieberle. New developments in methods for analysis of volatile flavor compounds and their precursors. In: AG Gaonkar, ed. Characterization of Food. Elsevier Science BV, 1995, pp 403–431.
12. P Schieberle. Primary odorants of pale lager beer: Differences to other beers and changes during storage. Z Lebensm Unters Forsch 193:558–565, 1991.

13. W Engel, W Bahr, P Schieberle. Solvent assisted flavor evaporation: A new and versatile technique for the careful and direct isolation of aroma compounds from complex food matrixes. Z Lebensm Unters Forsch A: Food Res Technol 209:237–241, 1999.
14. P Semmelroch, W Grosch. Studies on character impact odorants of coffee brews. J Agric Food Chem 44:537–543, 1996.
15. D Komarek. Key odorants in beer—influence of storage on the flavor stability. PhD dissertation, Technical University of Munich, Garching, Germany, 2001.
16. C Blockmans, CA Masschelein, A Devreux. Origin of certain carbonyl compounds formed during the aging of beer. Proc Congress—Eur Brewery Convention 17:279–291, 1979.
17. C Allen, OW Parks. Evidence for methional in skim milk exposed to sunlight. J Dairy Sci 58:1609–1611, 1975.
18. H Guth, T Hofmann, P Schieberle, W Grosch. Model reactions on the stability of disulfides in heated foods. J Agric Food Chem 43:2199–2203, 1995.
19. P Darriet, T Tominaga, V Lavigne, JN Boidron, D Dubourdieu. Identification of a powerful aromatic component of *Vitis vinifera L.* var. *Sauvignon* wines: 4-Mercapto-4-methylpentan-2-one. Flavour Frag J 10:385–392, 1995.

35

Extraction of Encapsulated Flavors by Accelerated Solvent Extraction

Gary W. Christensen
International Flavors & Fragrances, Inc., Union Beach, New Jersey, U.S.A.

I. INTRODUCTION

Liquid flavors are encapsulated in solid carriers to protect them from oxidation, volatilization, moisture, etc. There are numerous methods for encapsulation as well as types of solid carriers that are utilized. Spray drying and extrusion are the two most widely used techniques. Both methods typically encapsulate the flavor with some form of modified starch or sugar. It is necessary to monitor the performance of any encapsulation technique by determining the actual flavor loading versus the theoretical load.

Traditionally, refluxing-type extractions such as Clevenger distillation have been used for this purpose [1–3]. Completing a typical Clevenger extraction requires 2–3 hr in an entirely manual process. Also, this technique is limited to water-insoluble flavors. Similarly, Soxhlet extraction can also require several hours for completion and generates large amounts of waste solvent. Liquid/liquid extractions can be difficult as the carrier material can cause emulsions that necessitate further processing. Sorbent bed based techniques such as solid-phase extraction are an attractive alternative, but finding a single sorbent type that will function well for all the compounds in a typical compounded or natural flavor can be challenging.

Accelerated solvent extraction (ASE), also known as *pressurized solvent extraction*, is an instrumental technique that allows the extraction of solid or semisolid samples with organic solvents at temperatures above the boiling point of the solvent. In addition to the increased efficiency of extraction due to elevated temperature, the ASE instruments available today are fully automated and allow a much greater throughput of samples when compared to Clevenger distillation. The appropriate extraction conditions must be developed for each sample type. Temperature, time, and solvent must all be optimized to obtain the most efficient method.

II. MATERIALS AND METHODS

A. Materials

Samples of spray-dried and extruded flavors and neat flavor oils were obtained from the Delivery Sciences Group at the International Flavors & Fragrances Research and Development facility in Union Beach, NJ. High-performance liquid chromatography (HPLC)-grade methanol was used for all extractions and standard solutions. Ottowa sand and Celite 545 were obtained from Fisher Scientific.

B. Accelerated Solvent Extraction of Encapsulations

A Dionex ASE 200 was used for all work. A total of 1.0–1.3 g of sample was weighed into a plastic weighing boat and 8 g of sand (for extrusions) or 1.5 g Celite$^®$ (for spray-dried) was added and mixed with the sample. The sample was transferred to an 11-mL extraction cell and sealed. Methanol was used for all extractions. Extraction temperature ranged from 70°C to 150°C, depending on the test being conducted. Extraction time was also varied from 6 min to 30 min. The pressure was maintained at 1500 psi for all samples. The total extraction time for each sample was divided into three instrument cycles. For example, if the total extraction time was 30 min, there were 3×10 min cycles. After each cycle, a portion of the extraction cell volume was replaced with fresh solvent. After extraction, the volume of the extract was recorded and analyzed without further cleanup. Typical final volumes ranged from 25 to 30 mL. Each test was conducted using five replicates.

C. Gas Chromatographic Analysis

All gas chromatographic (GC) analyses were done using a Hewlett Packard 5890 Series II gas chromatograph with flame ionization detection. Detector temperature was 300°C. The column was a Restek Rtx-1 F&F, 100% methyl silicone, 50 m $\times$ 0.32 mm inner diameter (i.d.) with 0.5-μm film thickness. The oven program was typically 50°C to 280°C at 8°C/min. A split/splitless injector was used in the splitless mode (0.5 min) and maintained at 250°C. The carrier gas was helium set at 1.5 mL/min (constant flow).

D. Calculations

Standard solutions of the neat flavors were prepared in methanol and analyzed with sample extracts. The standard solutions were prepared to match approximately the theoretical concentration of the sample extracts. The integration parameters in the GC method were adjusted to include only those peaks attributable to the flavor. The total area of all these peaks was used in calculating flavor recovery. Sample extracts were compared directly to the standard solutions and percentage (w/w) loading was calculated after accounting for starting weight and extract volume. The five replicates for each test were averaged together.

E. Clevenger Distillations

For each sample 20 g was weighed into a 250-mL round bottom flask and 100 mL water was added. The flask was attached to the Clevenger trap and condenser and refluxed for 3 hr. The amount of oil evolved was read from the trap and the percentage oil was calculated.

III. RESULTS AND DISCUSSION

The primary goal of this work was to determine the optimal extraction conditions for a set of encapsulated flavors. Four separate encapsulations were tested: spray-dried orange, spray-dried peppermint, extruded orange, and extruded lemon. All flavors were stable beyond the highest temperature tested (160°C). The nominal flavor load was 20% for the spray-dried samples and 10% for the extruded samples. Clevenger distillations were performed on all samples as a comparison to the ASE results. These results are presented in Table 1. A secondary goal was to determine whether any patterns of flavor recovery emerged from varying extraction temperature

Table 1 Flavor Load of Encapsulations as Determined by Clevenger Distillation

Sample	Flavor, w/w, percentage
Spray-dry orange	18.0
Spray-dry peppermint	16.8
Extruded orange	6.13
Extruded lemon	6.51

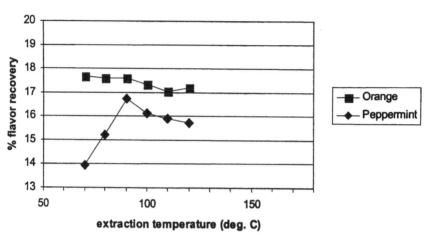

Figure 1 Recovery of flavor from spray-dried samples as a function of temperature.

and time. Ensuring that the method extracted all the components of the flavor was also an important consideration.

The effects of increasing extraction temperature are presented for the spray-dried samples in Fig. 1 and for the extruded samples in Fig. 2. Temperatures ranged from 70°C to 150°C and were increased in 10°C

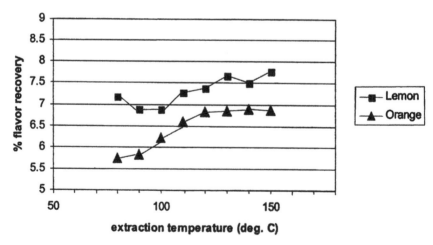

Figure 2 Recovery of flavor from extruded samples as a function of temperature.

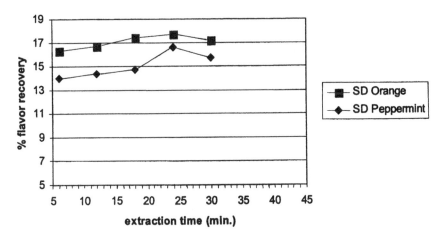

Figure 3   Recovery of flavor from spray-dried (SD) samples as a function of time.

increments. The extraction time was held constant (3 × 15 min cycles for a total of 45 min) for all samples.

The recoveries from the spray-dried samples drop off slightly after the optimal temperature is reached. The reason for this is not readily apparent. Loss due to volatility was considered, but since the extraction process is completely enclosed, volatility was not considered to be a significant source of loss. Since the spray-dried encapsulation carrier most likely remains

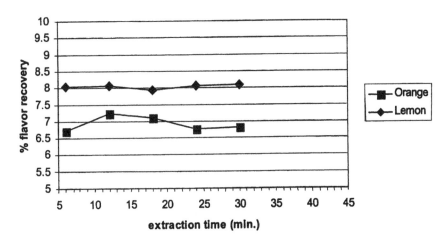

Figure 4   Recovery of flavor from extruded samples as a function of time.

intact during the extraction, it is hypothesized that after a certain temperature, flavor may partition back into the encapsulation, as opposed to staying in solution. Further tests are needed to determine whether this is occurring.

The recovery from the extruded samples plateaued after the optimal temperature was reached, and the dropoff seen with the spray-dried samples was not observed. The extruded product is a crystalline structure with the flavor immobilized within the glass, unlike the spray dry, which is more of a hollow sphere. In order to remove the flavor from the extrusion, the crystal must be disrupted. It is thought that once this occurs, the flavor cannot be reencapsulated and remains solvated.

The recovery of flavor as a function of extraction time is seen in Figs. 3 and 4. Extraction temperature was held constant for these tests: 150°C for the extruded samples and 90°C for the spray-dried samples. A pattern similar to that seen in Figs. 1 and 2 can be observed. The same reasoning can be applied here as before.

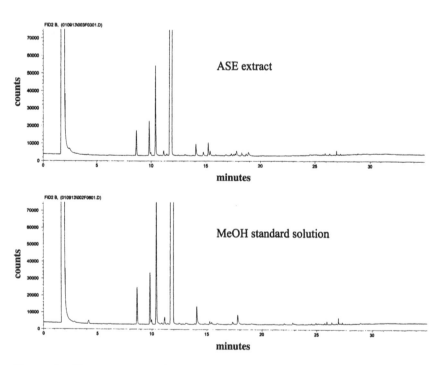

Figure 5   Chromatograms of orange flavor accelerated solvent extraction (ASE) extract and standard solution.

The relative standard deviation for each set of five replicates was less than 10% in all tests. It is also useful to note that the longest extraction time tested (30 min) was considerably less than the time for a typical Clevenger distillation (about 3 hr). Extract volumes were about 30 mL, which is also much less than in a typical Soxhlet or liquid/liquid method.

Chromatograms of the sample extracts versus standard solutions are shown in Figs. 5–7. The chromatograms are very similar in all cases. This demonstrates that ASE is capable of extracting all the components of the flavors in this work while not introducing any interferences that would hinder interpretation.

IV. CONCLUSIONS

Accelerated solvent extraction offers a viable alternative to Clevenger distillation as a method for determining the loading of flavor encapsula-

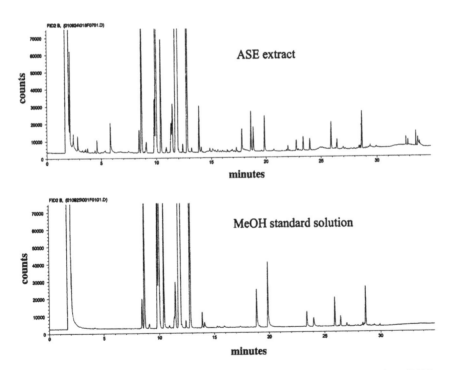

Figure 6 Chromatograms of lemon flavor accelerated solvent extraction (ASE) extract and standard solution.

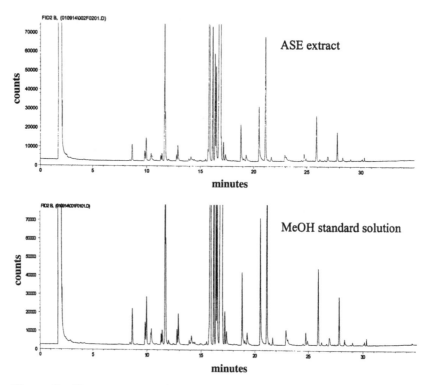

Figure 7 Chromatograms of peppermint flavor accelerated solvent extraction (ASE) extract and standard solution.

tions. The instrumentation adds the advantages of automation and reproducibility. Determining the optimal extraction conditions for each type of sample is critical for developing an accurate method.

REFERENCES

1. MH Anker, GA Reineccius. In SJ Risch, GA Reineccius, eds. Flavor Encapsulation. Washington, DC: American Chemical Society, 1988, p 80.
2. SJ Risch, GA Reineccius. In SJ Risch, GA Reineccius, eds. Flavor Encapsulation. Washington, DC: American Chemical Society, 1988, p 69.
3. LL Westing, GA Reineccius, F. Caporaso. In SJ Risch, GA Reineccius, eds. Flavor Encapsulation. Washington, DC: American Chemical Society, 1988, p 112.

Index